AMERICAN TANKS & AFVs OF WORLD WAR II

OSPREY
PUBLISHING

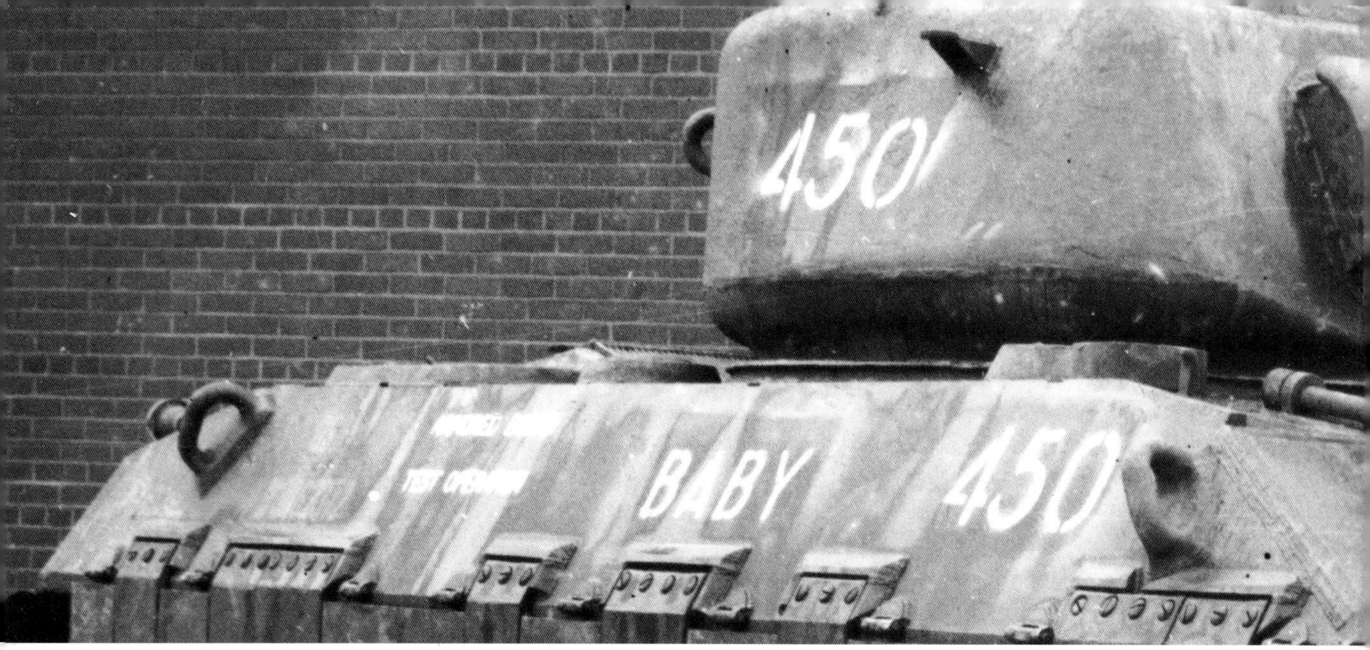
4501
BABY
450

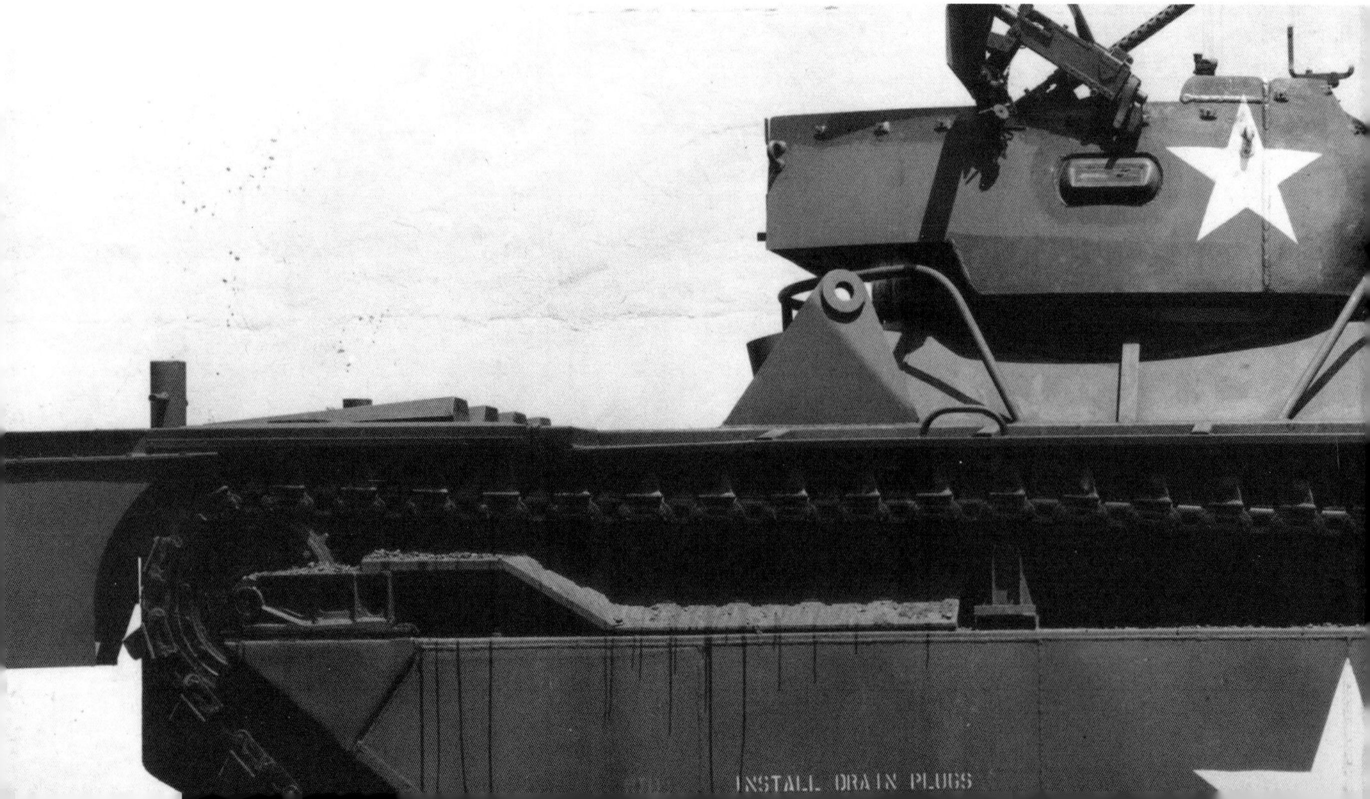
INSTALL DRAIN PLUGS

AMERICAN TANKS
& AFVs OF WORLD WAR II

MICHAEL GREEN

U.S.A.
9130950

OSPREY PUBLISHING
Bloomsbury Publishing Plc
PO Box 883, Oxford, OX1 9PL, UK
1385 Broadway, 5th Floor, New York, NY 10018, USA
E-mail: info@ospreypublishing.com
www.ospreypublishing.com

OSPREY is a trademark of Osprey Publishing Ltd

First published in Great Britain in 2014

This paperback edition was first published in Great Britain in 2018 by Osprey Publishing. Artworks variously by Tony Bryan, Richard Chasemore, Hugh Johnson, Jim Laurier and Peter Sarson © Osprey Publishing Ltd.

ISBN: HB 978 1 78200 931 3;
         PB 978 1 4728 2978 8;
         eBook 978 1 78200 980 1;
         ePDF 978 1 78200 979 5;
         XML 978 1 4728 3174 3

18 19 20 21 22   10 9 8 7 6 5 4 3 2 1

Index by Zoe Ross
Originated by PDQ Digital Media Solutions, Bungay, UK
Printed in China through C&C Offset Printing Co Ltd

Osprey Publishing supports the Woodland Trust, the UK's leading woodland conservation charity. Between 2014 and 2018 our donations are being spent on their Centenary Woods project in the UK.

To find out more about our authors and books visit **www. ospreypublishing.com**. Here you will find extracts, author interviews, details of forthcoming events and the option to sign up for our newsletter.

DEDICATION

The author would like to dedicate this book to World War II tanker veteran Tom Sator of the 4th Armored Division, 37th Tank Battalion, Company B, for his assistance on this book and many others.

ACKNOWLEDGMENTS

The author could not have completed this book without the support of many different people and organizations. The late Richard Hunnicutt supplied many historical pictures for this book as well as providing access to his research files. The individual quality of the historical pictures in this book varies reflecting the age of some of the pictures.

Other friends who contributed material, pictures, and often their time in reviewing the text for this book include Dean and Nancy Kleffman, James D. Brown, Joe DeMarco, Chris Hughes, Michel Krauss, Paul and Lorén Hannah, Kenneth Estes, Chun-Lun Hsu, Pierre-Olivier Buan and Christophe Vallier. A special word of thanks goes to Michael Panchyshyn for all his efforts on my behalf to ensure this book is as accurate as possible.

Institutions that assisted the author include the former Military Vehicle Technology Foundation and its former director, the late Philip Hatcher. Marc Sehring, curator of the Virginia Museum of Military Vehicles, also offered his support. David Fletcher (now retired) of the Tank Museum, Bovington, England, dug through the museum's photo files to help out the author. Randy Talbot, the TACOM Life Cycle Management Command (LCMC) staff historian, allowed the author access to their extensive historical photo collection.

Pictures credited to the Tank Museum, Bovington, are shortened to "Tank Museum" for the sake of brevity. Images from the TACOM LCMC historical office are shortened to "TACOM historical office" for the same reason. Lastly, the author wishes to thank the always helpful staff of the now-closed Patton Museum of Armor and Cavalry for their assistance through many years. Pictures credited to that museum are shortened to "Patton Museum."

CONTENTS

INTRODUCTION

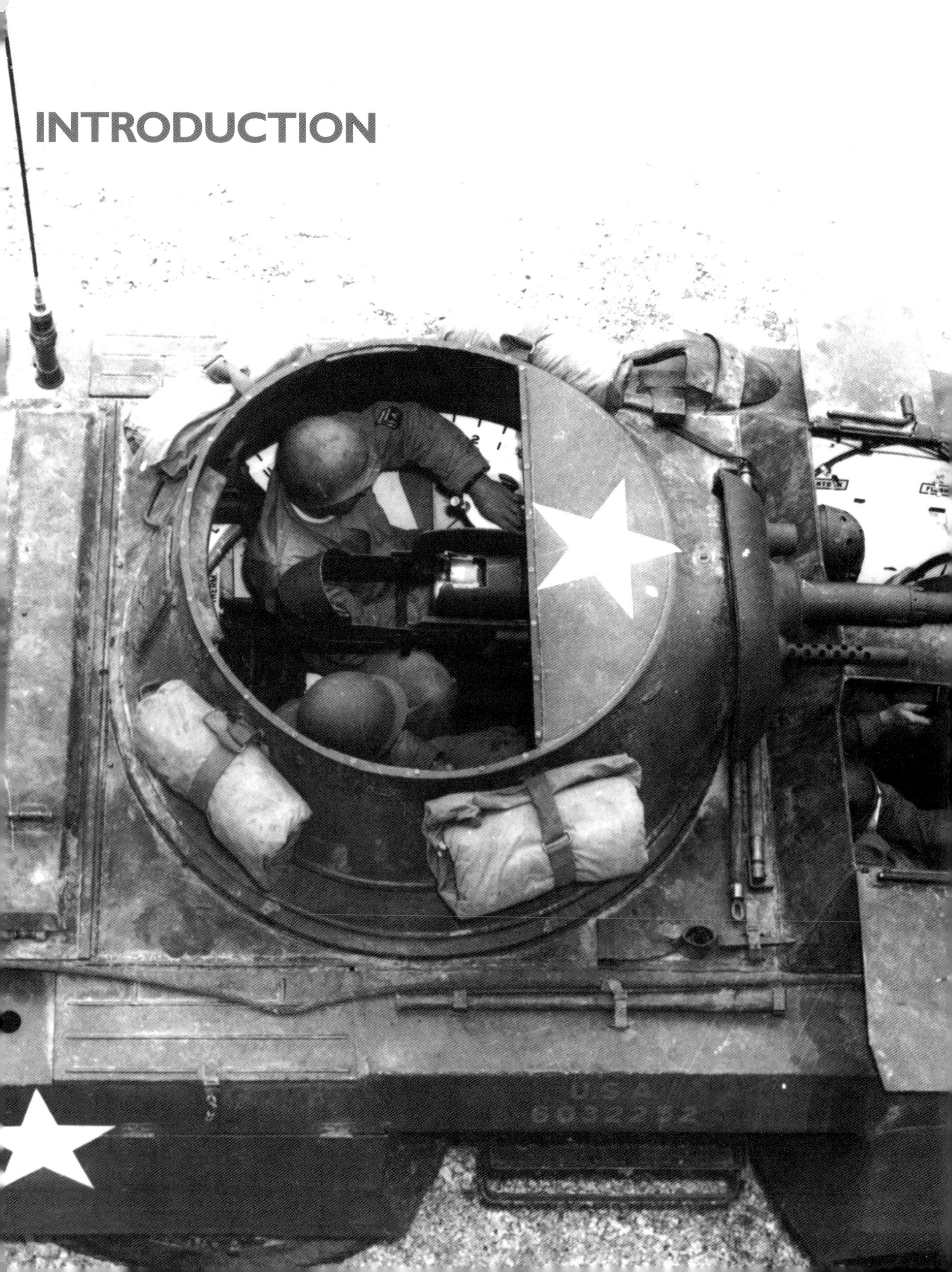

The equipping of the United States military with the full spectrum of weapons it needed to prevail during World War II was an unparalleled example of America's industrial might at the time. Among the many weapons produced by America's workers, tanks rate as an important example, with 88,140 built between 1939 and 1945. This was almost twice what Germany and Great Britain combined built during the same period, which numbered only 49,163 tanks. The factories of the Soviet Union built 76,827 tanks during the world conflict. When American wartime production was at its highest level in 1943, almost 30,000 tanks rolled off the factory floor in just that year alone. These tanks not only equipped America's ground forces but saw service with many Allied armies during World War II.

Beset by a faulty doctrine that insisted that tanks were not intended to fight other tanks – a role intended for specialized vehicles referred to as tank destroyers – American tankers had to learn the hard way that their German Army counterparts did not fight in the manner envisioned by the prewar senior leadership of the U.S. Army. It is to their credit that American tankers made do with the vehicles with which they were provided, until such time that American engineering talent could come up with a suitable tank that nearly matched the qualitative superiority enjoyed by late-war German tanks. Sadly, that tank showed up too late during the war in Europe, and in insufficient numbers, to have any effect on the battlefield.

In the Pacific Theater of Operations (PTO), the American medium tanks of the Marine Corps and U.S. Army dominated the opposition. However, the threat faced by these tanks in battle proved far different than what the U.S. Army faced in the European Theater of Operations (ETO). Instead of having to deal with well-armed and heavily armored enemy tanks and self-propelled guns, they often faced suicidal Japanese infantrymen carrying explosive charges who were willing to expend their lives to destroy an American tank, or heavily camouflaged and bunkered antitank guns.

In addition to the 18,620 tank-based variants, such as armored engineering vehicles, self-propelled artillery, armored recovery vehicles, and tank destroyers, American factories went on to design and build thousands of wheeled armored cars for reconnaissance purposes and armored half-tracks to transport the infantry into battle behind the tanks. Like the tanks, American armored half-tracks were modified to serve a wide variety of jobs including mortar carriers, self-propelled artillery, tank destroyers, and antiaircraft vehicles. So useful were these vehicles that many would remain in service with foreign armies for decades after World War II.

To complement its inventory of tanks and armored fighting vehicles, the American military industrial complex also designed and built over 18,000 amphibian tractors. Appearing in both unarmored and armored variants, they went into combat with a wide variety of armaments. Referred to as the Landing Vehicle, Tracked (LVT), they would serve not only with the U.S. Marine Corps, who often called them "Amtracs," but also the U.S. Army. They allowed the American military to take the fight to the far-flung Japanese Empire wherever it had established itself in the vast reaches of the PTO. These same vehicles would also see service in the ETO with the U.S. Army and Allied forces when it came time to cross various water obstacles, often used by the German military as defensive barriers.

CHAPTER ONE

EARLY MEDIUM TANKS

In the summer of 1919, the U.S. Army Ordnance Department began exploring the idea of fielding medium tanks that would be able to exploit any breakthroughs in enemy lines achieved by the heavy tanks. This was a concept pioneered by the British Army with great success with their fielding of the approximately 15-ton machine-gun-armed Medium Mark A "Whippet" late in World War I.

Up until the late 1920s, the U.S. Army often copied British Army advances in the field of tank development. The U.S. Army's embrace of this British Army concept of medium tanks resulted in a single prototype example of the approximately 23-ton medium tank M1921 showing up at Aberdeen Proving Ground (APG), Maryland, for testing in early 1922.

A prototype demonstrates the design and layout of a vehicle, but does not attempt to prove the manufacturing methods. Prototypes are therefore typically hand-built. All tons listed are in United States short tons (2,000 pounds per ton). Vehicle weights listed are when combat loaded with a full load of fuel and ammunition.

The M1921 was very similar to the British Medium D tank then under development and was armed with a turret-mounted, British-designed and built 6-pounder (57mm) main gun and two .30 caliber machine guns.

The M1921 was followed by a near-identical prototype, designated the medium tank M1922, which featured a British-designed suspension system. Both the M1921 and M1922 prototypes were built at the American government-owned and government-operated Rock Island Arsenal.

Problems with the original gasoline-powered engines in the prototype M1921 and M1922 tanks later caused the U.S. Army Ordnance Department to re-engine the M1921 with a specially designed and built gasoline-powered engine from the Packard Motor Company in 1925. Reflecting the new engine and some other minor changes, the vehicle was re-designated as the medium tank T1. The prefix "T" indicated a test vehicle or weapon.

Pictured is a World War I-era British Mark A Whippet medium tank. The vehicle was envisioned by the British Army as an exploitation tank that would assume the role once performed by the cavalry. This was a concept embraced by the U.S. Army following "The War to End All Wars." (Tank Museum)

The medium tank T1 was the follow-on to the medium tanks M1921 and M1922. Its outward appearance generally mirrored that of the M1921. In lieu of armored steel, it was constructed of soft steel plate as this cost a fraction of the former. Top speed of the T1 proved to be 14mph. (Patton Museum)

In January 1928, the U.S. Army's Ordnance Committee, which included representatives from that branch of the service intending to employ the vehicle and the technical sections of the Office of the Chief of Ordnance, recommended standardization of the four-man 21-ton medium tank T1 as the medium tank M1. By this time, the letter "M" indicated a model number of a weapon or vehicle standardized and approved for series production. This recommendation was approved, only to be rescinded a few months later.

Following the cancellation of the M1, the U.S. Army expressed a wish for another medium tank design not to exceed 15 tons in weight, the limit for its medium pontoon bridges and the majority of American highway bridges at the time. In response to that interest, the Ordnance Department had the James Cunningham, Son and Company build a prototype in 1929 of a four-man vehicle, designated the medium tank T2. It was based on an enlarged 7.5-ton light tank T1 that the same firm had submitted for consideration to the U.S. Army in 1927.

The T2 featured a turret-mounted experimental 47mm main gun and a coaxial .50 caliber machine gun. There was also a limited traverse 37mm gun mounted in its right front hull. As the front hull-mounted cannon interfered with the operation of the turret-mounted cannon, it was removed in 1931 and replaced by a single .30 caliber machine gun. The T2 looked very similar to the British Vickers medium tanks then under development.

Testing of the T2 between 1930 and 1932 produced positive results as the vehicle possessed a fair degree of mobility with its front hull-mounted gasoline-powered Liberty engine, which gave it a maximum speed of 25mph. The turret-mounted 47mm main gun provided an outstanding level of firepower for its time. Due to limited funding available to the U.S. Army during the Great Depression (which began in 1929) the tank was never proposed for series production. It was, however, retained for testing components and accessories for possible use in future tank designs.

U.S. Army Ordnance Department Terms

In the context of vehicle acquisition, a "standardized" or "standard" item is one for which initial issue, repair parts, and maintenance are available. The term "substitute standard" means a vehicle that fulfills all the capabilities of the standard item, but may differ slightly in details. When the term "limited standard" is used it means that an item is intended to be maintained by cannibalization or to be scrapped when no longer serviceable. Limited standard was seen quite a bit in World War II, when initial runs of items replaced by more modern equipment remained in use to prevent them from going to waste. Limited standard vehicles were often used for training. For vehicles intended solely for foreign military use, the Ordnance Department employed the term "limited procurement."

CONVERTIBLE MEDIUM TANKS

Besides those medium tanks proposed by the Ordnance Department, the U.S. Army also took a look at a number of experimental medium tank designs by prolific American automotive inventor J. Walter Christie following World War I. Testing of Christie's first concept vehicle, designated the medium tank M1919, began in 1921.

Christie's medium tank M1919 was soon superseded by an improved, rebuilt version, designated the medium tank M1921, which began undergoing testing in 1922. In contrast to the M1919 that had a weapon-armed turret, the M1921 had no turret and its main armament was mounted in the front hull. As with the M1919, the M1921 did not meet the U.S. Army's expectations and the design was never standardized and placed into production.

The medium tank T2. The 37mm gun barrel projecting from the right front of the hull identifies it as the original version of the vehicle, as the weapon was subsequently replaced with a .30 caliber machine gun. (Ordnance Museum)

Christie eventually came up with a vehicle design that generally met the U.S. Army's requirements and was awarded a contract on June 12, 1931, for seven examples of an 11-ton tank, designated the convertible medium tank T3.

The term "convertible" meant that a tank could either run on its tracks or on its large rubber-rimmed road wheels. Both the M1919 and M1921 had been convertible tanks and were referred to as "combined wheel and caterpillar" vehicles. The need for a tank to be able to operate with or without its tracks reflected the very short lifespan of tracks at the time.

Of the seven Christie medium tanks ordered, three were assigned to the U.S. Army Infantry Branch for test purposes and were armed with a turret-mounted 37mm main gun and a coaxial .30 caliber machine gun. The other four went to the Cavalry Branch of the U.S. Army for testing as light tanks and were armed with a turret-mounted .50 caliber machine gun in lieu of the 37mm main gun, and a coaxial .30 caliber machine gun. Even at this late date, the .50 caliber machine gun was still seen as an antitank weapon by the U.S. Army.

In a picture taken at Aberdeen Proving Ground, Maryland, during testing in June 1922, is J. Walter Christie's medium tank M1921 with its tracks fitted. The gun mount visible in the front of the vehicle's hull contained a 6-pounder (57mm) main gun and a coaxial .30 caliber machine gun. (Ordnance Museum)

The U.S. Army Tank Corps, formed in 1917, had been abolished by the American Congress by the Defense Act of 1920, leaving all future responsibilities for tank development with the Infantry Branch until 1931, when the Cavalry Branch was authorized to pursue its own line of development with "combat cars" (i.e. light tanks). As only the Infantry Branch of the U.S. Army was supposed to have tanks at this time – a reflection of the intense inter-branch rivalries then in play – the Christie medium tanks in service with the Cavalry Branch were actually designated the combat car T1.

The Christie-designed medium tanks were powered by government-supplied, gasoline-powered, liquid-cooled Liberty engines and had incredible mobility for their day, with a top speed on their rubber-rimmed road wheels of over 40mph. With their tracks on, they could reach speeds of over 25mph. By comparison, the light and heavy World War I-era tanks then in the U.S. Army inventory had a maximum speed below 10mph. These unheard of speeds were possible with the Christie suspension system because of the use of helically wound coil springs acting independently on the large road wheels.

Not everybody was enthralled with the Christie-designed medium tanks. U.S. Army Major-General Samuel Hof, the then-incoming Chief of Ordnance, not only doubted the true practicability of the convertible principle but also its reliability, stating that it had not been successfully demonstrated. He also went on to say in June 1931 that, "too much stress was being placed on speed at the expense of ruggedness and fighting ability."

The U.S. Army also had other concerns with Christie-designed medium tanks besides their speed with or without tracks. They felt the vehicle's two-man crew, limited firepower and armored protection were more suited for a light tank design than that of a medium tank. The U.S. Army therefore asked Christie for an improved version of his medium tank that better suited their needs. The two sides could not come to an agreement on the matter and parted ways.

In the collection of the former U.S. Army Ordnance Museum at APG is this nicely restored convertible medium tank T3 with its tracks stored on its upper hull fenders. The T3 was armed with a 37mm gun M1916 and a coaxial .30 caliber machine gun. An innovative feature of the vehicle was its sloped armor. (Michael Green)

CHRISTIE REPLACEMENT

The U.S. Army decided it made more sense to pursue its own line of development using the convertible principle with its medium tanks. It therefore went ahead and awarded a contract for five new medium tanks to American-La France that would incorporate their required specifications.

The new 14-ton vehicle that came forth from American-La France rode on a modified, Christie-designed suspension system, and was designated the convertible medium tank T3E2. It was wider than the Christie-designed vehicle by a few inches, which allowed for a crew of four, two in the front hull and two in the turret.

The prefix "E" following a vehicle designation meant that an experimental modification was made to the base vehicle, and E1, E2, and so forth indicated additional experimental modifications added to the base vehicle, but not significant enough to warrant a new model number.

The main armament of the T3E2 was a combination gun mount that incorporated a 37mm main gun and a coaxially mounted .30 caliber machine gun. There were also three more .30 caliber machine guns mounted in the vehicle's turret, one on each side and one in the rear. Testing of the T3E2

uncovered a series of minor problems that were addressed with modifications to the vehicle design. Reflecting these minor modifications, the T3E2 became the T3E3.

The biggest problem that could not be overcome with both the Christie-designed and the American-La France-designed tanks was the inability of their clutch-brake steering systems to deal with the high speeds the vehicles were capable of achieving. With this problem in mind, the Ordnance Department, between 1935 and 1936, decided to have a go at the idea of a convertible medium tank by coming up with a vehicle equipped with a controlled differential steering system.

The new four-man vehicle with a controlled differential steering system was designated the convertible medium tank T4, and 16 of these 13.5-ton vehicles were built, while three examples of a turret-less version, the T4E1, were also constructed. Power for these medium tanks came from a gasoline-powered, air-cooled radial aircraft-type engine. An advantage of air-cooled engines for tank designers was the elimination of the plumbing intricacies and the higher weight of water-cooled engines.

Despite testing showing the T4 and T4E1 to be underpowered, the U.S. Army wanted to standardize the design in 1936. This was rejected by the U.S. Army's Adjutant General's Office because the offensive power of these machine-gun-equipped tanks was not much more than the current machine-gun-equipped light tank M2 series just entering service, yet they cost twice as much to build.

With the fear of another major war in Europe breaking out, the U.S. Army went ahead in March 1939 and standardized the T4 and T4E1 as the convertible medium tank M1, limited standard. The 18 pilot vehicles would remain in service until March 1940, when they were declared obsolete. A pilot vehicle is intended to prove that the manufacturing line works and is able to actually produce the vehicle in numbers.

The American-La France convertible medium tank T3E2 and the follow-on T3E3 were hamstrung by their clutch and brake steering systems that proved unable to deal with high speeds. The convertible medium tank T4 seen here had formed part of the collection of the former U.S. Army Ordnance Museum and was fitted with a controlled differential steering system that was felt better suited for high-speed operation. (Neil Baumgardner)

LET'S TRY AGAIN

In late 1936, the Ordnance Department pushed ahead with development of the five-man medium tank T5 that would weigh no more than 15 tons. In the 1930s, the U.S. Army broke down its tanks by weight. A light tank was a two-man vehicle that could be transported by a tank carrier (truck). A medium tank was a vehicle weighing not more than 25 tons but too heavy or too large to be transported by a tank carrier. A heavy tank was any vehicle over 25 tons in weight.

The new medium tank T5 would be powered by a seven-cylinder Continental Motors gasoline-powered, air-cooled radial engine that developed 268hp at 2,400rpm. The vehicle's transmission was located in the front lower hull and was connected to the rear hull-mounted engine by a propeller shaft.

The T5 also had to meet the service requirements specified by the Infantry Board. Rather than design a new medium tank from scratch, the Ordnance Department decided to simply enlarge the chassis of the successful light tank M2 series, which featured the Ordnance Department-developed vertical volute spring suspension (VVSS), which had first appeared on the combat car T5 in 1934. The VVSS on the medium tank T5 consisted of three VVSS assemblies, each fitted with two small bogie wheels, on either side of the vehicle's lower hull rather than the two seen on the light tank M2.

Another important part of the suspension system that came with the decision to base the medium tank T5 on the light tank M2 series was the Ordnance Department's development of a very successful and durable track system made up of smooth rubber blocks (also known as pads) vulcanized around steel links held together by rubber-bushed track pins under tension. This type of track system is referred to as "live track." Untensioned track is called "dead track" and generally hangs loosely. The main advantage of dead track is that it is cheaper to manufacture and easier to repair than live track.

Besides the turreted convertible medium tank T4, there was a non-turreted version of the vehicle seen here that was designated the convertible medium tank T4E1. As with the T4, the T4E1 was armed only with .50 caliber and .30 caliber machine guns. (Patton Museum)

MEDIUM TANK T5

In order to determine the optimum mixture of weapons for mounting on the T5, a prototype appeared in 1937 that was referred to as the medium tank T5, Phase I. On the top of the lower armored hull of this vehicle was a box-like soft steel upper hull. Upon this soft steel upper hull a wooden mockup of another box-like level was installed, which featured a .30 caliber machine gun mounted in each corner. On top of this level was a wooden mockup of a small turret armed with a dummy 37mm main gun.

The Infantry Branch liked what they saw with the medium tank T5, Phase I and gave the green light for additional development. This eventually led to the replacement of the wooden superstructure and turret on the vehicle with a new soft steel superstructure and turret. The vehicle in this latter form now mounted up to eight .30 caliber machine guns.

The mounting of eight machine guns on the medium tank T5, Phase I, reflected the mindset – only two years before the start of World War II – that the tank would be employed in World War I-like scenarios, slogging across "No man's land" firing at German soldiers hiding in trenches in support of U.S. infantry, rather than fighting enemy armor.

The soft steel turret on the T5, Phase I retained a dummy 37mm main gun, as the Ordnance Department-developed 37mm tank gun M3 was not yet ready for production. At some point, the dummy 37mm main gun was replaced with twin 37mm guns M2A1 in mount T10 until the 37mm tank gun would be in production. Testing of the T5, Phase I began in February 1938 and went well enough that the vehicle was standardized as the medium tank M2 in June 1938.

Despite the standardization of the medium tank M2, the Ordnance Department continued to explore enhancements to the T5 design with the emphasis on improved armor protection and automotive performance that could be obtained without exceeding a weight limit of approximately 20 tons. Another slightly different

Shown is the medium tank T5, Phase I, with the mockup wooden upper hull and turret equipped with a dummy 37mm gun. It was the approval of the development of the T5 in May 1936 that eventually led to the production of the M4 series medium tanks beginning in 1942. (Patton Museum)

prototype was eventually built and designated the medium tank T5, Phase III. Due to its increased weight, the vehicle was fitted with a nine-cylinder Wright gasoline-powered, air-cooled radial engine that produced 346hp at 2,400rpm. There was never a medium tank T5, Phase II as that had been reserved for a design study.

The most interesting modification to the T5, Phase III prototype vehicle was "a proof of concept" mounting of a 75mm howitzer M1A1 in the right front of the vehicle's upper hull and the replacement of the original 37mm main-gun-equipped turret with another turret design equipped with a rangefinder and a .30 caliber machine gun. In this revised configuration the vehicle was designated the medium tank T5E2. Testing proved that it was a workable arrangement for a medium tank. The designation medium tank T5E3 was reserved for a proposed mounting of a diesel engine in the vehicle.

Part of the collection of the former Patton Museum was this example of the medium tank T5, Phase I. This particular vehicle was armed with twin 37mm guns. A curious feature of the T5 and later M2 series medium tanks was the installation of bullet deflectors on the rear corners of the hull. Presumably, the fire of the rear sponson machine guns could be deflected down and into any enemy trenches the tank might cross. (Richard Hunnicutt)

Even though the medium tank T5, Phase I, was standardized as the medium tank M2 in June 1938, the Ordnance Department continued to experiment with the design and eventually came up with an improved vehicle, designated the medium tank T5, Phase III. As a proof of concept it eventually mounted a 75mm howitzer M1A1 in its right front hull as seen here, and was referred to as the medium tank T5E2. (Patton Museum)

MEDIUM TANK M2 AND M2A1

Whereas the prototype T5, Phase I had been fitted with a turret mounting the twin 37mm M2A1 main guns, the 19-ton medium tank M2 had the turret-mounted 37mm tank gun, M3, installed. The machine gun arrangement on the T5, Phase I was retained with the M2. A more powerful, nine-cylinder Wright gasoline-powered, air-cooled radial engine was installed in the tank.

Unlike the non-ballistic soft steel construction of the T5, Phase I upper and lower hull, the M2 featured face-hardened armor (FHA) plates riveted together for both its upper and lower hull. FHA is normal steel armor plate put through an extra heating process to harden its outer surface while retaining the ductility of the original armor plate. Although standard, blunt-nose steel armor-piercing (AP) projectiles would often shatter on impact with FHA plate, the widespread introduction of armor-piercing capped (APC) projectiles by the various combatants during World War II would eventually lead to its falling out of favor.

Reflecting the budget constraints still in place at the time, the U.S. Army only ordered 18 units of the M2 with funds authorized for fiscal year 1939. Actual production of the tank began in the summer of 1939. Another 54 were authorized for fiscal year 1940. Tests conducted with the early-production examples of the M2 led the U.S. Army to order a number of improvements be included in the 1940 production vehicles.

The most noticeable external change to the improved M2 was the replacement of the original turret, which had sloping sides, with one that had vertical sides, to increase the interior working space for the crew. Armor thickness averaged 1.26 inches (32mm) with the gun shield being 2 inches (52mm) thick.

The improved M2 was fitted with a more powerful nine-cylinder Wright Aeronautical-designed, gasoline-powered, air-cooled radial engine designated the R975 EC2 that produced 400hp at 2,400rpm. The R975 EC2 engine was actually built by Continental Motors under license from Wright Aeronautical. Reflecting these changes and many others to the vehicle's configuration, the vehicle was re-designated as the medium tank M2A1. It was at this time that events in Europe would begin to influence the continued series production of the M2A1.

Two medium tank M2s undergo training at Fort Benning, Georgia, in early 1940. The M2 was armed with a single, high-velocity 37mm gun in place of the twin 37mm guns of the medium tank T5, Phase I. An identifying feature of the M2 was the sloped armor on the turret. Production of the vehicle began in the summer of 1939 at Rock Island Arsenal, and ended that same year after only 18 units. (Patton Museum)

In December 1940, the medium tank M2A1 replaced the medium tank M2 on the production lines at Rock Island. The M2A1 featured improvements, such as a larger turret, increased armor protection and better engine output. The M2A1's turret can be readily distinguished from the M2's by its vertical sides. Pictured are two 1st Armored Division M2A1s during a training exercise. (Patton Museum)

THE BIGGER PICTURE

The successful German invasion of Poland in the fall of 1939 followed by the speedy conquest of France and the Low Countries in the summer of 1940 was a massive shock to the U.S. Army, as France's army was considered the best in Europe at the time. It was this jolt that helped some of the more far-sighted senior officers of the U.S. Army to overcome the branch rivalries that had so hobbled the development of tanks during the interwar period, and allowed for the forming of a separate "Armored Force" on July 10, 1940, two weeks after the fall of France.

With the forming of the Armored Force, the term "combat cars" for the Cavalry Branch was done away with, all combat cars being referred to as what they had always been: light tanks.

The downside of the forming of the Armored Force was the fact that it was established as a "service test" and not officially authorized as a branch of service by the United States Congress. Although this provided the U.S. Army a great deal of flexibility in organizing it as it saw fit, it also planted the seeds of its demise. The more powerful bureaucracy of the Army Ground Forces (AGF), which was responsible for the organization and training of the U.S. Army's ground combat elements, would eventually take over its responsibilities and reduce it to nothing more than a powerless advisory board. The Armored Force was later renamed the "Armored Command" in July 1943, and eventually became the "Armored Center" in February 1944.

The medium tank M2A1 modified the original design of the medium tank M2 with the addition of gun shields to the sponson-mounted .30 caliber machine guns. Three horizontal bullet splash guards were also added to the glacis of the vehicle. (Patton Museum)

GETTING THEM BUILT

The fall of France helped overcome the reluctance of the U.S. Congress to become involved in the affairs of Europe and led them to authorize sufficient funding for the U.S. Army to rebuild after decades of stagnation in the interwar period. Among the many items of military equipment the U.S. Army believed would be of crucial importance for any future overseas conflict was a fleet of modern medium tanks to supplement the light tanks already in series production.

Aware that the Rock Island Arsenal could never build the anticipated number of medium tanks needed by the U.S. Army, William S. Knudsen – appointed by President Franklin Roosevelt in May 1940 as head of the government's military production program – met with K. T. Keller, President of Chrysler Corporation in early June 1940. Knudsen wanted to know if Chrysler could apply the mass production techniques it used to build cars to build medium tanks. When his request was met with an affirmative, he had Rock Island Arsenal, which was then building three pilot models of the medium tank M2A1, ship their blueprints for the vehicle to Chrysler's offices in Detroit, Michigan, where they arrived on June 17, 1940.

Following the delivery of the blueprints for the M2A1, a hard-working team of Chrysler employees labored around the clock for a month to figure out how to mass produce the M2A1. Their successful efforts resulted in the U.S. Army awarding Chrysler a contract on August 15, 1940, calling for the production of 1,000 units of the M2A1 to be delivered by August 1942.

To assist Chrysler in building the large number of M2A1s, the U.S. government sponsored the construction of the first plant built to mass produce tanks, the brand new Detroit Tank Arsenal, which was then the largest of its type in the world. While the plant would be owned by the U.S. Army, Chrysler would manage the day-to-day operations.

By the end of World War II, the Detroit Tank Arsenal had built 22,234 tanks. Following in second place was American Car and Foundry, which constructed 15,224 tanks. In third place was the Fisher Tank Arsenal, which managed to complete 13,137 tanks by 1945.

Shortly after the contract for 1,000 units of the M2A1 went to Chrysler, reports were received from overseas military observers which indicated that the M2A1 was already obsolete by European standards in both armor protection and firepower. The U.S. Army reacted by ordering production of the M2A1 to cease. Only 94 would eventually be built as training tanks.

What had come to the attention of the overseas military observers was the German Army's fielding of a medium tank, designated the Pz.Kpfw. IV, which featured a low-velocity, short-barreled, turret-mounted 75mm gun designated the 7.5cm Kw.K. 37 L/24. The initial versions of the Pz.Kpfw. IV rolled off the production line in 1936 and first saw combat during the German conquest of Poland in the fall of 1939.

Even as they trickled off the production lines at Rock Island Arsenal in 1940, it was recognized that the medium tank M2A1 with its thin armor and 37mm gun was already obsolete as a main battle tank. In June 1940, work began on a new medium tank design incorporating heavier armor and a 75mm main gun. Exactly one year later, the medium tank M3 seen here came off the production line. (TACOM Historical Office)

MEDIUM TANK M3

In answer to the threat posed by the German Army's Pz.Kpfw. IV, the Ordnance Committee and General Adna R. Chaffee, Jr., the first commander of the Armored Force, decided what they needed was a medium tank armed with a 75mm main gun. To meet that requirement as quickly as possible, a decision was made to redesign the M2A1 to accept thicker armor and the mounting of a limited traverse 75mm gun in its right front upper hull. As this had already been done experimentally with the medium tank T5E2, it would help shorten the development cycle for the redesigned M2A1, which became the M3 Medium Tank.

With the M3, the U.S. Army was also mimicking the general design layout of the French Army Char B1 heavy tank that featured a low-velocity 75mm gun in its right front upper hull. The tank had first come off the French production lines in 1935. With the fear of a German invasion, the French government had been in touch with American firms to see if the improved Char B1 bis could be built in the United States to supplement those being built in French factories.

In place of the 75mm pack howitzer M1A1 that was fitted into the right front upper hull of the T5E2, which normally fired a high explosive (HE) round, the redesigned M2A1 would mount the 75mm gun M2. This weapon was derived from a U.S. Army low-velocity antiaircraft gun designated the M6 that never entered service.

Rounds for the never-fielded antiaircraft gun M6 had been developed around the

Like its predecessors the medium tanks M2 and M2A1, the medium tank M3 featured a turret armed with a 37mm gun. However, the M3's turret was made of cast armor, and was in turn topped off with a small cast-armor rotating commander's cupola that incorporated a .30 caliber machine gun as pictured. (Patton Museum)

ammunition of the 75mm gun M1897, which was then the standard light artillery piece in the U.S. Army inventory. The 75mm gun M1897 was the American license-built copy of a famous French 75mm field gun widely employed in World War I that was commonly referred to as the "French 75."

The mounting of a 75mm gun M2 in the right front upper hull of the redesigned M2A1 was intended from the beginning to be only a stopgap measure until such time as the Ordnance Department and civilian industry could muster the skill set required to design and build a turret large enough to house a 75mm main gun.

The redesigned medium tank M2A1 with the 75mm gun to be mounted in its right front upper hull was designated the medium tank M3. The U.S. Army awarded a contract to Chrysler for 1,000 units at $33,500 each on August 28, 1940, 13 days after the contract for 1,000 units of the medium tank M2A1 was canceled. What made the awarding of the contract to Chrysler for the M3 so unusual was the fact that the design of the tank was not even finalized at the time, reflecting the haste with which the U.S. Army wanted to build up its medium tank numbers.

Design work on the new M3, the building of a full-scale wooden mockup, and the construction of the first pilot vehicle took place between August 1940 and January 1941 at Rock Island Arsenal. Unlike the M2A1, whose lower and upper hull was constructed of FHA plates riveted together, the Ordnance Department decided to use rolled homogeneous armor (RHA) plates riveted together for the lower and upper hull of the medium tank M3.

This medium tank M3 clearly emphasizes the height of the vehicle at 10 feet 3 inches. In contrast, the German Army's Pz.Kpfw. IV medium tank was only 8 feet 6 inches high. The Red Army's T34 medium tank that appeared in 1940 was 7 feet 10 inches tall. (Patton Museum)

The switch from FHA seen on the M2 and M2A1 to RHA plates on the new M3 is explained in this passage from the U.S. Army Center of Military History publication on The Technical Services, titled *The Ordnance Department: Procurement and Supply*: "For one thing, face hardened armor was so difficult to produce and hard to machine that its use in the expanded tank program of 1941–1942 was out of the question… Homogeneous armor was not only easier to produce but could be produced either by rolling or casting. The case for homogeneous armor was further strengthened when test firing showed that, if properly sloped, it had resistance to penetration substantially equal to face-hardened armor."

Unlike FHA, RHA is essentially uniformly hard throughout its depth and has a very high degree of ductility (the property of a material that allows it to withstand large amounts of deformation before fracturing). A disadvantage of RHA is its

Prior to America's official entry into World War II, the tanks and AFVs of the U.S. Army were finished in a factory-applied lusterless (flat) olive drab. The vehicle's registration numbers, also applied at the factory, were done in blue, referred to as "blue drab," because it had the same tonal attributes as the olive drab paint and blended together. There was no form of national symbol applied to prewar American tanks and AFVs. However, a modified version of the U.S. Army Air Corps' insignia, with the red and blue colors reversed, was applied to both the tanks and other AFVs of the 2nd Armored Division in late 1941, as seen on this M3 medium tank. On the turret of the tank pictured is another form of vehicle identification adopted by the 2nd Armored Division in late 1941. It consisted of a prefix code, with the letter indicating the company the tank belonged to and the numerals being the tactical number of the vehicle. (© Osprey Publishing Ltd.)

inability to form easily other than in flat plates. Because of this, the manufacturing of complex shapes with RHA plates raises costs.

To keep the manufacturing costs of the M3 at an acceptable level, the Ordnance Department decided to dispense with the 37mm gun M5 armed turret of the M2A1, which was constructed of FHA plates riveted together. In its place, the Ordnance Department decided to use a cast homogeneous armor (CHA) turret armed with the 37mm antitank gun M6 (an improved version of the 37mm gun M5 with a semi-automatic breech) and a coaxially mounted .30 caliber machine gun. The new CHA turret on the M3 was surmounted by a small CHA armored commander's cupola armed with a .30 caliber machine gun.

Unlike RHA, which is difficult to make in anything other than flat plates, CHA can be molded into almost any shape desired, such as tank turrets and gun shields. It can also be formed in varying thicknesses in a single pour, with weight distribution allocated to sections that are most vulnerable. A disadvantage of CHA is that it must often be made thicker and heavier because it lacks the toughness added by the work-hardening process of RHA. This disadvantage is partly offset by the rounded surfaces that mark CHA, which increase the chances of incoming projectiles glancing off. For the sake of brevity, RHA, be it welded or riveted together, will be referred to as "welded armor," and CHA as "cast armor" in some areas of the text.

The riveting of the welded armor plates together to form the lower and upper hull of the M3 was not without some disadvantages. This became readily apparent during

Belonging to the Virginia Museum of Military Vehicles is this medium tank M3 that lacks the cast-armor commander's cupola. It was provided to the British Commonwealth as Lend-Lease. The British disliked the commander's cupola, because it increased the already extreme height of the tank. Consequently, late-production Lend-Lease medium tank M3s were equipped with a simple split hatch. (Michael Green)

In a dramatically staged demonstration, a medium tank M3 is seen hurtling off a wooden embankment. Most tank crews of World War II would be loath to repeat such a maneuver with their vehicles due to the physical punishment inflicted on themselves and the chassis of their vehicle. (Patton Museum)

test firings conducted at APG by the Ordnance Department. These tests occurred shortly after the Japanese attack on Pearl Harbor on December 7, 1941, which pulled the United States into World War II.

The Aberdeen firing tests showed that heavy machine gun fire drove the rivets of tanks into the vehicle's fighting compartment, which would have caused great harm to the crew and the interior components of the tank. In addition, bullet splash would enter into the vehicle's fighting compartment through the riveted joints; even worse, AP projectiles as small as 37mm would part the riveted joints at the seams. Despite these known shortcomings, riveted armor construction was retained as a variant design to accommodate manufacturers (such as locomotive works) who had riveting experience, but could neither weld thick armor plates nor cast large armor hull structures.

INTO PRODUCTION

Completion of the final design work on the M3 took place in early 1941. There was no prototype stage for the tank. By the early summer of 1941, full-scale production began at the Detroit Tank Arsenal. At first, the M3 was authorized a crew complement of seven men. That number later dropped to six when the U.S. Army decided that the driver could double as the tank's radioman, as the radio was inside the vehicle's upper hull near the driver's position. There was an interphone system installed in the tank to allow the crew to talk to each other.

The welded armor front glacis of the M3 was sloped at 30 degrees and was 2 inches (50mm) thick. The cast-armor turret on the M3 was 2 inches (50mm) thick all the way around, with the front of the turret having a slope of 47 degrees. The sides and rear of the turret had only a 5-degree slope. The near-vertical side upper hull welded armor on the M3 was 1.5 inches (38mm) thick, with a slope of no more than 10 degrees. The thickest welded armor on the various versions of the German Pz.Kpfw. IV medium tank employed in the conquest of France was a mere 1.2 inches (30mm), sloped at 29 degrees on the turret gun shield.

To meet the somewhat unrealistic demands for still more tanks by both the American and British senior political leadership, three other civilian firms were awarded contracts by the U.S. Army to build the M3 and a somewhat different version for the British Army. These included the American Locomotive Company (Alco), the Pressed Steel Company, and the Pullman Standard Car Company. All told, the four companies completed almost 5,000 units of the M3 by August 1942.

While the various civilian firms were tooling up for the building of the M3, the U.S. Army still had a pressing need for training tanks, so a contract was also awarded to the Rock Island Arsenal to build 126 units of the M2A1. Only 94 units of the tank were built before the U.S. Army canceled the contract in August 1941, as the M3 was now coming off the production line in sufficient numbers to meet their needs.

An interesting picture of the VVSS at work on this medium tank M3 as it crosses a trench and plows into a tree. The large hull access doors were seen as ballistic weak spots and were eliminated in later production units. (Patton Museum)

Not everyone believed that the M3 was worth building in large numbers. The senior leadership of the Armored Force suggested that no more than a few hundred be built until such time that a turret large enough to mount a 75mm main gun might appear. Due to the pressing need of the British Army in North Africa for still more American-built medium tanks, the Ordnance Department wanted to continue M3 production despite the objections. One objection came from Major-General Adna R. Chaffee, Jr., who described the M3 as having "barely satisfactory performance characteristics."

It was the U.S. Army's 1st Armored Division that first saw action with the M3 on November 28, 1942, when a combined American and British force attempted to capture a small German-held town and airfield in Tunisia, North Africa. During the opening attack, the inexperienced American tank crews of the M3s suffered serious losses from hidden German antitank guns. Over the next few days of fighting, the American tankers suffered additional heavy losses at the hands of the more experienced German troops.

Despite the fielding of a replacement medium tank for the U.S. Army during the fighting in North Africa, the M3 would soldier on in ever decreasing numbers until the German surrender of their North African military forces in May 1943. Major-General Ernest N. Harmon, commander of the 1st Armored Division in North Africa, stated in a memo to the Commanding General, Allied Forces, on May 21, 1943, that he considered it "criminal to send an armored division into battle where it will be heavily engaged against antitank guns and tanks with the M3 tank."

The last combat action for the gun-armed M3 series in U.S. Army service occurred during the four-day assault on the Japanese-occupied island of Butaritari, part of the Makin Atoll in November 1943. Some would soldier on with the U.S. Army after being modified for other roles, such as an armored recovery vehicle (ARV) or prime mover.

The medium tank M3 was considered an interim design. The sponson-mounted 75mm gun, which harkened back to the medium tank T5E2, was the most serious deficiency. However, the urgency of the world situation necessitated the mass production of the M3. Pictured at the Israeli Armored Corps Museum is this M3 medium tank missing the cast-armor vehicle commander's cupola. (Vladimir Yakubov)

M3 MAIN ARMAMENT

The standard armor-piercing capped tracer (APC-T) round fired from the 75mm gun M2 was designated the M61 and weighed 19.92 pounds. The projectile weighed 14.96 pounds and had a muzzle velocity of 1,930ft/sec. It could penetrate 2.4 inches (60mm) of RHA sloped at 30 degrees at a range of 500 yards. For the sake of brevity, all penetration figures quoted from here on will be based on firing at an armor plate sloped at 30 degrees from the vertical.

At a range of 2,000 yards (over a mile), the M61 projectile could still penetrate 1.8 inches (46mm) of armor. The APC-T round made FHA obsolete on the battlefield for medium and heavy tanks. The German Army switched from FHA on their medium tanks to welded armor in 1943/44 due to the widespread introduction of the APC-T round.

There was also another AP round for the 75mm gun M2 that was designated the M72 Shot Tracer (AP-T), which had the same muzzle velocity as the M61 APC-T and matched its penetration ability at 500 yards, but at longer ranges the penetration dropped a bit. At 2,000 yards, the M72 projectile could only penetrate 1.5 inches (38mm) of armor. The term "shot" refers to a projectile that does not contain an HE element.

The standard high-explosive (HE) round fired by the 75mm gun M2 of the M3 had a projectile with a muzzle velocity of 1,470ft/sec with a normal propellant charge. The round was designated the M48 Shell (HE), Normal. Another version of the same round fitted with additional propellant, referred to as a "supercharge," had a muzzle

Taking part in a training exercise in 1942 at the U.S. Army's Desert Training Center, located in southern California, is this medium tank M3. Despite it being radio-equipped as evident by the large whip antenna, the vehicle commander is using a flag to signal other tanks in his unit. (Patton Museum)

velocity that topped out at 1,885ft/sec. It was designated the M48 Shell (HE), Supercharge. HE rounds are primarily intended for unarmored vehicles, buildings, and exposed enemy personnel.

A U.S. Army report dated June 12, 1951, by Major-General Gladeon M. Barnes, from the Chief of Ordnance Department Research and Development Division during World War II, says this about the effectiveness of 75mm HE rounds: "During 1939 the writer was assigned to the technical staff of the Ordnance Department and had conducted tests at the Aberdeen Proving Ground to determine the effectiveness against infantry of machine gun versus high explosive 75mm shells. These tests showed clearly that the 75mm gun was much more effective against personnel at all ranges than machine gun fire."

Pictured on display at the Virginia Museum of Military Vehicles is a British Army medium tank M3 commonly referred to as the "Grant." The British preferred to mount the radio in the turret, and as a consequence, the turret they designed for the Grant was much larger than the U.S. Army version of the M3. (Michael Green)

The M48 Shell (HE) came with point-detonating (PD) fuze which was selectable to provide either a super-quick (SQ) fuze or a delayed (0.05 or 0.15 seconds, depending on the model) fuze. As shipped, all M48 series main gun rounds came with their fuze set for SQ. To change their fuze setting for PD, the loader turned a setting screw with the screwdriver end of a wrench, so that the slot aligned with the PD setting, which was at right angles to the axis of the fuze housing. Loaders could do this in the dark by noting the position of the slot in the setting sleeve.

The super-quick fuze was prescribed against enemy personnel in the open, where the bursting projectile was more effective if it generated its lethal blast and fragments before burrowing into the ground. Delayed point-detonating fuze allowed the projectile to penetrate targets such as the light armor on antitank gun shields or buildings before bursting. It was up to the tank commander or gunner to inform the loader what type of fuze setting he wanted before the round was loaded into the main gun's breech.

There was authorized storage for 50 75mm main gun rounds in the M3. However, tank crews of all nations during World War II often carried more rounds than authorized for fear of running out at an inopportune time when engaged in combat.

The 75mm gun M2 was also provided with a smoke round designated the M89 Shell, Smoke, to assist in marking targets or obscuring friendly forces. This round was typically referred to as a "WP" (white phosphorus). The phosphorus in the WP rounds often ignited oil and grease deposits on enemy tanks, creating a surface fire, which could burn the target or cause smoke to enter into the vehicle itself through seams and joints, forcing the crew to abandon their vehicle, thus neutralizing it.

The 75mm gun M2 had a bore length of 7 feet and was fitted with an elevation gyrostabilizer beginning with January 1942 production vehicles. A War Department publication titled *Armored Force Field Manual: Tank Gunnery*, dated April 22, 1943, explains the function of the elevation gyrostabilizer: "The stabilizer will not lay [aim]

Pictured on the assembly line is an incomplete Grant. The three American companies that built the tank included the Baldwin Locomotive Works, the Pressed Steel Car Company, and the Pullman Standard Car Company. The first Grant was completed in July 1941. (Patton Museum)

the gun. It merely tends to keep the gun where it has been laid; that is, it eliminates extremely jerky movements caused by the movement of the tank. Even with the stabilizer, the gun does not hold constantly on the target. Watch the swing of the gun through the target and fire as proper sight setting crosses the target."

Because the 75mm gun in the M2 proved to be somewhat out of balance to operate the elevation gyrostabilizer properly it was decided to add a counterweight on the muzzle end of the barrel. The 75mm gun M2 also had a semi-automatic breech. This meant the spent cartridge case was ejected from the now open breech block upon counter-recoil and the weapon returning to battery, allowing the vehicle's loader to quickly insert another round into the breech. Inserting the fresh cartridge also tripped the closing springs, which automatically closed the breech. This was a welcome improvement over the manual breeches, such as the French 75, which required a separate action (and usually a second loader) to maintain a high rate of fire.

M3 SECONDARY ARMAMENT

The turret-mounted 37mm gun M6 in the cast-armor turret of the M3 was an improved and slightly longer version of the 37mm gun M5. Unlike the breech on the 37mm gun M5 that was manually operated, the breech on the 37mm gun M6 was semi-automatic. The weapon was fitted with an elevation gyrostabilizer beginning with January 1942 production vehicles. A counterweight was added just below the gun's barrel to improve the performance of the elevation gyrostabilizer.

The M3 originally had four .30 caliber machine guns fitted. Two were in the cast-armor turret, with one mounted to fire alongside the 37mm gun M6 and the other mounted in the rotating vehicle commander's cupola. The other two .30 caliber machine guns were located in the front of the upper hull in a fixed forward-firing position; one of these machine guns was eliminated from production tanks starting in June 1942 and the empty opening filled with a steel plug. There was authorized storage space on the M3 for 9,200 rounds of .30 caliber ammunition.

M3 ENGINE ISSUES

One of the biggest shortcomings of the M3 was its inheritance of the nine-cylinder Continental Motors gasoline-powered, air-cooled radial engine from the M2A1, which was designated the R975 EC2. The disadvantages with this type of engine were numerous. Test Engineers at Aberdeen Proving Ground stated in a January 14, 1942 report that the engine was underpowered, causing the horsepower to weight ratio to be too low to provide satisfactory performance.

Reports from U.S. Army training exercises in which the M3 took part at the same time also indicated that the radial engine was "unsatisfactory as to performance and life." Some officers of the Proof Department at Aberdeen Proving Ground recommended "that additional consideration be given to other power plants with a view to increasing the horsepower to weight ratio as well as improving the accessibility."

A further disadvantage of using an aircraft-derived radial engine is that it is inherently tall, with a drive shaft that is also higher above the engine bay floor than a comparable in-line or V-type engine. The overall engine bay height therefore drove up the height of the M3 upper hull and turret. Height over cupola of the M3 was 10 feet 3 inches.

The engine drive shaft on the M3 was connected to the five-speed synchromesh transmission located in the front lower hull of the vehicle. The synchromesh transmission powered the controlled differential in the M3, which supplied power to the two final drives that turned the drive sprockets at the front of each track. The differential and final drives were contained within a large three-piece cast-armor housing that bolted together to form the lower front of the vehicle hull.

The medium tank M3A5 Grant pictured resides at the Israeli Armored Corps Museum. The larger size of the 37mm main-gun-armed turret can be compared to the original American-designed turret on the right-hand side of the picture. The American military moved its tank radios to the turret with the next generation of medium tanks. (Vladimir Yakubov)

The entire housing could be removed to service the components contained within. Armor thickness of the housing was 2 inches (50mm) and was sloped from 0 to 45 degrees.

The lack of a suitable tank engine for the M3 was not the fault of the Ordnance Department. They had foreseen such a requirement long before the vehicle entered into U.S. Army service. However, during the 1930s money for such projects was not available and civilian industry had little interest in developing a specially designed tank engine that had no commercial applications.

Despite having more than enough funding in 1942 to develop a suitable tank engine for the M3 with a higher horsepower to weight ratio, it was clear to all within the U.S. Army that there was not enough time to do so. To delay production of the M3 until the perfect engine was available was not realistic. So, despite its shortcomings, the nine-cylinder Continental Motors, gasoline-powered, air-cooled radial engine would remain the most numerous in American medium tanks during World War II.

Following the declaration of war against Japan, the prewar national symbol adopted by the 2nd Armored Division for its tanks and AFVs was done away with, as it was felt that the red circle being employed was too similar to the Japanese national flag. In its place, a decision was made to adopt, U.S. Army-wide, a five-pointed white star as the new national symbol. The exception to that rule were the vehicles of the U.S. Army Armored Force that would have a yellow five-pointed star, as that color was deemed less visible to enemy observation. The yellow stars on either side of the tank turrets were joined by a horizontal yellow bar as seen in this artwork of a U.S. Army M3 medium tank belonging to the 1st Armored Division in Tunisia, December 1942. The yellow geometric markings on the front of the vehicle's hull were employed by the division to identify each tank company. The vehicle's registration number done in blue drab would sometimes be done in white, when yellow was not available. (© Osprey Publishing Ltd.)

BRITISH ARMY VERSION

The British Army lost the bulk of its tank fleet with the fall of France in the summer of 1940. Lacking the industrial infrastructure to build sufficient numbers of replacement tanks quickly, a British Tank Mission headed by Mr. Michael Dewar traveled to the United States to see if it was possible to arrange for British-designed tanks to be built in American factories, for which the British government was willing to pay cash.

Once in America, the U.S. Army quickly informed their British visitors that the only tanks for which they could contract were those acceptable to the U.S. Army. This was motivated by an unspoken fear that, should Britain fall to the Nazis, the Americans would be stuck with factories configured to manufacture tanks for which the U.S. Army had no use. However, there would be some modifications allowed to meet British Army requirements. The British Tank Mission therefore agreed to accept a modified version of the medium tank M3 and placed an order for almost 3,000 units with three different American manufactures.

Because British factories proved unable to build a sufficient number of tanks for their own forces, the British government received permission from the United States government to have a modified version of the M3 series of medium tanks built for use by their own army. The British-ordered M3 series medium tanks, named the "Grant," arrived in North Africa in the standard U.S. Army olive drab. As seen here, the British Army quickly repainted over the olive drab with a more appropriate light tan color known as Light Stone (British Standard Colour No. 61). There would be variations between units in the manner of its application. The tank's registration numbers, seen here in white, remain on their original olive drab background. Also visible on the vehicle pictured, belonging to the 22nd Armored Brigade, in May 1943, are the red tactical markings employed by the unit with which it was assigned to serve. (© Osprey Publishing Ltd.)

Two weaponless Grants being shown being prepared for rail shipment to a port from whence they will be shipped overseas for service with the British Army. The storage boxes on the upper-rear hull of the Grant differed in shape and size from those seen on their U.S. Army counterpart. (Patton Museum)

The expense of the M3s for the British Army was later offset by the "Lend-Lease Act" of March 11, 1941, which provided the British government military aid for its armed forces at no cost. The Lend-Lease Act was formally titled "An Act to Promote the Defense of the United States."

The biggest visual difference between the M3s constructed for the U.S. Army and the M3s being built for the British Army was their respective turrets. The British Army version of the M3 featured an American-made but British-designed cast-armor turret that was lower, larger, and roomier than the American version. While the British-designed cast-armor turret retained the American 37mm gun M6 and coaxially mounted .30 caliber machine gun, it did away with the machine-gun-armed commander's cast-armor cupola fitted to the U.S. Army's M3s.

The British Army felt the vehicle commander's cupola on the American M3 raised the already tall profile of the tank to an unacceptable height. In place of the commander's cupola, the British-designed turret featured a rotating two-piece split overhead hatch, one of the hatch portions having a rotating 360-degree periscope fitted. The height over the periscope of the split overhead hatch in the British Army version of the M3 was 9 feet 11 inches.

The British Army had been allowed to ask for a different turret design for their version of the medium tank M3 as it was their policy to have the vehicle's radio in the turret with the tank commander. The U.S. Army's policy at the time called for the vehicle's radio to be located in the superstructure of the tank. To make room for the tank's radio and crew the British-designed turret was much wider than its American-designed counterpart and 4 inches lower. Other differences between the American and British versions of the M3 were the interior stowage arrangements. As the British Army version of the M3 was going to North Africa they were also fitted at the factory with sand skirts on their fenders.

The British Army typically attached names to their tanks. They did the same with all of their American-acquired tanks, naming most of them after well-known American Civil War generals. The M3 with the British-designed turret was officially named the "General Grant I," normally shortened to just "Grant" by British tankers. As the British Army also received the standard U.S. Army version of the M3, the vehicle was assigned the name "General Lee I," typically shortened to just "Lee." It was British Prime Minister Winston Churchill's decision in August 1942 to drop the prefix "General" from the names of the American-supplied tanks and refer to them by the last names of the American officers only.

Pictured is a Grant tank recovered from a firing range. The gun shield on the front of the sponson-mounted 75mm gun is missing from this particular vehicle as is that for the turret-mounted 37mm gun. Despite the limited traverse of the sponson-mounted 75mm gun, when introduced into combat by the British Army in North Africa in May 1942, the Grant's firepower came as something of a shock to the Germans. (Christophe Vallier)

THE M3 IN FOREIGN SERVICE

The British Army M3s first saw action in North Africa during the Gazala battles of May 1942. Their front hull-mounted 75mm guns outranged the turret-mounted main guns of their German medium tank counterparts. The 50mm main gun on the majority of the Pz.Kpfw. III medium tanks encountered was designated the 5cm Kw.K. 38 L/42. It was 6.89 feet long and fired an AP projectile designated the Pzgr. 39 that had a muzzle velocity of 2,247ft/sec. It could penetrate 2.12 inches (54mm) of armor at 109 yards, and at 2,187 yards 0.87 inches (22mm) of armor.

A few of the Pz.Kpfw. III medium tanks that would encounter the British M3s in North Africa were armed with a longer-barreled 50mm main gun designated the 5cm Kw.K. 39 L/60, which was 9.85 feet long and fired a standard AP projectile designated the Pzgr. 39 with a muzzle velocity of 2,740ft/sec. The projectile could penetrate 2.64 inches (67mm) of armor at a range of 109 yards, and 1 inch (26mm) of armor at a range of 2,187 yards.

The early model Pz.Kpfw. IV medium tanks that also took part in these first encounters with the British M3s were armed with a 75mm main gun designated the 7.5cm Kw.K. 37 L/24, which was only 5.9 feet long. Such a short barrel was originally intended to deliver HE shell at relatively short ranges against fortified positions and personnel. Its only useful armor-defeating round was a shaped-charge projectile with a muzzle velocity of 1,476ft/sec. The shaped-charge projectile could penetrate 1.6 inches (41mm) of armor at 109 yards, and 1.2 inches (30mm) at 2,187 yards. The sloping of armor does not affect the terminal effects of shaped-charge warheads.

Making a bad situation even worse for the German tankers encountering the British M3s for the first time in May 1942, the main guns on their medium tank were

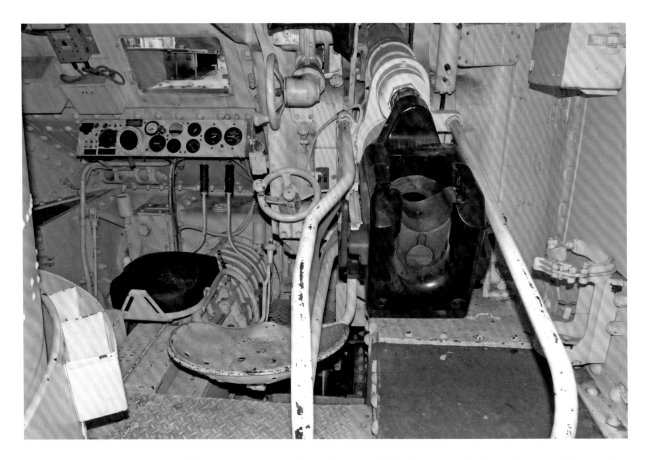

Looking into the front hull compartment of the Virginia Museum of Military Vehicle's Grant tank, one can see the gunner's seat with his elevation and traverse manual controls. Projected out behind and around the breech end of the 75mm gun is the recoil guard. (Paul Hannah)

unable to penetrate the frontal armor of the American-built tanks even when within normal combat ranges. In contrast, the APC-T projectiles fired from the 75mm gun M2 mounted in the British M3s easily penetrated the frontal armor on the German medium tanks they engaged at ranges up to 1,531 yards. German General Erwin Rommel, commander of the German ground forces in North Africa, wrote in his diary, regarding meeting the British M3s in combat for the first time that "the advent of the new American tank has torn great holes in our ranks. Our entire force now stood in heavy and destructive combat with a superior enemy."

Unfortunately, the combat domination of the British M3s did not last that long, as the German Army began fielding an up-gunned and up-armored version of their Pz.Kpfw. IV medium tank in the fall of 1942, known to the British Army as the Mk IV Special. The German Army designated it as the Pz.Kpfw. IV Ausf. F2. The German term "Ausf." refers to the version of the vehicle.

Armor thickness on the Pz.Kpfw. IV Ausf. F2 was beefed up to 2 inches (50mm) on the front of the vehicle's turret and hull. It was armed with a long-barreled 75mm main gun that was 10.6 feet long and designated the 7.5cm KwK. 40 L/43, which fired a standard APC-T projectile designated the Pzgr. 39 with a muzzle velocity of 2,428ft/sec. German Army reports from that time stated that this projectile could penetrate the frontal armor of the British M3s at a range of 1,640 yards – almost a mile. The British Army armored units fighting in North Africa in the fall of 1942 and spring of

Grants – Fire Now!!!

During World War II Lieutenant-Colonel George Witheridge served with the 3rd Royal Tank Regiment, 4th Armored Brigade, 7th Armored Division. In his letter to author Richard Hunnicutt he describes the events of May 27, 1942, when the M3 Grant tanks of his unit first engaged their German counterparts in North Africa:

"We ran for about ten minutes in the ordered direction, which was towards the Bir Hacheim area. The Stuarts (American M3 light tanks) well out ahead at about 2,000 yards; Cyril's Grant Squadron, on the left of the line, 'Pip,' who at all times was in the thickest of any show which ever it was, a tank battle or horse play in the officer's mess was in the center of the line and my squadron on the right. Except for the outposts of Bir Hacheim there was nothing between us and the approaching fast mobile enemy, who, as yet, was still out of sight to us. We believed that we would clash with a reconnaissance force in strength. There was no information to the contrary. Brigade told us that it would not be a serious threat to us.

"If there was nothing between us and the enemy, what had happened to the two Motor Brigades southeast of Hacheim, I thought? Surely there is something wrong. Some information is missing. Now as I reflect, I believe this earlier message of the day was correct in every detail and that at least one German armored division was turning our southern flank…

"Vast dust clouds were now to be seen ahead. Wireless messages came in fast, establishing that many enemy tanks were moving with purpose – never have I seen such confidence for they moved en mass[e] without protection front; not using the eyes and mobility of lighter armored vehicles ahead. They came 'flat out.' Here were the Panzer III and IV we knew so well. They came on without reducing speed and on a wide front at close interval of about 30 yards between tanks and in depth beyond vision. The latter was not entirely due to sand storms churned up by movement of hundreds of tanks, but to the enormous area these tanks occupied in depth. I could not possibly see them all.

Two British soldiers watch as a Grant tank is driven onto a railroad flatcar. It was the 75mm gun on the Grant that proved to be the most useful for the British Army in North Africa due to its ability to fire a very effective HE round against German-towed antitank guns. (Tank Museum)

"Gradually, I could see more clearly the leading tanks, each carrying infantry, clinging like flies on something sweet. Our light squadron by this time had been ordered not to engage the enemy but to protect the regiment's right flank. The desert was flat and wide open, with no hull down positions to be found anywhere. There was not even time to break to the right to take up positions on the enemy's flank. Heavens, a whole enemy armored division in tight formation was almost upon us…

"In our gunnery training we had zeroed our guns at 1,000 yards, and we knew that our shooting would be accurate and that we could penetrate most of the heavy frontal armor with our new 75's at that range; while we could also create havoc amongst all their tanks by rapid fire from our 37mm guns into their armored mass. Both the 75mm and 37mm were told to stand by to engage the targets as indicated. I thought to myself – thank heavens these are American guns as they can fire indefinitely, without the frequent necessity of 'topping up' the buffers with oil which was essential when firing the 2-pounder (40mm) gun with which British tanks had been equipped. As long as my tanks remained in action, they must, and I knew they would, keep those guns working smoothly-but-faster than ever before because the occasion called for extreme attention to danger from without and the need to destroy before being destroyed. We must stop this headlong German drive to the north.

"'Hullo Cambrai – Cambrai calling – Fire now!' 'Cambrai out.' This was the order for 75mm guns to open fire, the enemy then being within 1,000 yards of our guns. Immediately my small world seemed to vanish in the madness of the moment. We caused chaos in the German ranks, here plain to see were Panzers turning desperately to avoid the hail of death-dealing 75mm solid shot from the Grants, soon to be augmented with shot from the higher velocity 37mm guns. Some German tanks appeared to run into each other and the infantry clinging to their hulls were being thrown off their backs. Three German panzers were on fire, crews bailing out and great indecision seemed to reign. They were brought to a halt at about 900 yards. Now the famous Chestnut troop of 25-pounders, which had been close behind us, too close perhaps, joined in the slaughter of the panzers. Colored tracers from the several guns crisscrossed the space between the contestants, the air

was full of flying metal; noise and confusion and brave men were dying.

"A flash inside my tank – ammunition on fire – would explode any moment – I had experienced being 'burned up' in other battles and knew at most I had but about two seconds to get my crew out. 'Bale out' I yelled. Like lightning my crew were out and taking cover behind the tank, I then explained the bearing and direction they should take to the brigade's center line. Then singly they moved off trying to dodge the frightening tracers of armor-piercing shot which flew in all directions.

"Mounting the next tank I was just about to throw a leg over the cupola when there was a terrific crash and the tank burst into flame with greater fierceness than the one I just left. My other tank was already glowing red along its side. Bursting ammunition from within was popping out the cupola soaring into the sky like fireworks. Soon, this, my second tank would be doing the same. I then ran to the third tank where the crew was working at tremendous speed pumping shot from both guns. The gunner with the 75mm was having his own battle fight, with but an occasional correction from me, which was only necessary when I saw a particularly dangerous enemy – dangerous to me and my crews. The panzers had recovered from the first shock and were moving forward again but bearing away slightly to our right flank. Targets abounded.

"Wireless again. 'Hullo Cyril, Cambrai calling, why are you moving back, over.' Looking to my left I saw Cyril's squadron or what was left of it, moving slowly in reverse. No answer from Cyril. 'Hello Cyril, tell your chaps to switch off and stay where they are, over.' 'Hullo, Cambrai, Cyril answering. Sorry but I can't see a dammed thing for blood in my eyes. I'm out of bricks (ammunition) and my turret gun damaged – over.' 'Hullo, Cyril, Cambrai, O.K. well done – carry on.'

"Many of our tanks were silenced, some burning to my right and left. The whole situation seemed that we were about to be overrun and trampled by the mass of German armor, which by now appeared to be speeding up their move leaving many of their dead tanks behind.

"Now the shooting was even better for us as the enemy tanks gradually steered away exposing their thin flanks. We had just 19 Grants, 16 Stuarts at the start, since we were under strength in tanks due to the need to share the Grants with other deserving regiments. The German tanks, as they went past us were being hampered by our Stuart light tanks, like terriers against wild boars. The range was closed to 300–400 yards on our right as the Germans swarmed past on their drive to the north. They had now made a complete envelopment of the Bir Hacheim area, Eighth Army's southern end of the Gazala Line.

"The inter-com crackled again. '37mm traverse right, 400 right, right – steady fire.' Over the vane sight I had laid the 37mm gun in the turret onto the leading German tank which we hit – it stopped – then immediately began to move again and continued on its way. The heat and flame shot past my face and the next thing I remember was being pulled off the engine covers to the rear of the turret. Again, I told the men where to go to reach the brigade center line to our rear.

"How long it took me to recover I will never know, but as soon as the crew were away I looked for another tank to mount. That on my right had a shattered side plate. The tank on my left was on fire then for some unaccountable reason I started to look for Cyril Joly, the Commander of C Squadron, a young officer whom I looked on more as a father, although the years in age between us was no more than ten years. I ran to his tank and removed the Homolite fuel container which was on fire, left front of the turret. Then I saw that his 37mm gun had been knocked askew in its mount and that the tank had been hit by 50mm shot. It had received many hits, none of which had penetrated its frontal armor. Many had hit the turret, again without penetration.

"Cyril was out on his feet. A 50mm round had parted his hair putting him out of action for some time. I sat on the glacis plate and directed the driver to reverse the tank out of battle – then I got the crew together, directing them to what I had hoped would be safety. They carried Cyril with them. Doing this I had reversed some one hundred yards so I went back towards our other tanks. Then, either through confusion or tiredness or both I found myself trying to avoid the projectiles, whose tracers indicated their paths over the whole area, by attempting to jump over them. Of course this was a ridiculous action but I found it hard to stop as I forced myself to return to my tanks. All but five remained out of the total. I climbed up on a tank next to Pip's my C.O. and took up the fight again, but soon realized that Rommel's boys had either had enough of us or they decided to spare us for they flowed past on our right flank… For every penetration on a Grant there were twenty to thirty which did not. Crew members of dead tanks had either burned in their tanks or were taken away. Most of our wounded had got away but three still remained. After an injection of morphine and first aid we made them as comfortable as possible, while awaiting assistance from brigade."

1943 gradually had their M3s replaced with the next generation American-designed and built medium tank. Some of the British gun-armed M3s modified to serve as command vehicles during the fighting in North Africa went on to see service with British Army units serving in Italy.

The Australian Army would eventually receive 757 units of the M3 series, both the American and British versions, gasoline- and diesel-powered. According to Paul Handel of the Australian Army Tank Museum, the American version of the M3 that arrived in the country had their original cast-armor vehicle commander's cupola replaced with the two-piece overhead vehicle commander hatch seen on the British version of the M3. The Australian Army did not think the American version of the M3 was suitable for combat and reserved them only for training duties.

Both the American and British versions of the M3 series would also see service with the British Fourteenth Army during the Burma campaign. It was during this campaign that the American version of the M3 was first seen with its machine-gun-armed cast-armor vehicle commander's cupola removed and replaced with the same rotating two-piece split overhead hatch seen on the British Army version of the M3.

Besides the British Army and its Commonwealth counterparts, the largest user of the M3 series was the Red Army, who obtained 1,386 of them under a Lend-Lease program. The Brazilian Army would receive about one hundred M3 series tank under Lend-Lease but did not deploy them overseas when their forces supported the Western Allies during the fighting in Italy.

EXPERIMENTING WITH THE DESIGN

Even as the medium tank M3 rolled off the various production lines between 1941 and 1942, the Ordnance Department and civilian industry continued to tinker with the vehicle's design to see if they could come up with ways of improving the product as well as speeding up production and lowering costs. Improvements to the M3 were both internal and external, with almost 3,000 minor changes being made per month.

The biggest exterior change to the M3 was doing away with the large armored doors on either side of the vehicle's upper hull, which proved to be weak spots during ballistic testing. In place of these armored doors later production units of the M3 were fitted with an escape hatch in the floor of the tank's lower hull.

Another easily noticed external change to late-production units of the medium tank M3 was the replacement of the original 75mm gun M2 with the longer-barreled 75mm gun M3, which was 9.22 feet long. This changeover had been planned from the beginning of M3 production. The 75mm gun M3 was not a new tank gun but merely a lengthened version of the original 75mm gun M2.

On display at the Tank Museum in Bovington, England, is what is thought to be the first Grant made by the Pressed Steel Car Company. The main guns on the M3 medium tank series were gyrostabilized in elevation. This provided the crews a limited ability to fire accurately on the move. Since any change in the terrain or direction would throw off the stabilizer system, it was seldom employed in combat. (Tank Museum)

The longer-barreled M3 gun pushed up the weapon's muzzle velocity with the M61 APC-T projectile to 2,030ft/sec. The M61 APC-T projectile fired from the 75mm gun M3 could penetrate 2.6 inches (66mm) of armor at a range of 500 yards. At a range of 2,000 yards (over a mile), the projectile could still in theory penetrate 2 inches (50mm) of armor. The M61 APC-T was designed from the beginning to have a small HE element. However, production delays meant that it was not fitted until late in World War II.

When the HE filler started showing up in the M61 APC-T rounds, it came with a base-detonating fuze that was supposed to detonate after penetration of a target had occurred. Conventional point-detonating fuzes affixed to the nose of the projectile would have been destroyed during initial impact and penetration, so the base-detonating configuration was necessary to ensure survival of the fuze until penetration was complete. With the combination of the HE filler and base-detonating fuze, the M61 APC-T round became the M61 APC-HE-T. A later improvement, fitted with a different fuze, became the M61A1 APC-HE-T.

Because the 31-ton medium tank M3 suspension system was based on the suspension system of the 11-ton light tank M2 series, there were numerous suspension component failures. To correct this situation, the VVSS on the M3 was eventually redesigned and strengthened. The external features that reflect this change to a more heavy-duty suspension system appeared on some very late-production M3s, with the relocation of the track return roller from the top of the VVSS assemblies to a bracket extending rearward from the assemblies. In addition, a steel track skid was now fitted on the top of the VVSS assemblies in place of the track return roller.

In an effort to speed up production of the M3 series of medium tanks, a number of techniques were explored, including the replacement of the welded-armor hull with a cast-armor hull seen here. In this configuration, the tank was designated the medium tank M3A1. The example seen here was on display at the Aberdeen Proving Ground, Maryland. (Michael Green)

NEW VERSIONS

Based on their success with making the cast-armor turret for both the U.S. Army and British Army versions of the M3, the Ordnance Department and industry went ahead and came up with a version of the vehicle that featured a cast-armor upper hull and a riveted welded-armor lower hull. The development of a cast-armor upper hull did away with the ballistic vulnerabilities posed by using riveted construction in the upper hull of the M3. Reflecting this design change to the vehicle, this model was designated the medium tank M3A1. The prefix "A" followed by a number indicates the first model difference to the vehicle's design. Thus, medium tank M3A1 denotes the first design variant of the medium tank M3.

Firing tests conducted at Aberdeen Proving Ground with early-production units of the M3A1 showed excellent results and led to the Ordnance Department approving the construction of cast-armor upper hulls as an alternate type for the M3. However, the biggest problem turned out to be an insufficient number of foundries capable of making the large upper hull castings for the M3A1. This production bottleneck resulted in only 300 of the M3A1s being built between January 1942 and July 1942.

Of the 300 M3A1s built, 272 units retained the same nine-cylinder Continental Motors gasoline-powered, air-cooled radial engine found in the M3. The other 28 had been fitted experimentally with a Guiberson T-1400-2 diesel engine that sadly did not live up to expectations and was soon dropped from consideration.

The Ordnance Department had conducted ballistic tests with a proposed welded armor hull for the M3 and liked what they saw, so they contracted with Rock Island Arsenal to build two pilot units of an M3 with a welded-armor lower hull and upper hull, one of which was fitted with the standard cast-armor turret. The Ordnance Committee had this version of the vehicle designated the medium tank M3A2 in August 1942. It retained the same gasoline-powered, air-cooled radial aircraft-type engine found in the M3.

DIFFERENT ENGINES

Almost as quickly as the M3A2 appeared, its production was canceled, with only 12 having been built between January and March 1942. This was due to the fact that the Ordnance Department decided to power its M3s that had a welded-armor lower hull and upper hull with two 12-cylinder General Motors G-71 liquid-cooled diesel truck engines mated together, the arrangement was designated the 6046. This version of the M3 series was designated the medium tank M3A3, with 322 units built between March and December 1942. Forty-nine of the M3A3s went to the British Army, in whose service it was referred to as the "Grant II."

Because the two joined liquid-cooled diesel engines took up more room than the single air-cooled gasoline engine, the entire rear hull of the M3A3 had to be lengthened and redesigned. Since not enough M3A3s were coming off the production lines, it was also decided to mount the twin-diesel engine arrangement in a modified version of the original M3 with the riveted lower and upper hull. These vehicles were designated the M3A5, and 591 units were built between January and December 1942. One hundred eighty-five of these tanks were provided to the British Army and were lumped together with the M3A3 and referred to as the "Grant II."

The twin-diesel engines in the M3A3s and M3A5s were joined at their fan ends with a heavy junction plate and at the flywheel ends with a double clutch housing. A transfer unit then transmitted power to a single propeller shaft. The drivers of the tanks had a single pedal with adjustable linkage providing uniform engagement between the two clutches. The two engines operating in conjunction had a gross horsepower of 410 at 2,900rpm.

If one engine went down, the drivers on the M3A3s and M3A5s had clutch lockout cables leading to the instrument panels that allowed them to disengage either engine in case of a failure, so that the surviving engine might operate without the drag of the other. On paved level roads, the twin diesel engines of the M3A3s and M3A5s provided the vehicles a top speed of 30mph, while on a single engine a top speed of 20mph was possible.

Tests conducted between December 1941 and April 1942 at APG between gasoline-powered M3 tanks and those versions equipped with the twin diesel-powered engine arrangement favored the diesel engines. An Ordnance Department report from 1942 stated the following regarding the twin diesel-powered engine arrangement: "The power plant has very good performance. Mileage and cruising range are better than those with gasoline engines. The cooling and starting characteristics of these engines are equal to or better than those of the standard production M3 tanks."

IMPROVISED ENGINE ARRANGEMENT

The need to explore improvised engine arrangements for the M3 series occurred because of a growing shortage of gasoline-powered, air-cooled radial engines and diesel engines due to the demands of the other services, a problem which lasted until the early months of 1943. The strangest engine arrangement for the M3 series involved coupling five Chrysler Motors commercial liquid-cooled, gasoline-powered automobile engines to a single drive shaft, with a combined total of 30 cylinders.

Referred to as the Chrysler A57 Multi-bank, this massive power plant required the lengthening of the vehicle's engine compartment by 11 inches and the rear portion of the upper hull 15 inches. To compensate for the increase in vehicle length and weight (25,000 pounds), the designers also increased the spacing between the center and rear VVSS assemblies units of the suspension system. The modified vehicle was designated the medium tank M3A4.

U.S. Army tests of the A57 Multi-bank in October 1942 demonstrated that it furnished adequate power for the M3A4 tank, with 410hp at 2,900rpm. However, continual failures of a minor nature made maintenance on this vehicle very difficult. Nonetheless, Chrysler received a contract to build 109 units of the M3A4, which it completed between June and August 1942. All these tanks remained in the United States as training vehicles during World War II.

In an extract from a July 1943 U.S. Army report by Brigadier-General John K. Christmas, the assistant chief of the Tank-Automotive Center (T-AC) in Detroit, he talks about the advantages incurred by the Ordnance Department in exploring so many different engine concepts for the M3 series tanks: "But in the use of this diversity of engines, not only were production goals for the year 1942 fully met, but a great reduction in the amount of new facilities required were attained. The extensive use of tanks equipped with these various engines now in service should demonstrate conclusively to the using forces which types of tank engines are preferable."

The Ordnance Department would reclassify the entire M3 series of medium tanks as limited standard in April 1943 and obsolete in April 1944.

M3 SERIES TANK VARIANTS

The most numerous variant of the M3 series tanks to see service in World War II was a tank recovery model. Between October 1942 and December 1943 various firms converted a total of 807 units into tank recovery vehicles using the chassis of the M3, M3A3, and M3A5, which respectively received the designations M31, M31B1, and M31B2.

To perform their tank recovery role the M31 series was equipped with a 30-ton capacity winch and a Gar Wood crane that could lift 5 tons without boom jacks being fitted. With the boom jacks fitted the Gar Wood crane on the M31 series could lift 15 tons.

As the British Army replaced the M3 series of medium tanks with the next generation of American-designed and built medium tanks, supplied under Lend-Lease, the M3 series of medium tanks remaining in British Army service were pushed into secondary theaters of operation. The M3 medium tank pictured belongs to C Squadron, 3rd Carabiniers (Prince of Wales Dragoon Guards) engaged in combat with Japanese ground forces in occupied Burma in 1944. Reflecting the threat from close-in attack by Japanese infantry, the rear engine deck of the tank has been covered with raised screening wire to prevent the placement of mines or other explosive devices. The tank, which has the original American-designed turret, the British Army using both versions, is missing the vehicle commander's cupola, and has been fitted with additional storage bins. In place of the original base olive drab, the tank has been repainted in a dark, drab olive green, which had been approved in early 1943 for British tanks in the Far East. It was officially known as SCC13 Jungle Green. Also visible is a very large American national symbol applied to the vehicle even though it was in British Army service, a not uncommon occurrence.
(© Osprey Publishing Ltd.)

During the process of converting M3 series gun-armed tanks into tank recovery vehicles, the vehicles were de-gunned, retaining only two .30 caliber machine guns for self-defense. To hide the fact that these tank recovery vehicles were almost completely defenseless they featured dummy 37mm and 75mm guns.

Some of the M31 series of tank recovery vehicles were provided to the British Army under Lend-Lease and went on to receive the designation "Grant ARV II." The letters "ARV" stood for armored recovery vehicle. The British Army also converted some of its inventory of M3 series tanks into makeshift ARVs. Unlike the models used by the U.S. Army, the British Army dispensed with the 37mm gun-armed turret of the M3. The open-topped vehicle, designated the "Grant ARV I," could be fitted with a jib crane with a chain block hoist.

On display at the Tank Museum in Thun, Switzerland, is this M31B1 tank recovery vehicle. Conversions based on the M3A3 and M3A5 diesel-powered medium tanks were designated M31B1, while those based on the gasoline-powered M3 medium tank were designated M31.
(Pierre-Olivier Buan)

This Ram II medium tank is on display at the Base Borden Military Museum in Ontario, Canada. Based on the power train and suspension system of the M3 medium tank, this Canadian/British design mounted a 6-pounder (57mm) gun in a rotating turret. The British preferred the Ram over the Grant. Production of the Ram II did not begin until early 1942. Although 1,899 were built, only a few conversions, such as personnel carriers, were used in combat.
(Paul Hannah)

SEARCHLIGHT TANKS

The strangest variant of the M3 series tanks to be built during World War II was a searchlight-equipped version intended to be used during nighttime combat engagements. It was pioneered by the British Army, which referred to this top-secret vehicle as the "Canal Defense Light" (CDL) to confuse anybody as to its real purpose. It involved removing the 37mm gun-armed turret of M3 series tanks and replacing it with a new manually operated turret containing a high-intensity carbon arc lamp in one half of the turret and an operator in a separate compartment on the side of the turret. The British Army had approximately 110 of their M3s converted to this purpose. However, they never employed them in combat.

The U.S. Army had the CDL demonstrated to them in early October 1942 by the British Army. The American observers left most impressed with what they saw and had 497 units converted using both the M3 and M3A1 between June 1943 and sometime in 1944. As a cover name they were referred to as the "Shop Tractor T10." The term "Leaflet" was another codename for this hush-hush program.

The extreme secrecy surrounding the CDL meant those within the U.S. Army who might have considered using the vehicle in their military operations in Northwest Europe were unaware of its potential. As a result, the vehicles were eventually placed into storage and their crews transferred to other duties.

The only combat use of the CDLs occurred when a small number of the British and American versions were reactivated to assist in protecting the various Rhine River crossing points in the spring of 1945 from German floating mines, barges, swimmers, etc. that might have been employed to destroy the Allied temporary bridges. After the Rhine crossing the CDL vehicles continued to be used at various bridge sites behind the front lines to assist in facilitating the around-the-clock maintenance and repair of the bridges due to a lack of ordinary searchlights.

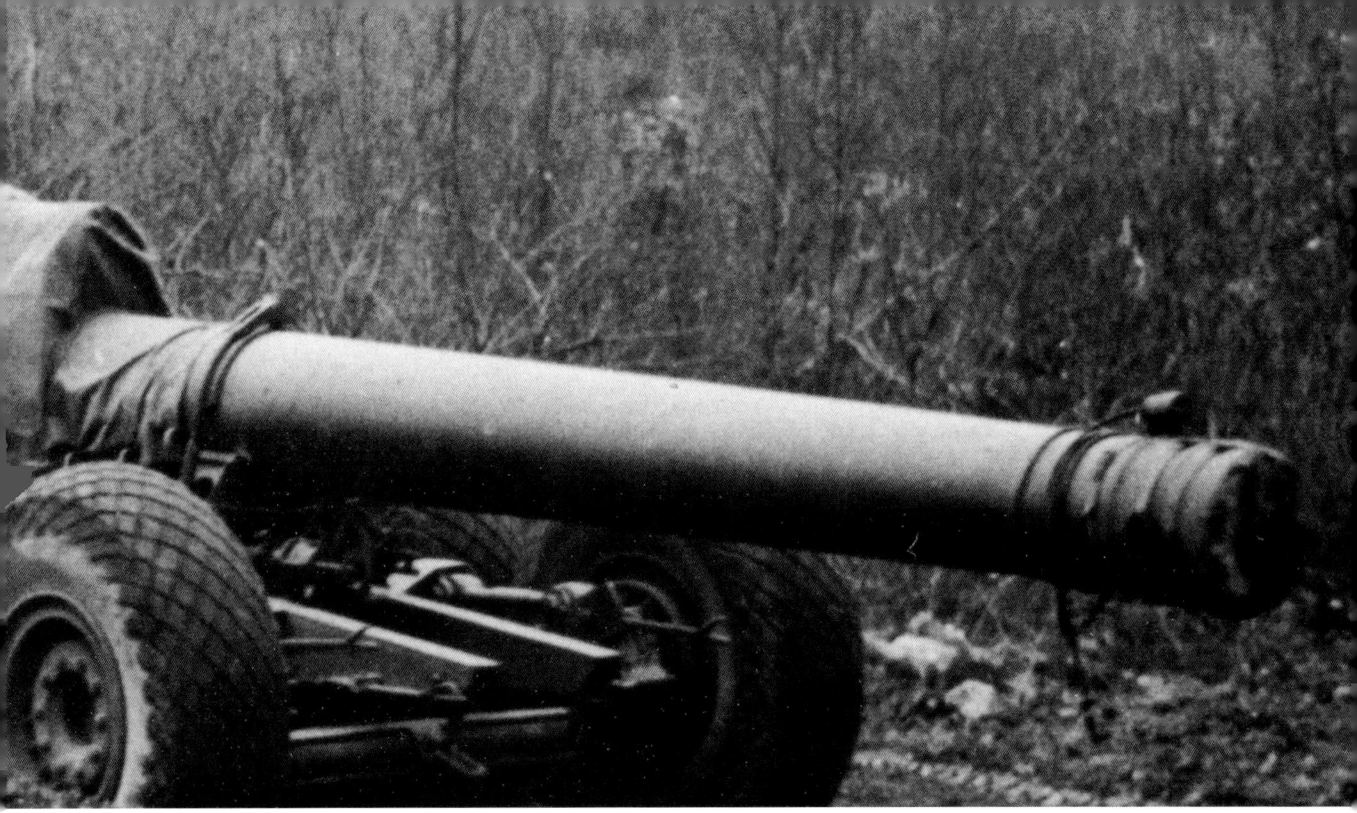

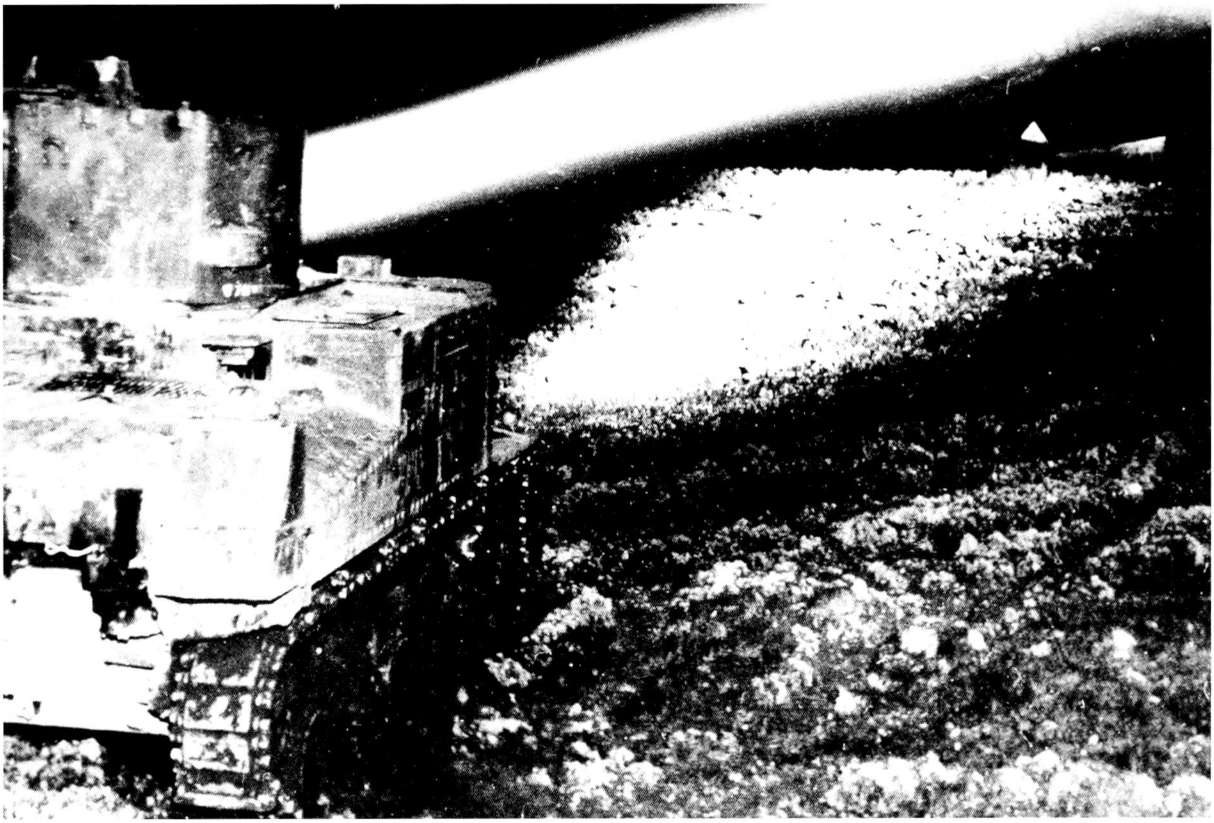

CHAPTER TWO

M4 SERIES TAN

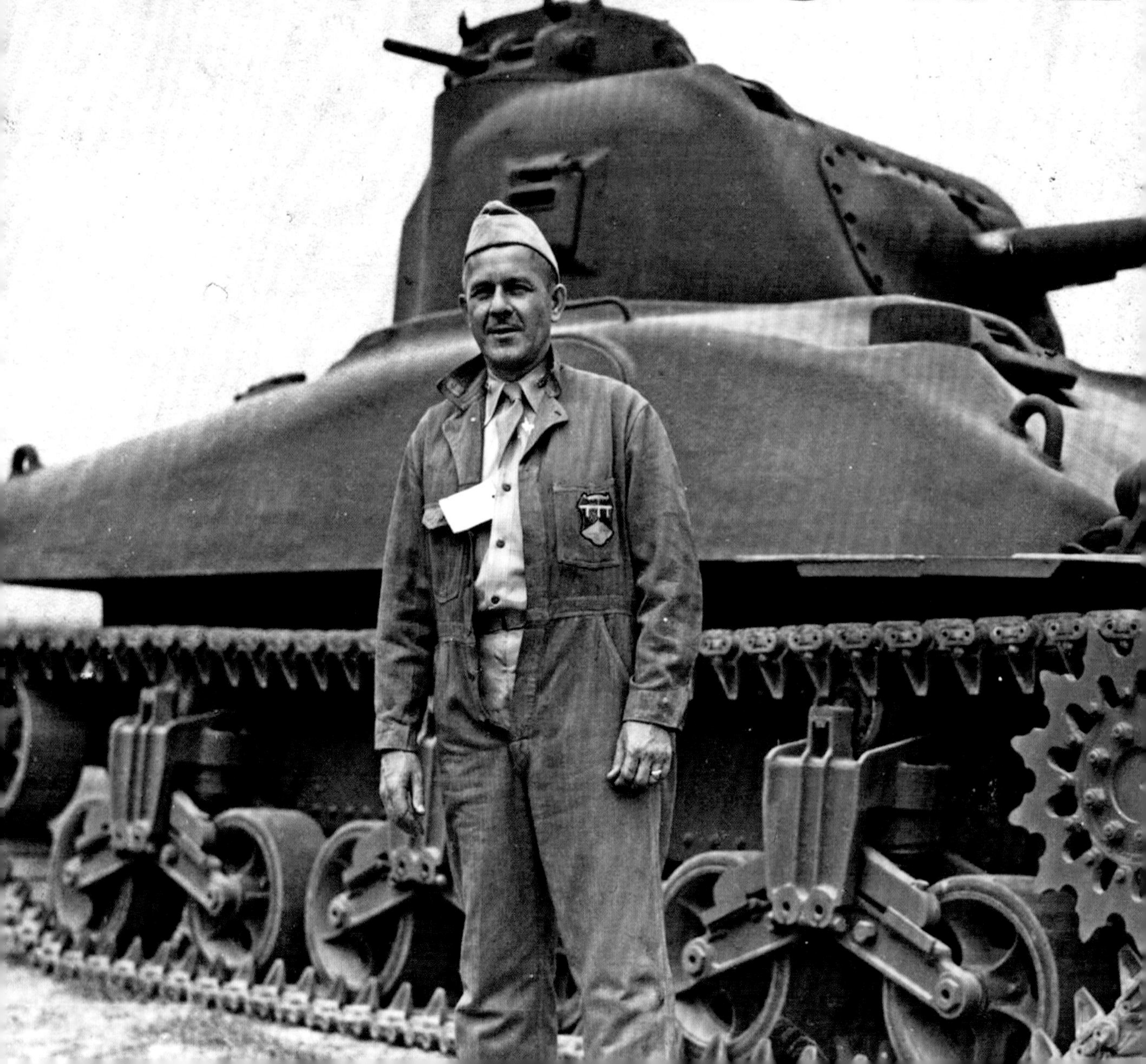

On August 31, 1940, the Armored Force released detailed characteristics for its desired new medium tank to be armed with a turret-mounted 75mm gun M3. However, the new tank design team brought together at APG by the Ordnance Department was then still working on the production drawings of the medium tank M3 and had no time to spare until that project was completed at the end of January 1941. On February 1, 1941, the Ordnance Department directed that its tank design team begin work on the medium tank M3 replacement as quickly as possible.

There was a conference regarding the new tank held at APG on April 18, 1941. Out of that conference came the confirmation of key features of the new tank, one of the most important being the retention of the basic chassis of the M3 powered by a nine-cylinder, Continental Motors gasoline-powered, air-cooled radial engine. This would speed up the design and eventually production of the new tank, while not disrupting the existing production of the M3 series tanks.

It was also decided at the April conference that the upper hull of the new tank would be made out of either cast or welded armor. The lower hull on all of the new tanks would be made of welded armor. The cast-armor turret of the proposed new tank would be surmounted by the same machine-gun-armed cast-armor vehicle commander's cupola seen on M3s built for the U.S. Army.

In place of the seven-man crew of the M3 tank series, the new medium tank would have a crew of only five men, with the vehicle commander and gunner sitting on the right side of the turret and the loader on the left side of the turret. The driver and bow gunner sat in the vehicle's front hull, with the driver on the left side and the bow gunner on the right side. The bow gunner had a manually operated .30 caliber machine gun mounted directly in front of his seat, as well as two remote-controlled, fixed forward-firing .30 caliber machine guns to his left. To allow the crew to talk to each other, an interphone was to be installed.

Presentation Day, September 3, 1941. Major-General Jacob Devers, head of the Armored Force, poses with the medium tank T6, the prototype of the M4 series medium tanks. Devers played a significant role in the development of American medium tanks. The T6 was armed with the short-barrel M2 75mm main gun. (National Archives)

Thinking outside the box, the designers of the next-generation medium tank envisioned that various turret cast-armor gun shields could be made to mount a variety of different weapons besides a 75mm main gun and a coaxial .30 caliber machine gun. These would include a 105mm howitzer with a coaxial .30 caliber machine gun, two 37mm guns M6 with a coaxial .30 caliber machine gun, a British 6-pounder (57mm) gun with a coaxial .30 caliber machine gun, or three .50 caliber machine guns for antiaircraft purposes.

The wooden mockup of the medium tank T6 is shown at Aberdeen Proving Ground in August 1941. It inherited the hull side access doors and the machine-gun-armed commander's cupola from the M3. Both of these features were dropped from the vehicle before it was standardized. (Patton Museum)

MEDIUM TANK T6

In June 1941, the Ordnance Committee ordered the building of a full-sized wooden mockup and a pilot model of the new tank, designated the medium tank T6. Upon completion of the wooden mockup followed by a few design changes, APG started to construct a pilot tank with a cast-armor upper hull and turret, with a riveted welded lower hull. The cast-armor turret on the T6 built by APG resembled nothing more than a scaled-up version of the cast-armor turret on the M3 series tank. Rock Island Arsenal built a T6 pilot tank with a welded-armor upper hull, but without a turret.

Like the M3 series tanks, the T6 differential and final drives were contained within a large, three-piece cast-armor housing that bolted together to form the lower front of the vehicle hull. The T6 also appeared with the original M3 series tank suspension system that is identified by the track return roller being located on the top of the VVSS assemblies.

APG rolled out its 30-ton T6 pilot tank on September 2, 1941. The vehicle's turret was armed with the 75mm gun M2, as the intended 75mm gun M3 was not yet ready. Because the elevation gyrostabilizer system for the tank's turret was supposed to be for the 75mm gun M3, it proved necessary to add counterweights to the muzzle end of the 75mm gun M2 to permit the elevation gyrostabilizer system to work properly. The elevation gyrostabilizer in the turret of the T6 was a modified version of that fitted to the 37mm gun M6 in the turret of the M3 series tanks.

VEHICLE DETAILS

The T6 turret could be rotated manually by the vehicle's gunner by squeezing a lever on a vertical drive handle located on the top of a gear mechanism to the right of his seated position. It could also be turned by the gunner with a pistol-grip control handle that operated an electro-hydraulic power traverse unit that could rotate the turret 360 degrees in just 17 seconds. On the roof of the T6 turret was a cast-armor cupola armed with a .30 caliber machine gun for the vehicle commander. The height of the tank with the vehicle commander's cupola was 9 feet 6 inches.

The turret traverse system offered the innovation of speed control, by which turret speed was determined by how far the vehicle's gunner turned his control. Only a few tanks of the day offered powered operation, which usually featured ON/OFF power, to be followed by fine adjustment using manual control. To stop the turret from turning, the gunner needed only to release the control handle and the turret stopped automatically. Vision for the gunner on the T6 came from an overhead fixed forward-facing periscope device that sat on the upper front slope of the turret and was referred to as the "protectoscope." It had an adjustable armored shutter to protect the upper prism of the periscope.

There was a pistol port on either side of the turret fitted with a protectoscope and armored doors on either side of the vehicle's upper hull. The tank's driver had a direct vision slot protected by an armored visor in the front of the upper hull and the bow gunner a protectoscope, later replaced by a direct vision slot protected by an armored visor.

There was an adjustable seat for the driver on the T6 that allowed him to operate the tank with the overhead hatch closed, or when the overhead hatch was open he could drive the vehicle with his head projecting out from the top of the front upper hull. Strangely enough, there was no overhead hatch for the bow gunner on the T6, or for the loader in the turret.

Welders are pictured at work in the unfinished hull of a medium tank M4A1 at Lima Locomotive Works in Ohio. At this point the vehicle's hull is being moved from station to station around the factory floor on its bogie wheels. When completed the tank weighed 33 tons combat loaded. (Patton Museum)

The second medium tank M4A1, built by the Lima Locomotive Works in March 1942, was shipped to England bearing the name "MICHAEL" in honor of Michael Dewar, head of the British Tank Mission to the U.S. It is seen here on display at the Tank Museum in Bovington, England. The first ten M4A1s made by Lima were built with non-ballistic (unarmored) steel, for test purposes. This tank is one of the ten, with the other nine transferred to APG. (Tank Museum)

SUGGESTED CHANGES TO THE T6

After representatives of the Armored Force and the Ordnance Department inspected the T6, some major and minor changes were ordered. The first major change to the revised T6 was to do away with the armored doors on either side of the upper hull, as they were a serious weak spot in the tank's armor protection. In their place, a single escape hatch was installed in the bottom of the hull floor just behind the bow gunner's seat.

The second major change requested to the revised T6 was the elimination of the cast-armor vehicle commander's cupola. It was replaced with a rotating, two-piece flat circular split hatch with one of the hatches fitted with a 360-degree rotating periscope, designated the M6. This brought the revised T6's height down to 9 feet.

Also asked for on the revised T6 was an overhead hatch for the bow gunner in the front hull. Instead of a pistol port fitted with a protectoscope on either side of the T6 turret, there was only a single pistol port, minus the protectoscope, on the left side of the turret.

NEW DESIGNATIONS

On September 5, 1941, the Ordnance Committee recommended that the revised T6 be standardized as the medium tank M4. That recommendation was approved a month later by the Ordnance Department, with a couple of added requirements for the new tank that included a ball mount for the manually operated .30 caliber machine gun in the front upper hull, and if possible, the addition of a .50 or .30 caliber machine gun on the turret for antiaircraft purposes. Eventually, it was decided in February 1943 that the .50 caliber machine gun was a better choice.

By November 1941, the manufacture of M4 pilot tanks had begun. The following month, the Ordnance Committee decided to attach slightly different designations to the new medium tank M4. Those built with a cast-armor upper hull were referred to as the medium tank M4A1, and those with an upper hull constructed of welded armor were designated as the medium tank M4. APG continued to operate and test the pilot T6, now classified as an M4A1, for further development.

Cast-armor upper hulls were quicker and cheaper to manufacture for the M4 series tank than welded-armor upper hulls. However, warships and aircraft were at a higher priority than tanks and there was not enough leftover foundry capacity in the United States to construct a sufficient number of cast-armor upper hulls for all of the M4 series tanks being ordered.

The fixed machine guns were eliminated from the M4 series medium tank design almost immediately, and this early medium tank M4A1 has had the holes covered over. The gunner's sight rotor was also eliminated early on, and was replaced by an overhead periscope as seen on this example. This vehicle has the original M34 gun mount that featured a very narrow rotor shield. (TACOM Historical Office)

The Ordnance Department took delivery of the first M4A1 from the Lima Locomotive Works in February 1942. The second vehicle went to Great Britain, bearing the name "MICHAEL" on its side, in honor of Michael Dewar, head of the British Tank Mission in the United States.

In March 1942, the Pressed Steel Car Company started building M4A1s, with Pacific Car and Foundry Company following in May of that same year. Production of the M4A1 continued at the three firms until December 1943, with 6,281 units built.

Following the production of 30 M4A1s by the Lima Locomotive Company, and ten units by the Pressed Steel Car Company, the overhead gunner's protectoscope, inherited from the T6, was replaced by a forward-looking periscope, designated the M4. The M4 periscope sat further back on the very top of the turret, making it far

A very early-production medium tank M4A1 can be identified by a number of features. These include the direct vision slots for the driver and bow gunner, the two fixed .30 caliber machine guns in the front hull, and the sight rotor in the top-right front of the turret. (Patton Museum)

less vulnerable to direct projectile strikes than the protectoscope. Unlike the protectoscope, the M4 periscope could be retracted into the turret if damaged and replaced if the need arose. There was only a simple unarmored lid to cover the opening on the top of the turret roof when the M4 periscope was removed.

In July 1942, the Pressed Steel Car Company started production of the M4. Four other manufacturers soon joined in, including Baldwin Locomotive Works, the American Locomotive Company, the Pullman Standard Car Company, and the government-owned Detroit Tank Arsenal.

By the time production of the M4 ended in January 1944, a total of 6,748 units had been completed. A directive issued in March 1942 had dropped the two fixed, forward-firing hull-mounted .30 caliber machine guns from the M4 and M4A1. During World War II, the M4 and M4A1 were the types most widely employed by the U.S. Army.

Due to the more angular shape of the M4 welded-armor upper hull, it had room for 97 rounds of 75mm main gun ammunition versus only 90 rounds of main gun ammunition in the more rounded cast-armor upper hull of the M4A1. It was not uncommon for some units to increase the onboard main gun ammunition storage arrangement in their M4 series tanks during World War II.

An April 27, 1944 War Department document titled *Report of the New Weapon Board* details the U.S. Army's impression of the ammunition storage arrangement in the M4 series tank in the Italian Theater of Operations: "When going into combat,

Pictured is the first medium tank M4A1 made by Pacific Car and Foundry. The VVSS was inherited from the M3 medium tank and can be identified by the return rollers mounted on the top of the bogie assemblies. Notice the pistol port on the left-hand side of the turret. (TACOM Historical Office)

There are many M4 series medium tanks on display as monuments throughout Europe. The M4A1 shown above rests in Patton Square in Ettelbruck, Luxembourg. The vehicle was 19 feet 2 inches long with the main gun forward, and 8 feet 7 inches wide. Its height was 9 feet. (Pierre-Olivier Buan)

the crews invariably put a full complement of ammunition in the floor of the turret basket because they are anxious to carry a very large quantity of ammunition. Tank crews are very little concerned with protection of ammunition and consider accessibility and quantity [main gun rounds] of primary importance."

Other than a few minor differences resulting from the welded-armor upper hull of the M4 and the cast-armor upper hull of the M4A1, the two tanks were almost identical. The engines, power trains, and suspension systems were the same. They both featured a nine-cylinder, Continental Motors gasoline-powered, air-cooled radial engine that generated 400hp at 2,400rpm. The engine in the M4 and M4A1 was designated the R975 C1 as it had a slightly lower compression ratio than found on the R975 EC2 engine in most of the M3 series tanks and the T6.

Both the M4 and M4A1 inherited the differential and final drives arrangement from the T6, which was contained within a large three-piece cast-armor housing that bolted together to form the lower front of the vehicle's hull. Eventually, a one-piece cast-armor housing was produced for the entire M4 series of tanks, which could vary in contours and thickness based on the foundry that made it.

Being installed in the rear hull of a medium tank M4A1 is an air-cooled, gasoline-powered, Continental R975 C1 radial engine, which weighed 1,137 pounds. The maximum sustained speed of the M4A1 on a level road was 21mph. It could reach a top speed on a level road of 24mph for short periods. (Patton Museum)

MACHINE GUNS

Authorized storage for the .30 caliber machine-gun ammunition on the M4 series tanks was 4,750 rounds. Many tank crews loaded their vehicles with additional .30 caliber machine-gun ammunition. A rationale for this behavior appears in the "After Action Report" (AAR) of the 741st Tank Battalion. They expended approximately 100,000 rounds of .30 caliber machine-gun ammunition during a single day's combat engagement near Etouvy, France, during the breakout from the Normandy beachhead.

A War Department document titled *Report of the New Weapon Board*, dated April 27, 1944, describes the U.S. Army's impression of the usefulness of the .30 caliber machine gun located in the front glacis of the M4 series tank: "The bow machine gun is much used in combat operations, although all crews agree that this weapon is extremely hard to fire accurately and would like some means of sighting to be provided. As it is known that the tanks draw fire and that the bow machine gun is never used unless the position of the tank is known to the enemy, it was suggested

Early combat experience in the Pacific Theater of Operations quickly demonstrated to the U.S. Marine Corps that light tanks lacked the bulk to push through jungle undergrowth, or the needed firepower to penetrate Japanese bunkers. This realization encouraged the Marine Corps to ask the U.S. Army for a supply of M4 series medium tanks to supplement their inventory of light tanks. What the U.S. Army had available at the time was the first-generation M4A2 medium tanks, armed with a 75mm main gun. Pictured is a Marine Corps first-generation M4A2 named "China Gal" of Company C, of the 1st Tank Battalion (Medium), on Tarawa in November 1943. The vehicle is in olive drab, with the registration numbers in blue drab. There is no national symbol on the tank pictured. (© Osprey Publishing Ltd.)

that, to improve accuracy of fire, the percentage of tracers be increased to fifty or one hundred percent."

Authorized storage for .50 caliber machine-gun ammunition on the M4 series tank was 600 rounds. The .50 caliber machine gun mounted on the turret roof of the M4 series tanks was far less popular with American tankers than their onboard .30 caliber machine guns, as is evident in this March 1945 U.S. Army report titled *United States vs. German Equipment*: "The American .30 cal. MG is considered one of the best weapons we have. Its rate of fire is sufficient. It is a well-built weapon and very dependable under the toughest conditions. The American .50 cal. MG is excellent for the same reasons. It is a good weapon for aircraft defense, but the mount on tanks makes its use for ground targets impractical."

U.S. Army tanker Tom Sator remembers that it normally fell to the loader to operate the .50 caliber machine gun on his M4 series tank, as the tank commander was typically too busy to perform that job. Tom believed it was certain death to expose himself outside the confines of his tank to fire the .50 caliber machine gun in the face of stiff enemy resistance. He only did it once and swore he would never do it again.

The late Jim Carroll (a Marine Corps tanker) remembered that they never bothered to even mount their .50 caliber machine gun on their M4 series tank as it was not safe to expose oneself to Japanese fire for even a moment in the close-in fighting that was so characteristic of combat in the Pacific Theater of Operations (PTO).

A Grizzly I medium tank on display at Base Borden in Canada. The Grizzly was essentially an M4A1 medium tank assembled by the Montreal Locomotive Works, and factory equipped to Canadian/British requirements. Due to production cutbacks in the M4 series medium tank program, only 188 units were produced from October through December, 1943. (Paul Hannah)

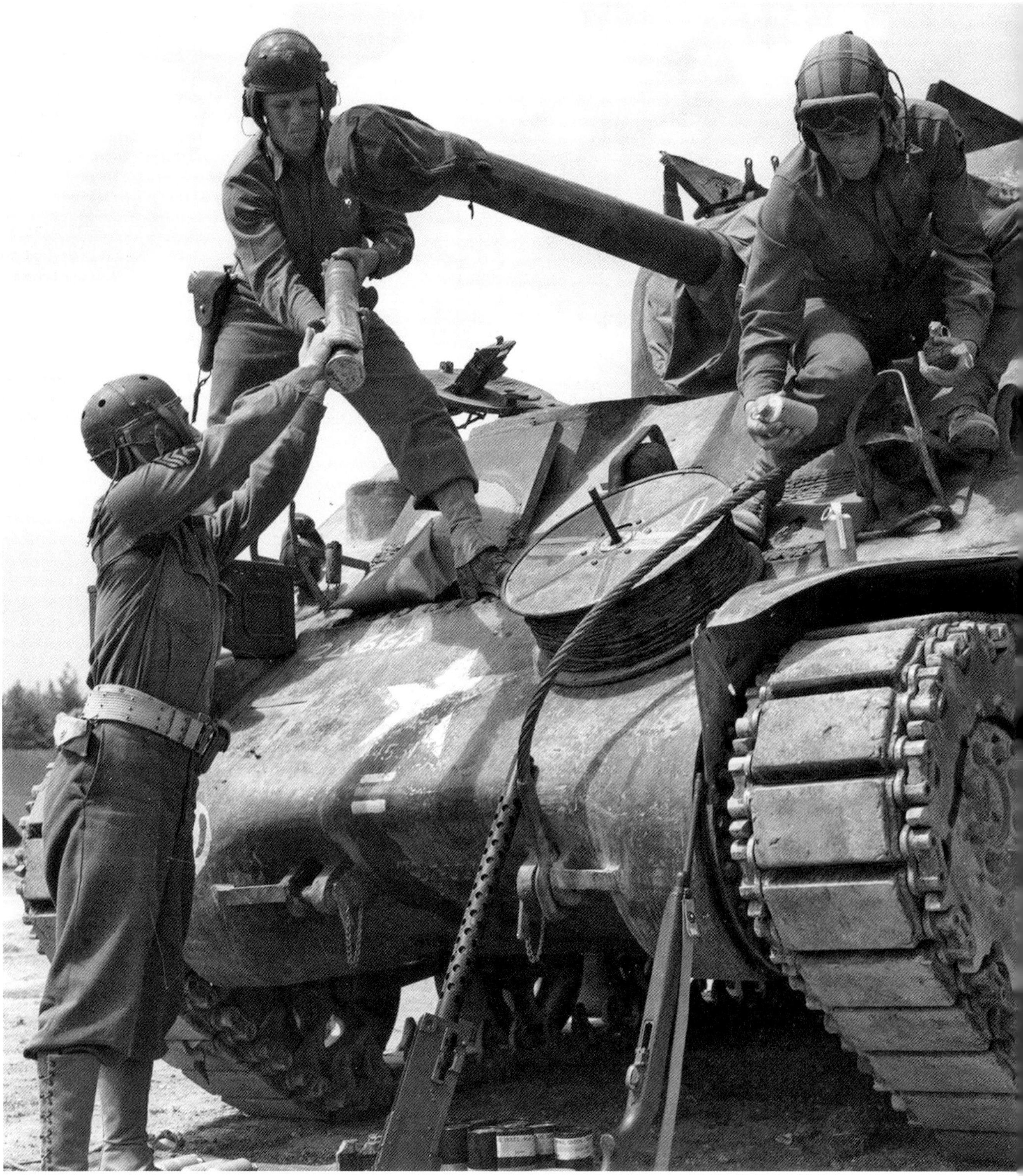

ARMOR PROTECTION LEVELS

Armor thickness on the welded armor glacis of the M4 was 2 inches (51mm) with a 56-degree slope. The armor thickness on the glacis of the M4A1 was also 2 inches; however, due to its more rounded upper cast armor hull, the slope varied from 37 to 55 degrees. The steep slope on the welded armor glacis of the M4 and M4A1 inherited from the T6 caused the designers of the tanks to come up with very narrow overhead hatches for the driver and bow gunner that made ingress and egress difficult. This was a design feature that made everybody unhappy and wasn't corrected until the introduction of the second generation of M4 series tanks that received a new, larger hatch design.

The steep slope of the welded armor glacis on the M4 accounted for the driver's and bow gunner's hatch protrusions on the vehicle, which were not as pronounced on the M4A1 due to its smoother cast armor contours. Both the driver's and bow gunner's overhead hatches featured a 360-degree rotating replaceable periscope designated the M6. Eventually, the direct vision slots (protected by armored visors) in front of the driver and bow gunner positions were eliminated from the entire M4 series of tanks, because they proved to be ballistic weak spots, and were replaced by replaceable, fixed forward-looking M6 overhead periscopes. Baldwin Locomotive Works continued to build M4s with the direct vision slots until production ended in 1944.

Armor thickness on the upper hull sides of the M4 and M4A1 was 1.5 inches (38mm). The cast-armor turret that was the same for both vehicles had an armor thickness of between 3 inches

The vehicle pictured is a welded-hull medium tank M4, probably made in September or October 1943 by Alco Production of the welded-hull M4 medium tank began in July 1942. It differed only slightly from the medium tank M4A1 in its interior arrangements, as its more angular-shaped hull allowed for an increase in main gun ammunition storage. The vehicle crew is bombing up with main gun ammunition, smoke grenades, and their small arms. (Patton Museum)

(76mm) and 3.5 inches (89mm) on the cast-armor gun shield, which was not sloped. The side walls of the rounded cast armor turret were 2 inches (51mm) thick with little or no slope. The turret roof was an inch thick (25mm).

In August 1943, the Chrysler Corporation began production of M4s. In an effort to save welding time and labor, their tanks consisted of a cast-armor front end married to the rear two-thirds of a typical M4 welded-armor hull. This somewhat unusual arrangement was generally referred to as the "composite hull" variant.

Late-production medium tank M4s built by the Detroit Tank Arsenal differed from those made by other firms. A single-piece cast-armor front hull section that extended past the driver's position was combined with a welded-armor hull as is seen on this medium tank M4. Such vehicles are now commonly referred to as having a "composite hull." The example pictured is on display in southern California. The vehicle was heavily damaged as it was previously used as a range target, thus the penetration holes in its armor. (Chris Hughes)

Of the 1,676 units produced of the M4 composite hull variant, perhaps 70 of the early examples featured armored front castings with small hatches, while the majority were equipped with large hatch castings very similar in shape to the second-generation M4A1.

MEDIUM TANK M4A2

As with the M3 series, the lack of enough gasoline-powered, air-cooled radial aircraft-type engines proved to be a serious problem for the M4 series tanks. The Ordnance Department was forced to use whatever motive power was available and therefore

Pictured is a late-1943-production medium tank M4A2 powered by the General Motors 6046 diesel engine. This tank is equipped with just about every modification that had been added to the original design by fall 1943. Note the later M34A1 gun mount that featured a full-width rotor shield that provided protection to both the new direct telescopic sight on the right side and the coaxial .30 caliber machine gun on the left. (Ordnance Museum)

took the twin, liquid-cooled GM diesel truck engine arrangement designated the 6046 and installed it in the M3A3 and M3A5, and mounted it in a welded-hull M4. Reflecting the various modifications made to the tank, primarily in the engine compartment, they designated it the medium tank M4A2. Production began in April 1942 at a number of different manufacturers and continued until May 1944, with 8,053 units built.

U.S. Army testing of the M4A2 resulted in both some positives and some negatives. Everybody liked the vehicle's excellent horsepower to weight ratio as well as its low fuel consumption. The maximum operational range of the M4A2 tank on roads was about 150 miles as compared with the M4 and M4A1 being approximately 120 miles due to the greater thermal efficiency of diesel fuel. However, everybody felt that the twin-diesel arrangement was too sensitive to dirt, which would result in shorter engine life and poorer reliability than the other types of engines used in the M4 series tanks.

·In March 1942, the U.S. War Department issued a policy which stated that only gasoline-powered M4 series tanks would go to American forces serving overseas. This would ease the burden on the logistical units as only one type of fuel would be required for both the U.S. Army tanks and its massive inventory of unarmored, gasoline-powered wheeled vehicles.

Pictured is a first-generation M4A2 of Company A, 3rd Tank Battalion, on Iwo Jima in February 1943. The base color of the vehicle is olive drab. However, in this case, it has been over-painted with irregular patches of black. As factory-built add-on armor kits for the first generation of M4 series medium tanks arrived in the field, they were quickly welded onto the hulls of the vehicles and painted in whatever was available at that moment, in this case a dull green. The tactical and registration numbers on the tank shown are in yellow. The lusterless (flat) olive drab paint applied to American tanks and AFVs in World War II actually dated back to official use with the U.S. Army beginning in 1917. It consisted of black and ochre mixed together. U.S. Army tanks and AFVs often employed a gloss (shiny) olive drab during the 1930s. (© Osprey Publishing Ltd).

Diesel-powered M4 series tanks were employed for training purposes within the United States only, or for supplying under Lend-Lease to Allied nations also fighting the Axis. Despite this ban on sending the M4A2 overseas with American fighting men, when the Marine Corps initially requested gasoline-powered M4 series tanks, they eventually ended up with an inventory of about 200 M4A2s in late 1943, as that was what the U.S. Army could spare from its inventory.

The M4A2 was initially disliked by some Marine Corps tankers as can be seen from this November 1942 quote from a memo written by Lieutenant Colonel William R. "Rip" Collins, commanding the Marine Corps Tank Battalion School at Jacques' Farm, located near San Diego, California: "From all that I can gather the M4A2 is a stinker which is probably the reason why the Army is so liberal with them." However, combat use of the M4A2 soon convinced the majority of Marine Corps tankers of the desirability of their diesel-powered, twin-engine tank. When informed they had to trade in their M4A2 tanks for single-engine, gasoline-powered, M4 series tanks in March 1944, they strongly objected.

The Marine Corps tankers' fondness for the M4A2 came about for a couple of important reasons. Combat use had quickly demonstrated to all that having two engines meant that damaged tanks could make it to safety even on one engine. Disembarking from a damaged tank in a combat zone is always fraught with danger and is never relished by any tanker. Watching gasoline-powered tanks quickly turn into raging infernos after being struck in combat by enemy weapons, Marine Corps tankers saw with their own eyes how diesel fuel was less flammable compared with highly volatile gasoline.

The advantages of the twin-engine, diesel-powered arrangement of the M4A2 is seen in this extract from a letter written by Marine Corps Major Richard Schmidt and his 4th Tank Battalion staff after the Tinian operation concluded on August 1, 1944:

> On at least three different occasions on Saipan, tanks of Company "C" were partially damaged when magnetic mines exploded over the engine compartments. In each instance the result was one engine destroyed and several fuel lines shattered with fuel [diesel] blown over the hot engines. But on neither of these occasions did a tank catch fire. Never was a vehicle destroyed or a crew lost. And always the damaged tank was able to return to safety powered by its one good engine.

Among the many different types of M4 series tanks supplied to the British Army during World War II, the M4A2 was the preferred version due to its twin-engine arrangement since it was the easiest to work on. In total, the British Army would receive 5,041 units of the M4A2. They would refer to the M4A2s they took into service as the "Sherman III." The Red Army received via Lend-Lease a total of 1,990 units of the M4A2 during World War II. Almost 400 M4A2s went to the Free French Army.

The majority of medium tank M4A2s were provided to the British and Soviets as Lend-Lease. Pictured is a restored example in British Army markings taking part in a yearly event held at the Tank Museum, Bovington, referred to as "Tankfest." The Free French Army also received a great many M4A2s under Lend-Lease. (Christophe Vallier)

Pictured is the Ford Motor pilot of the medium tank M4A3. This was made using a medium tank M4A2 hull. The angle of the upper-rear hull plate as well as the configuration of the engine deck does not accurately reflect the standard appearance of the production medium tank M4A3. (TACOM Historical Office)

Shown is a late-production M4A3 with the full-width M34A1 gun mount. The sand shields fitted to this vehicle seldom appeared on M4 series tanks employed by the American military during World War II. As with the M4 and M4A2, the glacis plate on the M4A3 was composed of both cast and welded armor. The .30 caliber bow machine gun in the tank's front hull was normally assigned to the lowest-ranking member of the crew. Since there was no sights for the bow machine gun, unlike their German medium tank counterparts, the bow machine gunner, or BOG, was forced to employ the fall of the weapon's tracers to direct his fire onto a target or targets. Many tankers thought it was a useless position, while other felt that it was useful in keeping enemy infantry at bay. (Patton Museum)

"Sherman" was the official British designation for the M4 series tanks. The name was in honor of the American Civil War general, William Tecumseh Sherman. It was never an official U.S. Army designation, although it did become popular with some American tankers fighting in Europe during World War II, and shows up in some U.S. Army reports from the period. In the decades after the war, it became the primary name for the tank, which continues to this day.

Under Lend-Lease, the British received a total of 2,096 M4s, which they designated as the "Sherman I." A good number of their M4s were of the composite hull type which they designated "Sherman I Hybrid." A total of 942 units of the M4A1 entered into British and Commonwealth service during World War II, and were referred to as the "Sherman II."

While the production of M4 series tanks in the United Kingdom was contemplated in 1942, it was never implemented, most likely for political reasons. The only foreign manufacturer of the M4 series tanks was Montreal Locomotive Works in Canada. Using over 90 percent American-made components, they assembled a slightly modified version of the M4A1, outfitted to suit Canadian and British preferences (different internal storage arrangements). Officially designated the "Grizzly I," only 188 units were completed between October and December 1943, when production was terminated, as it was then thought that American factories had sufficient capacity to meet all future requirements.

MEDIUM TANK M4A3

In August 1941, the U.S. government asked officials of the Ford Motor Company if they would be willing to manufacture tanks. Ford agreed to produce medium tanks, and for a power plant, offered a liquid-cooled, gasoline-powered V-8 design it had developed in-house. Ford-engine M4 series tanks were designated the "Medium Tank M4A3." In outward appearance, the welded-armor hull M4A3 differed from the welded-armor hull M4 only in its rear engine compartment arrangement and some other minor features. As with the M4 and M4A1, the M4A3 weighed about 33 tons. Production of the M4A3 began at Ford in June 1942 and continued until September 1943 with 1,690 units built.

More must be said about what was designated the "Ford GAA engine." It started out in June 1940 as an experimental 12-cylinder aircraft engine with a very high output power for its size and weight. Reduced to eight cylinders as a tank engine, maximum gross power was 500hp at 2,600rpm. The net power as installed in the M4A3 was 450hp at 2,600rpm.

A memorandum titled "Subject: Medium Tanks, M4A3," addressed to the Commanding General, Army Ground Forces Washington, DC, June 1943, shows that the Ford engine exceeded the performance of other medium tank engines at the time: "Based on experience in operations and after exhaustive comparative tests, the

Belonging to the Virginia Museum of Military Vehicles is this medium tank M4A3. As seen here, a modification was introduced in which armor plates were welded to the front of the driver's and bow gunner's protruding hatch hoods in an attempt to increase the protection levels on the front of the vehicle's glacis. (Michael Green)

Pictured is a production medium tank M4A3 undergoing a dynamometer test at APG. The M4A3 can be distinguished from the earlier medium tank M4 by the squared-off upper-rear hull plate, which extended below the level of the upper track run. To reduce the amount of dust thrown up by the two engine exhaust vents, a deflector grill was fitted as seen here. (National Archives)

Armored Force considers the Ford tank engine superior to any other medium tank power plant now available. It is therefore recommended that the Ford-powered medium tank be declared suitable for overseas supply and placed in active theaters of operation at the earliest practicable date."

It was at that point, if sufficient production capacity had existed, that the U.S. Army would have completely replaced the Continental Radial engine with the Ford GAA engine. As it was, the decision was to reserve M4A3s for U.S. Army armored units, while progressively replacing all other types of M4 series tanks with the M4A3 both at home and overseas.

Ford was the only automobile company dropped from the M4 series tank program in the production cutbacks of late 1943. Reasons cited were to free up Ford facilities for other war work, as well as high cost and low productivity. However, Ford continued to produce engines up until the end of World War II. Most of these were supplied to Chrysler and Fisher Body, who took over the manufacture of M4A3s (and M26 heavy tanks) in 1944 and 1945. All told, Ford's Lincoln Plant produced 26,954 GAA engines.

During World War II, most of the M4A3s actually produced by Ford served as training tanks in the United States. However, period photos and historic documents from 1945 show that a small number served in both Europe and the Pacific. On the other hand, with progressive replacement, Chrysler and Fisher-built second-generation M4A3s made up over half of the U.S. Army's inventory of the tank by the end of the war.

While the M4A3 entered into Marine Corps service in late 1944, only a very small number were Ford-built. The British Army received six Ford-built M4A 3s as Lend-Lease, and used them only as test vehicles. They were designated the "Sherman IV."

MEDIUM TANK M4A4

The Ordnance Department eventually approved the mounting of an improved version of the Chrysler A57 Multi-bank gasoline-powered engine that had been installed in the M3A4 medium tank into the M4 series. This configuration of the M4 series tank was designated the M4A4. Like the M3A4 medium tank with the same engine arrangement, the large and bulkier power plant mandated a lengthened hull and suspension system for the M4A4. Ordnance Department tests showed the A57 Multi-bank was the least satisfactory of the engines selected for the M4 series tanks and recommended that when a sufficient number of other types of engines became available, production should come to an end.

An early-1943-production medium tank M4A4 is shown on display in France. To make room for the larger and bulkier Chrysler A57 Multi-bank engine, the M4A4 hull had to be lengthened 11 inches. Note the increased distance between the road wheels as compared with other M4 series medium tanks. The approximate operating range of the M4A4 on level roads was 150 miles. (Pierre-Olivier Buan)

With no interest in the A57 Multi-bank engine-powered M4A4 except for stateside training purposes, the U.S. Army decided to give most of them to the British Army. Automotive engineers from the British Tank Mission initially looked with absolute horror on the complex A57 Multi-bank-powered M4A4 tanks being offered to them. However, they took them into service anyway and eventually rated them as being more reliable than the nine-cylinder Continental Motors gasoline-powered, air-cooled radial engines in the M4 and M4A1. This was partly accomplished by doing away with the individually mounted carburetors on each of the five engines and combining them into one bank on top of the power plant for easier access by the crew.

Chrysler completed 7,499 units of the M4A4 between July 1942 and September 1943. Of that number, 7,167 units went to the British Army under Lend-Lease. In British military service, the M4A4 became the "Sherman V." Of the remaining M4A4s, two went to the Red Army and 274 to the Free French Army in North Africa in 1943. Of the remaining 56 M4A4s remaining, 22 went to the Marine Corps, who employed them for training duties only. The tank was reclassified as limited standard in May 1945.

The lengthening of the M4A4, and the larger and heavier power plant, drove the vehicle's weight to almost 35 tons. Due to a serious U.S. Army shortage of M4 series tanks during the Battle of the Bulge (December 1944 through January 1945), the British Army returned a number of M4A2 and M4A4 tanks to the U.S. Army. Most went on to see service with the American Third Army, under the command of Lieutenant-General George S. Patton.

THE M4A6 TANK

The only other M4 series tank to see the mounting of an alternate power plant was the composite-hull M4A6, which featured a Caterpillar Tractor Company modification of a Wright G200 air-cooled, gasoline-powered radial turned into a diesel engine with a fuel injection system. Originally referred to as the Caterpillar D200A, it later became the Ordnance Engine RD-1820. Unlike the other engines in

The medium tank M4A6 was fitted with a Caterpillar Tractor Company radial diesel engine designated the RD1820. Due to the size and bulk of the engine, the rear ¾ hull of the M4A4 was mated to the front-hull casting used on the late-production composite hull M4 medium tank. Only 75 M4A6s were produced before the project was terminated due to production cutbacks. (TACOM Historical Office)

the M4 tank series, the new power plant in the M4A6 could operate on a variety of petroleum products ranging from crude oil to gasoline, making it the first example of a multi-fuel tank power plant.

Chrysler began production of the M4A6 in October 1943. However, the Ordnance Department decided to discontinue production of the vehicle in February 1944 after only 75 units were completed. The reason for this cancellation had to do with changing military requirements and a decision to concentrate on building more of the gasoline-powered M4A3 tanks. The M4A6 tanks all came with the same unusual combination of a cast-armor hull front mated to a rear welded-armor hull, as did late-production M4 tanks built at the Detroit Tank Arsenal. The tank was reclassified as limited standard in May 1945.

BRITISH ARMY COMBAT DEBUT

The British Army suffered some serious battlefield reverses in North Africa in June 1942, which included the loss of a great many tanks. Unable to make up those losses quickly, British Prime Minister Winston Churchill asked American President Franklin Roosevelt if he could make available as many of the new M4 series tanks as possible to help stem the German advance in North Africa.

Originally, it was envisioned that an entire American armored division under the command of then-Major-General George S. Patton would be sent to Egypt to assist the British Army. However, the logistics of such a move would have taken too long to be of much use to the hard-pressed British Army then fighting for its life. Instead, it was decided to take 300 M4 series tanks out of the U.S. Army inventory and ship them by sea directly to the British Army in Egypt.

Despite one transport ship carrying tanks being sunk en route to Egypt, by September 1942 almost 300 M4 series tanks had arrived in that country, with most of them being M4A1s and a smaller number of M4A2s. Modified by the British Army for use in a desert environment, which included the addition of sand shields and British stowage arrangements, these tanks would play an important role in helping the British Army eventually stop the German advance toward Egypt. As time went on, additional deliveries of M4 series tanks to the British Army in North Africa made it their most commonly used tank in that theater of combat.

British Army opinion of the M4 series tank ran very high during its time in combat in North Africa. That sentiment appears here in a statement from a senior British Army officer that showed up in a July 1943 U.S. Army report titled *Tankers in Tunisia*: "In my opinion the Sherman is the finest tank in the world, better than anything else we have and also better than anything the Germans have. It will be the best tank for the next five years."

U.S. ARMY COMBAT DEBUT

Various units of the U.S. Army were employed in the invasion of French North Africa (which includes the current countries of Morocco, Algeria, and Tunisia) in November 1943. The endeavor was referred to as Operation *Torch* and was spearheaded by the U.S. Army's 1st and 2nd Armored Divisions. The inexperienced American tankers had their first encounter with the more experienced German forces in North Africa in December 1942, when a platoon of five M4 series tanks from the 2nd Armored Division went up in flames from a combination of accurate German tank and antitank gunfire.

The M4 series medium tanks first saw combat with the British Army at El Alamein in October 1942. Its U.S. Army combat debut was in November 1942 with Operation *Torch*, the invasion of Northwest Africa. This camouflaged M4A1 of the 1st Armored Division was photographed in Tunisia in February 1943. (Patton Museum)

Pictured is an Ordnance Department display of the onboard stowage of the M4 series medium tank showing the many items carried within the vehicle. Many tankers blamed the gasoline fuel used by most M4 series tank as the primary cause of tank fires. However, battlefield studies revealed that the main gun rounds were the primary culprit. Ammunition stowage was positioned high in the hull and poorly protected. (Ordnance Museum)

Being winched onboard a semitrailer, designated the M15, is a trackless, burned-out medium tank M4A1. The tractor truck with the armored cab was referred to as the M26. By summer 1944 it was clear to all American tankers in the ETO that they were vulnerable to almost every antitank weapon in the German arsenal. (Patton Museum)

On February 19, 1943, American tankers felt the full wrath of the German Army in North Africa when its tank-led spearhead punched a 2-mile-wide hole through American lines at Faid Pass in Tunisia, pushing them back 50 miles in what is best described as a rout. In the process, German Army units inflicted heavy losses on the still inexperienced U.S. Army units in their path during a series of engagements known as "the battle of the Kasserine Pass." Over 100 American tanks, most of them belonging to the M4 series, littered the barren North African landscape in its aftermath. It took until February 25, 1943, before the American lines were restored to their original positions.

Despite their losses in North Africa, many American tankers considered the M4 series a fine tank and more than sufficient for any job at hand. In the U.S. Army report *Tankers in Tunisia*, Sergeant Becker, who saw combat in North Africa, remarked, "I like the M4. I look at the German tank and thank God I am in an M4."

Not everybody who served on M4 series tanks in North Africa considered the tank free of flaws. Sergeant James H. Bowser, a tank commander in the 1st Armored Division, is quoted in the same report: "Yes sir, this is my third tank, but I've still got all my original crew with me. We were burned out of our other two tanks under fire. If they were diesel, it wouldn't have happened, but these gasoline jobs blaze up on the first or second hit and then you've got to get out. None of us like 'em, sir. We'd rather have the diesels."

Despite the belief of many American tankers who served on M4 series tanks during the fighting in Northwest Africa that it was the gasoline fuel their vehicle used that burned so easily, this was not the case. Instead, it proved to be the main gun ammunition stored within them that caused the fires. A British officer commented in *Tankers in Tunisia*: "Self-sealing gasoline tanks for tanks are nice, but they are not vital. It is the ammunition, not the gasoline that burns. German tanks burn too if ammunition is hit. I think that the Germans aim to hit our ammunition. In one battalion fifteen tanks were penetrated; eleven of them burned, ten because of ammunition… In another battle fifteen tanks were penetrated; seven burned, all but one by ammunition fire."

Another U.S. Army report, *Observations on Problems in Armored Units*, dated June 6, 1945, from the Office of the Chief of the Commanding General Army Service Forces, confirms the British officer's opinion on the cause of tank fires: "Fires in Tanks: Recent experience reinforces earlier data, from sixty to ninety percent of M4-75mm go by burning and almost all burn when hit by a *Panzerfaust* [German late-war infantry single-shot antitank weapon firing a shaped-charged warhead]. The greater proportions of fires originate in the ammunition, a few slow fires in oil of power train, and stowage the remainder in the engine compartment."

ARMOR PROTECTION ISSUES

Once a projectile of almost any type penetrated the armor of an M4 series tank upper hull, it would often strike the main gun ammunition racks primarily located in the sides of the upper hull (referred to as "sponsons" in the American military and "panniers" in the British Army), igniting the propellant charges contained within the metal cartridge cases, which would begin to burn within a few seconds. In many cases, this happened before the crew could evacuate their vehicle. On the original M4 series tanks, the majority of the 75mm main gun rounds were stored within light-sheet steel boxes, with another 12 rounds stored vertically in clips along the inside wall of the turret basket for easy access by the loader.

There were numerous reasons for the lack of heavier armor not appearing on the M4 series tanks. Armor is the heaviest element of any tank design. Steel armor plate 1-inch thick weighs 40 pounds per square foot. When the M4 series tanks came about, the engineers designing the vehicle were constrained because existing cranes fitted to the majority of transport ships could not load or unload anything heavier than 35 tons. Portable bridges employed by the U.S. Army in the early days of the war also had an upper weight-bearing limit of no more than 35 tons.

Due to the need to produce a tank with a turret-mounted 75mm main gun as quickly as possible, the M3 chassis and engine became the basis for the M4 series tanks. While this decision certainly sped up the M4 series tank design and production, it saddled the

To increase the armor protection on their M4 series medium tanks, it was common for crews to weld spare track links onto the hulls as seen here. Sometimes the spare track links would also be welded onto the turret sides. (Patton Museum)

engineers with a number of design problems, such as figuring out the minimum amount of armor protection needed by the tank to carry out its intended battlefield roles.

The Ordnance Department looked to see what the main antitank weapon of the German infantry was in 1941. That turned out to be a towed 37mm antitank gun, designated the 3.7cm PaK 35/36 L/45, which fired an AP-T projectile with a muzzle velocity of 2,444ft/sec. In theory, it could penetrate 1.34 inches (34mm) of armor sloped at 30 degrees at a range of 109 yards. At 546 yards, it could only penetrate 1.14 inches (29mm) of armor. These penetration figures illustrate that the M4 tank series was well protected against this weapon.

What the American tank designers failed to foresee when laying out the armor protection levels on the M4 series tanks were the amazing advances in antitank weapon technology made by the German military prior to, and during, World War II. These weapon advances quickly left the M4 series tanks badly under-protected for the roles they assumed during the conflict. While many stopgap measures appeared, both officially and unofficially as field modifications, none could really meet the expectations of the American tankers who served in the M4 series tanks.

As the Marine Corps inventory of M4A2 medium tanks was depleted, they returned to the U.S. Army for a resupply of additional M4 series medium tanks. Available at that time was the second-generation M4A3 medium tanks. The vehicle pictured in this artwork belongs to Company A, 4th Tank Battalion on Iwo Jima in February 1945. It has a very elaborate camouflage scheme consisting of a combination of the base olive drab, with field-applied splotches of black, sand, and dull green. The vehicle also has raised wire mesh screening on the crew hatches, and sandbags on the rear engine deck, to reduce the effectiveness of enemy hand-placed explosives. The wooden planking on the hull side was intended to counter the use of magnetic mines by Japanese infantry tank-hunting teams. (© Osprey Publishing Ltd.)

While some M4 series tankers believed that cast-armor upper hulls were superior in deflecting armor-piercing projectiles than their more box-like, welded-armor upper hull counterparts, tests between the two types of hulls showed without a doubt that the welded-armor hull M4 series tank was better at deflecting projectiles.

The high losses suffered by American tanks, such as the M4 series, is addressed in Belton Y. Cooper's book *Death Traps: The Survival of an American Armored Division in World War II*. In his book, he states the 3rd Armored Division had 648 tanks completely destroyed and another 700 knocked out, but eventually repaired and placed back in service again. This took place between the time they went ashore in France in late June 1944 and the end of World War II in Europe.

In the ETO, a widely used method of adding extra protection was placing sandbags on the front, as seen here on this medium tank M4A1. To prevent the sandbags from sliding off the sloped glacis of the tank, they have been secured by a steel crossbeam at the bottom of the glacis. In reality, the sandbags did little, other than add weight, straining the drivetrain components. (Patton Museum)

THE SHAPED-CHARGE THREAT

Besides the threat from German kinetic energy rounds, a new danger for the M4 series tanks and other armored fighting vehicles came late in World War II from the widespread employment in the German military of hand-held rocket launchers equipped with shaped-charge warheads. These included the *Panzerschreck* ("tank terror") and *Panzerfaust* ("tank fist"). Kinetic energy refers to the energy an object has in motion.

To counter the threat from German kinetic energy rounds and shaped-charge-armed rocket launchers, American tankers began covering the outside of their vehicles with a variety of materials, including spare track links. Other items employed included sandbags followed by cement. Initially, the most common method of holding sandbags onto the tanks was with communication wire or common chicken wire. As time went on, some units built elaborate metal brackets to support the sandbags in place.

Some units would apply as many as 170 sandbags to each of their tanks. It was felt that the application of these materials to the exterior of tanks might deflect kinetic energy projectiles and prematurely detonate shaped-charge rocket warheads, thereby reducing their penetrative abilities. The actual effectiveness of these field expedient modifications varied depending on whom you talked to. The commanding officer of the 756th Tank Battalion noted in an After Action Report (AAR) of his unit that "sandbagging the front of the tank greatly improves the morale of the crews."

Some M4 series tankers decided to take the add-on protection to the next level by covering the sandbags on their vehicle's glacis with a layer of cement as seen in this photograph. Other tankers dispensed with the sandbags and added only a layer of cement to the glacis of their tanks. (Patton Museum)

Some units swore that the addition of sandbags to their tanks was worth the effort, while other units dispensed with them altogether due to the strain placed on the vehicle's automotive components. Lieutenant-General George S. Patton frowned on the use of sandbags in his Third Army due to the toll they took on suspension system components. This belief often caused problems with armored divisions being temporarily assigned to his command, coming in from either the First or Seventh Armies, whose commanders had no objections to the use of sandbags on tanks.

The addition of cement to the exterior of M4 series tanks was not as common as the application of sandbags in the European Theater of Operations (ETO). To verify the effectiveness of cement in repelling rockets armed with shaped-charge warheads, the U.S. Army's 709th Tank Battalion conducted a test that involved firing a number of rockets from a *Panzerschreck* at the cement-covered glacis of an M4 series tank. The cement did not stop the shaped-charge rocket warheads from penetrating the glacis of the vehicle, but it did reduce the damage inside the tank. Depending on the point of entry, shaped-charge warheads can ignite fuels, fluids, and ammunition within a struck vehicle.

A popular answer to the threat from hidden German soldiers armed with the *Panzerschreck* or the *Panzerfaust* was referred to as "reconnaissance by fire." A description of the merits of that practice appears in an article by an unnamed American tanker, titled "Technique of the Tank Platoon as the Point in an Exploitation," in *Armor in Battle*, a March 1986 publication by the U.S. Army Armor School:

> We come to that highly controversial subject: reconnaissance by fire. On this subject the writer had two complete changes of opinion. During his first days in combat, he employed it extensively. Later it seemed distracting, to destroy the element of surprise. Then he gaily rode into a neat ambush just across the Rhine. From there on he fired on everything remotely suspicious on the ground that it was German in any case. Of course, the life span of tractors and other farm vehicles of suspicious silhouette was short indeed.
>
> More seriously it should be said that reconnaissance by fire is almost a necessity if moving steadily. It is sometimes a waste of ammunition, but it has a decided morale factor. It is good for your morale and decidedly disturbing to the other fellows. However, it should be carefully controlled and done intelligently.

THE JAPANESE ANTITANK THREAT

The threat in the PTO to the M4 series of tanks was different from that found in the ETO. The Japanese Army deployed very few tanks and self-propelled guns against the American military during World War II. What they did have was an excellent towed 47mm antitank gun that was introduced into field service in 1942. It was a scaled-up copy of the German-towed 37mm antitank gun. It could penetrate 2 inches (51mm) of armor at 1,000 yards. This meant it could punch a hole through the M4 series tank in any area except the glacis or the front of the turret most of the time.

A Marine Corps document dated May 1945, titled *Iwo Jima, 4th Tank Battalion Report*, describes Japanese antitank measures involving their 47mm antitank gun:

> Antitank gunnery was generally excellent. The gun positions were usually well constructed and well concealed. Guns had good fields of fire and good alternate fields of fire. Most guns had alternate positions, in some cases several, and this allowed the enemy to shift guns, so that located positions would be empty the next day, and our tanks were surprised by fire from previously undetected guns. Fire discipline of 47mm gun crews was excellent, and few erratic or waste rounds were observed.

To make up for the shortfall of tracked and towed antitank weapons to deal with American tanks, the Japanese Army often depended on the fighting spirit of their infantrymen. Lacking the hand-held anti-armor rocket launchers provided to the German infantry in the latter part of World War II, the Japanese infantrymen evolved their own method of tackling enemy tanks.

During one combat encounter on the Japanese-occupied island of Saipan on July 8, 1944, a Japanese soldier tossed a hand grenade onto the turret of a Marine Corps tank commanded by Sergeant Frederick Grant Timmerman who was operating the .50 caliber machine gun on the roof of the vehicle. Rather than allow the grenade to drop within the turret and no doubt kill his crew, according to his Medal of Honor citation, "Sergeant Timmerman unhesitatingly blocked the opening with his body holding the grenade against his chest and taking the brunt of the explosion."

This passage from the U.S. Army Center of Military History publication, The Technical Services, *The Ordnance Department: On Beachhead and Battlefront*, describes how Japanese infantrymen dealt with tanks:

> The Army's 193rd Tank Battalion noted a pattern of Japanese infantry attacks in 1945 on Okinawa. Japanese squads of three to nine men attacked individual tanks. Each man in the squad filled a role. One man threw smoke grenades to blind a targeted tank. The next man threw fragmentation grenades to force the tank's crew to close their hatches. Another man placed a mine on the tank's tracks to immobilize it. A final man placed a mine or explosive charge directly on the tank to attempt to destroy the tank.

AMERICAN COUNTERMEASURES

To counter Japanese towed antitank guns and infantry tank-hunting teams armed with satchel charges, magnetic mines, regular mines, and on some occasions, with explosives strapped to their bodies, American tankers in the PTO would adopt some of the same countermeasures employed by their counterparts fighting in the ETO. This included adding spare track links, sandbags, or cement to the exterior of their vehicles. The downside of this extra protection meant an increase in weight and sometimes vehicle width that would cause problems for the U.S. Navy when attempting to transport these vehicles by landing craft.

As Japanese infantrymen would often attempt to set off explosives over the engine deck of tanks in order to immobilize them, Marine Corps tankers would sometimes place a layer of sandbags over that portion of their vehicles. Because Japanese infantry tank-hunting teams would also target the hatches on M4 series tanks with hand-emplaced mines or explosive charges, some Marine Corps tankers installed steel mesh covers over their hatches or even went as far as welding on nails points up on the hatches of their tanks.

The Marine Corps tankers of C Company, 4th Marine Tank Battalion, attached 2-inch by 12-inch wooden boards to the sides of their vehicles with a 4-inch gap between them and the vehicle's exterior hull in which they poured cement. The intention was that the wooden planks would defeat magnetic mines, and the cement would reduce the effectiveness of the AP round fired from Japanese-towed 47mm antitank guns.

American tankers in the Pacific discovered that canvas or neutral materials applied on the tanks made the magnetic mines slide off. The U.S. Army's 713th Flame Thrower Tank Battalion mixed sand with paint to stop magnetic mines from adhering to the surface of their vehicles, in a manner similar to that of the German's *Zimmerit*. However, since they did not encounter any magnetic mines in combat, the 713th's concept remained unproven.

In summer 1944, Chrysler Engineering, part of the Chrysler Corporation, did a trial installation of the heavy tank T26E3's 90mm main-gun-armed turret onto the hull of a second-generation M4 medium tank as shown here. Had this been available for immediate shipment in quantity, it might have been accepted. However, the powers-that-be in the U.S. believed that this arrangement would not be ready for delivery before January 1945. In their minds it therefore made more sense to await the production of the heavy tank T26E3, which later became the M26 heavy tank. (National Archives)

In response to the threat from enemy infantry tank-hunting teams, a number of countermeasures were explored by the Ordnance Department. These included everything from a special flame-thrower device mounted externally on the four corners of an M4 series tank, to a variety of explosive devices, be it antipersonnel mines or hand grenades on the exterior of the tanks. It was intended that these devices be detonated from within the vehicle by the driver when enemy tank-hunting teams approached. Due to the danger of these devices going off accidentally around friendly troops, none of these proposed arrangements was adopted.

As with the German Army, the Japanese Army employed standard antitank mines to stop enemy tanks. The Japanese Army would also use improvised antitank mines. Some comments on the use of antitank mines by the Japanese Army appear in a U.S. Marine Corps report titled *Armored Operations on Iwo Jima*, dated March 16, 1945:

> The enemy practice of tying aerial bombs to a yard stick mine is an expensive method of mining. Their policy is to completely demolish the tank with the mine as distinguished from the German policy of merely stopping the tank with the mine and destroying it by gunfire.
>
> Japanese mining was erratic. In one case the mines were marked with stakes. In another case a complete field beside a road was mined while the road was left untouched. Tanks on the beach were able to pick their way through [mine] fields because the mines were spaced too far apart.

FIREPOWER ISSUES

Following the U.S. Army landings in France on June 6, 1944, there was a quick push inland after the German defensive positions along the French coast were overcome. It was shortly thereafter that the American soldiers had a shock when they discovered what an incredible terrain obstacle the Normandy farmers' hedgerows created. The thick and tall centuries-old hedgerows began right behind the American landing beaches, codenamed "Omaha" and "Utah," and extended up to 50 miles inland in some areas. They provided excellent cover and concealment for the defending German units.

To make matters even worse, U.S. Army M4 series tankers who were assigned to assist the infantry in overcoming the Normandy hedgerows began encountering the 50-ton German Pz.Kpfw. V Panther medium tank in ever increasing numbers beginning in July 1944. The frontal armor on this tank proved impervious to the M61 APC-T projectiles fired by the 75mm gun M3 even at pointblank range.

The welded-armor glacis on the Panther tank was 3.15 inches (80mm) thick and sloped at 55 degrees, with the front of the Panther tank welded-armor turret being 4 inches (100mm) thick, fitted with a rounded cast-armor gun shield also 4 inches thick. How was it that the U.S. Army did not have an M4 series tank in service that possessed a main gun that could penetrate the Panther tank's frontal armor?

While the primary effectiveness of main guns on modern tanks comes from their ability to defeat the armor protection of other tanks, when the M4 series tank first appeared in 1942, its main purpose was envisioned by the U.S. Army as a deep attack (exploitation) weapon and not as a tank-versus-tank combat vehicle. This flawed concept is seen in a memorandum by Lieutenant-General Lesley J. McNair, commander AGF, dated January 23, 1943, in which he states the following: "It is believed that our general concept of an armored force – that it is an instrument of exploitation, not greatly different in principle from horse cavalry of old – is sound."

To allow the M4 series tanks to fulfill their primary mission as a weapon of exploitation, the U.S. Army fielded highly specialized vehicles for combating enemy tanks, known as tank destroyers, supposedly mounting weapons of sufficient power to penetrate any opposing armored vehicle. The idea that it took a specialized vehicle to destroy enemy tanks was championed by McNair, a man who had a great deal of influence with the then-U.S. Army Chief of Staff, General George C. Marshall, who had appointed him to the job.

American tankers engaging late-war German tanks in Northwest Europe beginning in the summer of 1944 and into early 1945 soon saw the fallacy of McNair's strong belief in the value of tank destroyers. Colonel S. R. Hinds, of Combat Command B of the 2nd Armored Division, stated in the March 1945 report *United States vs. German Armor* that "[i]n spite of the often quoted tactical rule that one should not fight a tank versus tank battle, I have done it almost invariably, in order to accomplish the mission."

Lieutenant-Colonel Wilson M. Hawkins, commanding the 3rd Battalion, 67th Armored Regiment of the 2nd Armored Division, was quoted in the same report: "It has been stated that our tanks are supposed to attack infantry and should not be used tank vs. tank. It has been my experience that we have never found this an ideal situation for in all our attacks we must of necessity fight German tanks. Therefore, it is necessary for a tank to be designed to meet adequately this situation."

PROBLEMS IN OVERCOMING THE PANTHER TANK

The Panther medium tank was first confronted by the U.S. Army during the fighting in Italy in early 1944. Its 75mm main gun was designated the 7.5cm Kw.K. 42 L/70 and fired an APC-T round known as the Pzgr. 39/42, which had a muzzle velocity of 3,035ft/sec. The projectile portion of that round could penetrate 3.5 inches (89mm) of armor at a range of 2,187 yards. Despite the outstanding attributes of the Panther tank's 75mm main gun, the vehicles did not make much of an impression upon the American frontline forces fighting in Italy because they were used in small numbers and restricted by poor terrain.

Unfortunately, the Italy experience led to the false belief among the U.S. Army senior leadership that the Panther tank, like the Pz.Kpfw. VI Tiger Ausf. E heavy tank, was only going to be fielded in small numbers. Hence, there was no need to spend the time and effort in developing tank or tank-destroyer main guns able to destroy these late-war German vehicles. At the lower levels of command, not everybody was convinced the Panther tank would only be deployed in small numbers. However, their voices would not be heard.

The M4 series tank problems with the Panther tank caused General Dwight D. Eisenhower (Supreme Allied Commander of the Allied Expeditionary Force) to send one of his subordinates back to the tank designers in the United States in July 1944 to look for a quick solution.

Eisenhower's subordinate was shown an M4 mounting a 90mm gun M3 armed turret from the future heavy tank M26. This arrangement seemed to be the answer to the Panther tank problem. However, he was told it could not be made ready for at least six months and it made more sense to wait for the M26 supposedly due out in that time frame. Sadly, while the M26 did make it to Northwest Europe in February 1945, it was fielded in such small numbers initially that it would have no real influence on the battlefield before the war in Europe ended in May 1945.

THE BRITISH ARMY SOLUTION

The British Army did not make the same mistake the U.S. Army did regarding the widespread fielding of the Panther tank. They had grasped right away that the vehicle was not going to be formed into a small number of independent battalions like the Tiger Ausf. E heavy tank, but was intended as a replacement for the Pz.Kpfw. IV medium tank in the Panzer Divisions.

The British Army therefore set out to field a tank main gun that could penetrate the frontal armor of anticipated German late-war tanks and had it in service before the Allied invasion of France on June 6, 1944. As a tank main gun, the weapon was designated the 17-pounder (76.2mm), quick-firing Mark IV. It was mounted in the M4 or M4A4 and became known as the "Firefly." It began appearing in British Army service in early 1944. By the time the British and American armies had landed in France there were almost 350 in British service. When the war in Europe ended in May 1945, 2,200 units of the Firefly had been constructed.

The 17-pounder main gun on the Firefly possessed the penetrative abilities to punch holes in the frontal armor array of late-war German tanks at normal combat distances. This ability was not appreciated by the U.S. Army until after the invasion of France, and large numbers of Panther tanks were encountered in June and July 1944. Desperation drove American General Omar N. Bradley to inquire in August 1944 from his British Army counterpart if some Fireflies could be made available to the U.S. Army. This request was denied as the British Army felt that they did not have enough Fireflies in service at the time to spare.

In September 1944, the British Army informed their American comrades in arms that they could now spare at least enough 17-pounder quick-firing Mark IV tank guns to re-arm up to 100 M4 series tanks for them. This time it was the U.S. Army which declined, as there was then a temporary shortage of 75mm main-gun-armed M4 series tanks in the ETO due to extremely high combat losses.

The program to up-arm U.S. Army M4 series tanks with the 17-pounder, quick-firing Mark IV tank gun bore no fruit until near the end of the war in Europe when 100 units out of 160 planned were completed. Twenty were transferred to the British Army, with the other 80 vehicles never issued to units in the field. The ultimate fate of these vehicles is lost to history.

The Chrysler Corporation was requested to design a spacer system that would allow extended end connectors to be attached to both the inner and outer track links of M4 series medium tanks to improve off-road mobility. The modifications kits were not available until early 1945, too late to see much service during World War II. The M4A1 tank pictured might be referred to informally as the "M4E9," although it lacks the extended end connectors that would normally be fitted in this configuration, but the space for the inner connectors between the track and the hull side can be seen here. (Michel Krauss)

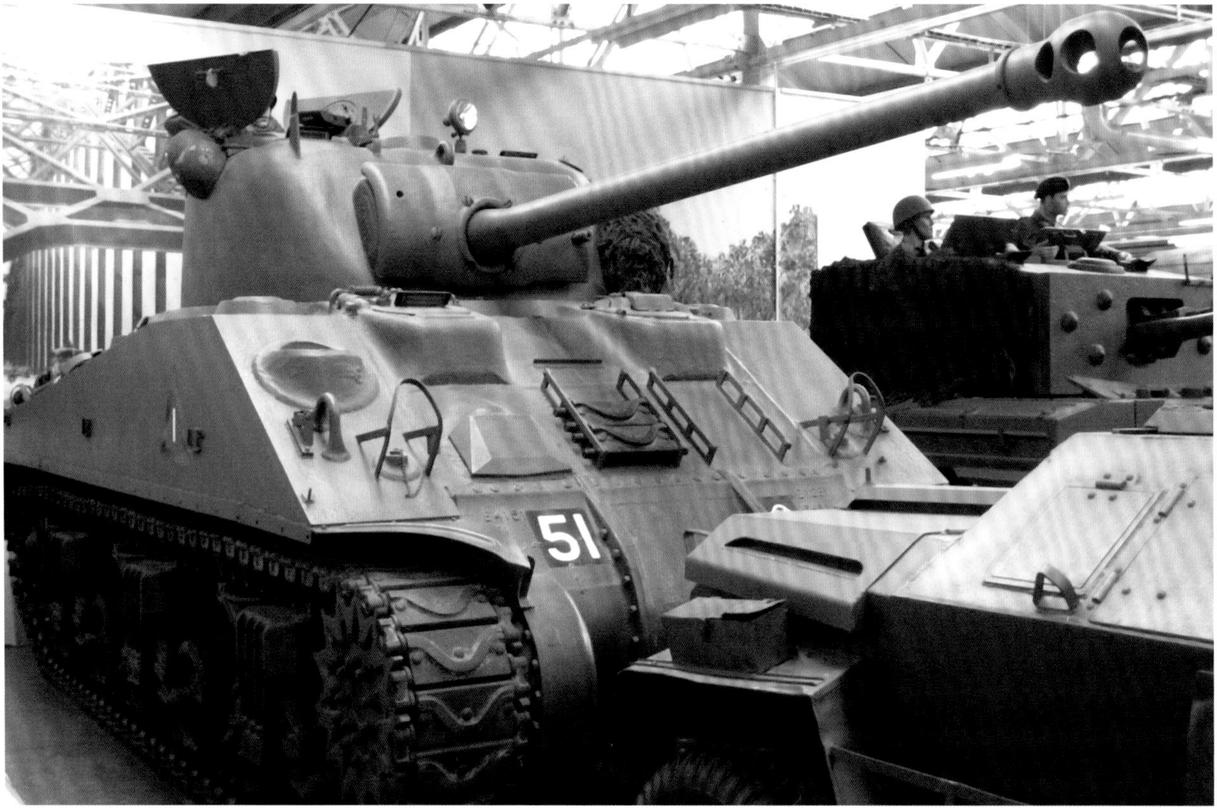
51

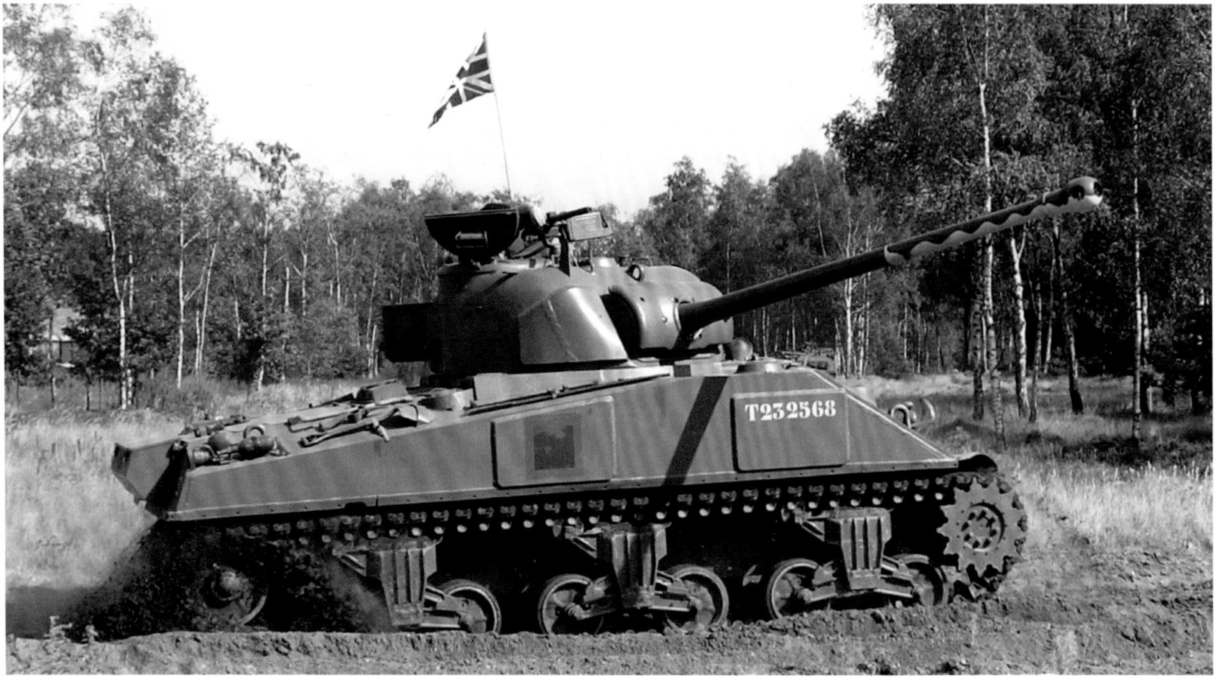
T232568

UPGRADING EARLY-PRODUCTION M4 SERIES TANKS

Beginning in spring 1943, the turret armor of the M4 series tanks was thickened. The original M34 combination gun mount, which included a 3-inch-thick cast-armor gun shield and a narrow 2-inch-thick cast-armor rotor shield in front of it, was replaced in production by the M34A1 combination gun shield that had a 3.5-inch-thick cast-armor gun shield and a full-width, 2-inch-thick cast-armor rotor shield behind it.

Experience gained by American tankers in North Africa convinced them that the German Army adoption of telescopic sights for their tanks was a superior arrangement than the periscope gun sight mounted in the early-production M4 series tanks. The addition of the telescopic sight M55 for the gunner, who sat on the right side of the 75mm main gun, created a small opening for the telescopic sight on the front of the tank's cast-armor turret that introduced a ballistic weak spot and was compensated for by the wider M34A1 cast-armor rotor shield.

Prior to the introduction of the telescopic sight, the M4 series tank gunner relied solely on an overhead, fixed forward-looking periscope designated the M4, or the improved M4A1, both of which incorporated the telescope M38 fitted with a ballistic reticle.

A modification was conceived in early 1943 to have early-production M4 series tanks with the original M34 cast-armor rotor shield upgraded to approximately the M34A1 standard by the addition of a welded-on cast-armor section on the right side of the rotor shield. However, as best as researcher Joe DeMarco can determine, something went wrong and the contractors tasked with the project were unable to obtain the necessary parts to complete the work until early 1945, with none making it into the field before World War II ended.

FIRE-CONTROL IMPRESSIONS

The War Department's *Report of the New Weapon Board*, dated April 27, 1944, describes the U.S. Army's impression of the effectiveness (or lack thereof) of the M4 series tank fire-control system:

> Because of the excessive dispersion which occurs with the M4 periscope, firing of the main guns is confined almost entirely to the artillery method of sensing and locating burst and giving corrections in mils to the gunners. The average dispersion which occurs as a result of slack in the periscope holder and linkage extends four mils in both planes. This dispersion is so great that guns do not stay bore-sighted with the telescope after any operation… There is little use of the coaxially mounted telescope; the dispersion which results from its use is even greater than that experienced with the M4 periscope. In addition, the optics of the M55 telescope are unsatisfactory, resulting in unsatisfactory light-transmission characteristics. Furthermore, most gunners report that it is very difficult for them to get their heads into proper position for sighting through the coaxial telescope.

The March 1945 report *United States vs. German Armor* mentions further problems with sighting, as First Lieutenant Coulter M. Montgomery of the 66th Armored Regiment points out:

On display at the Tank Museum, Bovington, is a Sherman VC "Firefly" based on the medium tank M4A4. It is armed with a 17-pounder (76.2mm) main gun that proved to be one of the few Western Allies' tank-mounted weapons that could penetrate the frontal armor of the German Panther tank at combat distances. (Tank Museum)

Belonging to the Belgian National Military Museum is this Sherman VC "Firefly." Due to the recoil length of the 17-pounder main gun in the Firefly turret, the radio, normally mounted in the rear of the turret bustle, was relocated to an armored box at the rear of the tank, as seen on this vehicle. (Michel Krauss)

Our sight reticle is okay, but our sights are not nearly powerful enough. These new telescopic sights are an improvement over the old periscope sight, but are still not powerful enough. The Germans seem to have better glass in theirs. Couldn't a ten or twelve power scope be devised to fit in the periscope? We shouldn't want to sacrifice the field of vision that our periscope gives us, but the telescope in it isn't nearly powerful enough. The position of the telescope is not satisfactory. The gunner has to cramp himself too much to use it.

Tom Sator of the 4th Armored Division remembers looking through the gun sight of an abandoned Pz.Kpfw. IV during his time in the ETO and swears he could see blades of grass a mile away. While he admits that might be a bit of an exaggeration, the obvious superiority of the German gun sight he examined that day compared with that found in his tank was extremely depressing and made him want to take the next boat back to the United States.

MORE MODIFICATIONS

In spring 1943, the Ordnance Department embarked on a program to upgrade the thousands of very early-production first-generation M4 series tanks then employed as training vehicles in the United States. During the rebuilding process, the tanks went through a modernization program, which incorporated the addition of many of the improvements developed after they first rolled off the assembly lines. These upgraded first-generation M4 series tanks then went off to various war zones with both the American military and other countries under Lend-Lease.

Combat experience had shown that the pronounced driver's and bow gunner's hatch hoods, which protruded from the glacis of the welded-armor M4 series tanks, were a ballistic weak point. To correct this problem, the vertical surfaces on the front of these hoods had 1.5-inch-thick welded armor plates welded onto them at an angle of approximately 35 degrees from the vertical. This armor-upgrading process occurred both on the assembly line and in the field and brought the protection afforded by the two hatch areas up to approximately the rest of the welded hull glacis.

While M4A1s were not officially specified as needing the addition of welded-armor plates on the driver's and bow gunner's hatch hoods, period photographs show that some had the plates added to their less-pronounced hatch protrusions.

In early-production M4 series tanks, part of the cast armor interior turret wall had been machined away in front of the gunner's position to make more room for the gunner when operating his electro-hydraulic power traverse unit control or electric turret power traverse system. Combat experience in North Africa had quickly demonstrated that this had been a serious mistake. As a result, early M4 series tanks were refitted with add-on armor plates on the right front of their cast armor turrets beginning in spring 1943 through summer 1943.

In summer 1943, new cast armor turret castings entered the production pipeline. On them, the pistol port was eliminated, and the area of the "thin spot" was thickened, so that the welded-on armor turret patches were no longer required. This thickening can be detected in close-up pictures of M4 series tanks so modified.

So as not to interrupt the production flow, in early 1943 the Ordnance Department came up with a "Quick Fix" to the problem of main gun ammunition fires in earlier and then-current production M4 series tanks. The 12 75mm rounds clipped to the wall of the turret basket were eliminated. The thin steel ammunition containers were retrofitted with 0.25-inch (6.35mm) armor plates. One-inch (25.4mm) welded armor plates were welded to the exterior of the upper hulls opposite the 75mm ammunition racks, one on the right side and two on the left side.

In the case of very late-production M4A1s, the upper hull armor casting was thickened by an inch in the necessary areas, obviating the need for the welded-on appliqué plates. Finally, the thin perforated sheet metal that originally enclosed the vehicle's turret basket was removed to improve access between the vehicle's fighting compartment and turret. More importantly, this provided the crew with more avenues of emergency escape.

The necessary components began to enter the pipeline in mid-1943, and manufacturers began installing the various modifications in new production tanks as they were able to procure them. Tank depots were also mandated to install modifications when the supply of the various kits became available to them. The U.S. Army desired to bring the 1,300 odd M4 series tanks in the United Kingdom up to the latest standards. Thus a sufficient number of modification kits were shipped to Great Britain, so that just about every U.S. Army M4 or M4A1 already there that needed modifications had them retrofitted before D-Day (June 6, 1944). The troops training there could not spare the time, so the upgrade program was contracted out to British firms.

MOBILITY IMPROVEMENTS

Field reports had indicated that the existing 16.56-inch-wide tracks on the VVSS of early-production M4 series tanks did not always provide sufficient off-road flotation. To rectify this situation it was decided to use duckbill-shaped devices referred to as "extended end connectors" to increase the track's ground contact area and thereby lower the tank's ground pressure. As there was no space between the inside of the suspension system and the lower hull on M4 series tanks to attach extended end connectors, Chrysler was tasked by the Ordnance Department in February 1944 to design a spacer system for the sprockets and other suspension components in order to provide the necessary room for the addition of extended end connectors on the inside of the track.

Testing of the Chrysler-designed spacer arrangement proved successful, and the Ordnance Department ordered 1,000 units for fitting on remanufactured early-production M4 series tanks. They also ordered another 1,000 units of the Chrysler-designed spacer arrangement in kit form to be added to selected M4 series tanks in the

Both the welded-hull M4A2 and M4A3 medium tanks were adapted to mount the new "T23 turret." Reflecting the changes to these second-generation tanks, the M4A2 was designated M4A2(76)W, and the M4A3, as seen here on display in Luxembourg, became the M4A3(76)W. The "W" in the vehicle's designation is shorthand for "wet stowage," which featured ammunition bins that included liquid-filled inserts intended to extinguish propellant fires in the event the bins were penetrated. (Pierre-Olivier Buan)

field. With the extended end connectors fitted to both the inside and outside of the tracks, the total width was just a bit over 23 inches and lowered the vehicle's ground pressure to 10 pounds per square inch. Vehicles equipped with the Chrysler-designed spacer arrangement had "E9" added to their designation to reflect the upgrade.

NEW HATCHES AND PISTOL PORT ISSUES

Beginning in November 1943, a small overhead oval hatch was fitted in the roof of M4 series tanks to assist the loader in entering and leaving the vehicle. Combat experience had shown that it was extremely difficult for the three men in the turret to exit their vehicle in a timely manner when forced to use only the vehicle commander's overhead hatch. This loader's hatch was provided with a hatch lock and featured counterbalanced springs to make it easier to open and close.

The driver's and bow gunner's hatches on the early-production and then-current production M4 series tanks were also fitted with hatch locks and counterbalanced springs beginning in 1943. In place of the original circular split hatch for the vehicle commander of the M4 series tanks, the remanufactured tanks sometimes received a newly designed vehicle commander's cupola that was developed for an improved second generation of M4 series tanks.

Because the Ordnance Department felt that the pistol port on the left side of first-generation M4 series tank turrets was a ballistic weak spot, it ordered the factories building the tank to delete it from production units of the vehicle beginning in April 1943. Those vehicles on the assembly line when the decision was made would retain

Efforts by the Ordnance Department to keep the M4 series medium tanks viable on the battlefield led to a fairly significant redesign of the hull front sections, along with the introduction of numerous improvements. Most "second-generation" M4 series medium tanks were armed with a more powerful 76.2mm main gun mounted in the turret developed for the abortive T23 medium tank. The first units were produced in January, 1944 and designated M4A1(76)W, an example of which is seen here. (TACOM Historical Office)

the casting impression of the pistol port but would never have it installed. Those M4 series tanks that had already rolled off the assembly line with the pistol port and not yet been issued to field units had their pistol ports welded shut.

Strong user feedback on the usefulness of the pistol port resulted in the Ordnance Department backing off on its original decision. New turrets with pistol ports, as well as loader's hatches, were introduced into production starting in November 1943.. Despite its misleading name, the pistol port was actually used by crews to pass ammunition into the tank during resupply. This method required one less man, because no one was needed to stand on the hull to pass ammo down through the loader's hatch. Ammunition could be passed by a man standing on the ground directly into the turret. Photo research by Joe DeMarco and Pierre-Olivier Buan have led them to conclude that the majority of M4 series tanks used in the ETO in 1944 had no pistol ports or had them welded shut.

THE SECOND GENERATION OF M4 SERIES TANKS APPEARS

Even before the M4 series tanks began rolling off the assembly lines of American factories in early 1942, the Ordnance Department was considering ways of improving it. By early 1943, the Armored Force concluded that it would have to make do with the M4 series tanks for the rest of the war and set out to modify the vehicle's existing design to improve its combat effectiveness. The M4A4 and M4A6 would not be part of this upgrading process.

To improve the off-road mobility of the second-generation M4 series medium tanks, many were built with a greatly improved suspension system incorporating tracks widened to 23 inches. This was referred to as the Horizontal Volute Spring Suspension (HVSS) system. Pictured is an M4A3(76)W HVSS at the former Patton Museum of Cavalry and Armor. (Michael Green)

Very soon, however, the number of changes requested for inclusion on the M4 series of tanks had grown to such an extent that the need for a major redesign of the entire series became apparent. This effort began in July 1943, and would result in a much improved second generation of M4 series tanks.

Chrysler received permission in September 1943 to build a number of pilot models incorporating numerous second-generation changes to M4 series tanks. These vehicles included the latest manufacturing modifications. However, due to delays in acquiring drawings necessary to begin the project, actual work did not start until December 1943, with the completed and semi-completed second-generation pilots showing up in February 1944.

SECOND-GENERATION MODIFICATIONS

The biggest change to the majority of second-generation M4 series tanks would be the addition of a more powerful 76mm main gun, designated the M1A1, M1A1C, or M1A2, which was mounted in a new, larger, cast-armor turret. Rather than take the time to design a new turret with space for a larger 76mm main gun, the Ordnance Department decided to use the pre-series production version of a cast-armor turret originally developed for the never-fielded T23 medium tank.

Unlike the first-generation M4 series tank with a cast-armor rotor shield mounted in front of the cast-armor gun shield, the second-generation M4 series tank armed with a 76mm main gun did away with the rotor shield and relied solely on a full-width vertical cast-armor gun shield for protection, designated the combination gun mount M62, which was 3.5 inches (89mm) thick. Not everybody was pleased with the new gun mount. Many tankers felt the mount was too light for the 76mm gun and the extra play in the mount caused inaccuracy in firing.

Another major modification for the second-generation M4 series tanks included the incorporation of a new vehicle commander's cupola developed by the Libby-Owens-Ford Glass Company. It was fitted with six laminated bullet-resistant glass vision blocks uniformly spaced around a central 21-inch diameter hatch. There was also an M6 periscope installed in a 360-degree rotating mount in the overhead hatch cover. Due to production shortfalls, some second-generation M4 series tanks left the factory with the original rotating two-piece circular split hatch from the first-generation M4 series tanks at the vehicle commander's position.

The loader on the second-generation M4 series tank armed with a 76mm main gun was originally provided with the same rotating two-piece circular split hatch for the

A key identifying feature of the second-generation M4 series medium tanks was the redesigned, slightly thicker (2.5-inch) glacis plate shown here on an M4A3(76)W HVSS. It was sloped at 47 degrees as opposed to the 60 degrees of first-generation M4 series medium tanks. The modified angle permitted the designers to do away with the protruding driver's hatches of first-generation M4 series medium tanks. (Michael Green)

vehicle commander from the first-generation M4 series tanks. Eventually, this hatch was replaced by the same small oval hatch applied to the modernized first-generation M4 series tanks beginning in November 1943.

An easily noticed exterior spotting feature that identifies second-generation cast- and welded-armor upper hull M4 series tanks is the disappearance of the very pronounced driver's and bow gunner's hatch protrusions seen on first-generation upper hull M4 series tanks. This came about because the slightly thicker 2.5-inch-thick (64mm) welded armor glacis of the second-generation M4 series tanks (M4, M4A2, and M4A3) was set at a less steep angle of about 47 degrees compared with the approximately 57-degree slope seen on the first-generation cast- and welded-armor upper hull M4 series tanks.

The reset of the glacis on second-generation M4 series tanks allowed for the inclusion of new larger, counterbalanced, angled overhead hatches for the driver and bow gunner, which made ingress and egress for the driver and bow gunner a much easier process. These new overhead hatches were designed by the Fisher Body Corporation of General Motors and were copied from those developed for the M10 tank destroyer.

As with the original first-generation small, narrow overhead hatches, the new, much larger overhead hatches contained a 360-degree rotating M6 periscope. Both the cast- and welded-armor front upper hulls of the second-generation M4 series tanks also retained the fixed forward-facing M6 periscopes. Rather than having the fixed forward-facing periscopes directly in front of the periscopes in the driver and bow gunner overhead hatches, they were offset to the center of the front upper hull.

With the introduction of the larger overhead hatches for the driver and bow gunner on the second-generation M4 series tanks, there also appeared in the design a slightly raised rear turret bustle. These new high bustle turrets can be recognized in pictures because there is almost no downward slope on the top portion of the turret bustle, whereas the low bustle turrets for first-generation M4 series tanks has a pronounced downward slope on the top of the turret bustle.

The raised turret bustle's intended purpose was to clear the new, larger upper hull hatches that appeared with the introduction of the second-generation M4 series tanks armed with the 75mm main gun. However, when 300 surplus early-production first-generation M4 series tank turrets became available because the chassis of the tanks were being converted into tank recovery vehicles, Fisher Body was granted permission to mount them on the chassis of second-generation M4A3 tanks after upgrading them with the raised vehicle commander's cupola and an overhead oval hatch for the loader. To make sure that the low bustle turret would clear the new, larger driver and bow gunner hatches, Fisher Body flame-cut some of the lower rear corners of the recycled turrets, which can be seen on some close-up pictures.

An important improvement to the second generation of the M4 series medium tanks was the addition of a new raised commander's cupola, as seen here. It featured six laminated glass vision blocks around its circumference, and an overhead periscope in the hatch cover. It was also retrofitted to some first-generation M4 series medium tanks as a field modification or during remanufacture. (Michael Green)

SECOND-GENERATION SUSPENSION IMPROVEMENTS

The first-generation M4 series tanks rode on the VVSS combined with 16.56-inch-wide (42.1cm) tracks. The shortcomings of this system were apparent to all. As early as December 1941, the Ordnance Department looked at other types of suspension systems. The system showing the most promise was the Horizontal Volute Spring Suspension (HVSS) system first considered for fitting on the medium tank M2. While the HVSS never appeared on the M2, it did make it onto the medium tank M3 for test purposes only.

In early 1943, the Armored Board began testing two first-generation M4 series tanks fitted with the HVSS system that retained the 16.56-inch-wide tracks of the VVSS currently in use. While it showed promise, there was insufficient improvement to warrant the costly retooling needed to change over the first-generation M4 series tanks from the VVSS to the new, narrow track HVSS. Work, therefore, began on designing an improved HVSS system in September 1943 utilizing more bogie wheels and wider tracks. Tests proved this new wide track HVSS system to be far superior to the VVSS and the earlier narrow track HVSS system in durability, flotation, riding qualities, and bogie wheel life.

As the name indicates, the volute springs in the HVSS system were mounted horizontally on the bogie assembly instead of vertically. Elimination of friction from sliding shoes, improved geometric design, addition of shock absorbers (VVSS-equipped M4 series tanks did not have shock absorbers) and bogie wheels, and, together with the use of center-guides instead of outside track guides, decreased the possibility of track throwing. The wide track required the use of dual bogie wheels, doubling their number and distributing the wear more uniformly than in the VVSS design.

In September 1943, the Ordnance Committee recommended development of the improved HVSS system with 23-inch-wide center-guided tracks for application to a first-generation M4 pilot vehicle, which was designated the M4E8. As other M4 series pilot tanks underwent conversion to the new type of suspension system, they also saw their designations change. The M4A1 became the M4A1E8, the M4A2 the M4A2E8, and the M4A3 became the M4A3E8.

Early testing showed that the 23-inch-wide version of the HVSS system improved performance of the pilot tanks, but needed some additional refinement. In November 1943, the Ordnance Committee recommended acquiring ten more sets of the HVSS system to be mounted on M4A3E8 pilot tanks. These vehicles were completed in early summer 1944 and sent out for testing to a variety of military bases.

HVSS IS APPROVED

Even before the testing of the M4A3E8 pilot tanks was completed, the Ordnance Committee recommended in early March 1944 that the HVSS system be fitted onto 500 second-generation M4A3 tanks equipped with a 76mm main gun and having the "Wet" main gun ammunition storage arrangement. On tank data plates this arrangement is abbreviated to "M4A3 76 Gun, Wet." For the sake of brevity the author will convert this to M4A3(76)W.

In total, Chrysler would build 4,542 units of the M4A3(76)W between March 1944 and April 1945. They began construction of these tanks fitted with the HVSS system in August 1944. As they were the only firm that kept track of how many second-generation M4 series tanks they built with the HVSS system, we now know courtesy of researcher Joe DeMarco that 2,167 of the 4,542 units built of the M4A3(76)W were equipped with the HVSS system.

Most U.S. Army documents from World War II do not distinguish between a second-generation M4 series tank riding on the original VVSS or having the HVSS system. A few documents mention "wide tracks" or "twenty-three inch tracks." There are also a few period documents that list the M4A3(76)W incorrectly as the "M4A3E8," which was intended solely for designating the pilot tanks, although it might have been adopted for the sake of brevity in documents or by the tankers and supporting troops in the field.

The addition of the HVSS system increased the weight of the second-generation M4 series tanks somewhere between 3,000 to 5,000 pounds depending on the type of track used. It also increased the tank's width to 9.8 feet with sand shields fitted. The first-generation of M4 series tank without sand shields were 8.7 feet wide. On the positive side, the increased width of the new tracks that came with the HVSS system reduced ground pressure from 14.5 pounds per square inch to about 10 pounds per square inch.

Belonging to a private collector is this M4A1(76)W HVSS. The effectiveness of the "wet" system is somewhat doubtful. That the ammunition bins were armored, and moved from the sponsons to the floor of the tank, were the real factors that increased crew safety. (Michael Green)

M4A3(76)Ws with the HVSS system fitted reached Northwest Europe in December 1944. Experience had already shown U.S. Army tank units that a track 23 inches wide was still not enough to keep a tank from bogging down in deep mud. In response, the Ordnance Committee approved production of kits for field installation with 39-inch grousers. These were still in development when World War II ended and therefore never went into production. The M4A3(76)W with the HVSS system fitted also eventually reached U.S. Army armored units fighting in Italy near the end of the war in Europe.

A total of 3,426 units of the M4A1(76)W were built between January 1944 and July 1945 by the Fisher Body Corporation of General Motors. Research at the National Archives undertaken by Joe DeMarco showed that 1,255 units were built with the HVSS system. Additional findings uncovered by DeMarco showed that out of a total of 2,915 units of the M4A2(76)W that rolled off the assembly lines between May 1944 and May 1945, 1,321 units were fitted with the HVSS system.

The American First and Third Armies received an allotment of M4A1(76)Ws with the HVSS system starting on April 6, 1945. Efforts by many tank history buffs have yet to uncover any pictorial evidence showing them in troop service before the war in Europe officially ended in May 1945. The majority of M4A2(76)W tanks, with or without the HVSS system, were reserved for Lend-Lease shipment to the Soviet Union. Pictorial evidence shows that the Red Army used the M4A2(76)W with the HVSS system in combat against the Japanese Army in Manchuria in June 1945.

MAIN GUN AMMUNITION RELOCATION

One of the most important, though not externally visible, second-generation modifications to the M4 series tank involved the placing of most of the main gun rounds in the tank's lower hull in storage racks located below the rotating turret basket. The purpose of this arrangement was to reduce the potential for fires caused by projectiles penetrating the tank's exterior armor and puncturing the metal cartridge cases formerly stowed high up in the hull sponsons and the turret basket.

Along with the relocation of the main gun rounds to the bottom of the tank's lower hull, second-generation M4 series tanks came with provisions for protecting those same rounds by placing liquid-filled metal insert containers called "Ammudamp" within the main gun storage racks. The rounds themselves sat in liquid-free cavities that formed part of the storage racks. The liquid within the metal inserts only engulfed the main gun rounds when penetrated by a projectile. On top of the liquid-filled metal inserts were removable plugs used to refill them if the need arose.

Second-generation M4 series tanks equipped with these liquid-filled metal inserts were designated "wet" stowage. Tests generally showed that wet stowage did not effectively quench the ignition of the propellant charges within tanks struck and penetrated by overmatching weapons. Realizing that the wet stowage system was not the perfect solution, some in the U.S. Army recommended that armored containers of higher ammo stowage capacity take its place. There was, however, no follow-through of this recommendation on the M4 series tanks before World War II ended.

U.S. Army tanker Tom Sator recalls that nobody ever informed him that his M4A3(76)W had water-protected main gun ammunition storage racks in the hull bottom of his vehicle or how it was supposed to work.

SECOND-GENERATION DESIGNATION CONFUSION

According to an Ordnance Department report, the term "wet" represented not only the fitting of liquid-protected main gun ammunition racks, but also indicated that the tanks so designated incorporated all the other second-generation modifications as well. This report is somewhat misleading, as the incorporation of features found on second-generation M4 series tanks was phased in at different times and with different manufacturers over the course of their introduction into field service. This meant that many so-called M4 series "wet" tanks did not have all the second-generation features, such as the HVSS system.

It was the early-production units of the M4A1(76)W tanks without the HVSS system that first reached U.S. Army units stationed in Great Britain in April 1944. However, those tanks didn't make it to Northwest Europe until July 1944. Their first combat employment would take place with Lieutenant-General Omar Bradley's First Army during Operation *Cobra*, which lasted from July 25 to July 31, 1944.

It took until September 1944 for the first of the M4A3(76)W tanks without the HVSS system to show up in Northwest Europe. They would see action with Lieutenant-General George S. Patton's Third Army that same month. Other than the original shipment of M4A1(76)W tanks minus the HVSS system that arrived in Great Britain back in April 1944, the bulk of the second-generation M4 series tanks shipped to Northwest Europe without the HVSS system consisted of M4A3(76)W tanks. One hundred and forty M4A3(76)W tanks without the HVSS system fitted were allocated to U.S. Army armor units fighting in Italy. They first went into the field in August 1944. Another 70 of the vehicles were provided to the U.S. Seventh Army that invaded Southern France on August 15, 1944.

The British Army eventually received 1,330 units of the M4A1(76)W without the HVSS system fitted under Lend-Lease. These tanks were designated the "Sherman IIA" and were supplied in lieu of first-generation M4A4s. The British Army was also provided five units of the M4A1(76)W tanks with the HVSS system, which they designated the "Sherman IIIAY." None saw operational use with the British Army. The British Army was never supplied the M4A3(76)W with or without the HVSS system.

Not impressed with the armor-penetration abilities of the 76mm main gun fitted to the M4A1(76)W they were supplied under Lend-Lease, the British Army decided to ship them off to British and Commonwealth armor units fighting in Italy, as there was a less serious threat from late-war German tanks in that theater of operations. The British Army would also supply the 1st Polish Armored Division fighting in Northwest Europe with about 180 units of the M4A1(76) W without the HVSS system. The Free French Army in the ETO were supplied with the M4A1(76)W and the M4A3(76)W without the HVSS system from U.S. Army stockpiles during the latter portion of World War II.

SECOND-GENERATION TIMELINE

Upon standardization of second-generation M4 series tanks with the 76mm main gun and the wet ammunition storage racks into the armored force, production of vehicles armed with the 75mm main gun was limited to just the M4A3 tank. These M4A3 tanks, which also had wet ammunition storage racks and other improvements, would be designated M4A3(75)W. At this point in the war, those first-generation M4 series tank without the wet storage system, such as the M4 and M4A2, came to be called "dry" stowage tanks. For example, the original M4A3 tank was now referred to officially as the M4A3 75mm, Dry Stowage.

In February 1944, the Fisher Tank Arsenal began building the first of the M4A3(75)W tanks. Production of this tank would continue until March 1945, with 3,071 units completed. The second-generation M4A3 tank was not fitted with the cast-armor turret from the T23 medium tank as were all the second-generation M4 series tanks armed with a 76mm main gun. Rather it retained the original cast-armor turret from the first-generation M4 series tank. However, it was eventually upgraded with the new raised vehicle commander's cupola common to most second-generation M4 series tanks armed with a 76mm main gun. None of the M4A3(75)W tanks would be exported under Lend-Lease during World War II.

The need to up-arm the M4 series of medium tanks was realized early on by the U.S. Army Ordnance Department. The result of their labor was a second generation of M4 series medium tanks fitted with a new turret design, mounting a 76mm main gun. The artwork shows a second generation M4A1(76) belonging to Company D, 66th Armored Regiment, 2nd Armored Division. It is taking part in Operation *Cobra*, launched in France in July 1944. To better blend in with the surrounding foliage, the base olive drab has been over-painted with splotches of black. The vehicle's tactical number, the registration numbers, and the tank's nickname appear in white, yellow having been discontinued by the U.S. Army in December 1942. Blue drab registration numbers would last on some vehicles into the postwar era. (© Osprey Publishing Ltd.)

A feature not found on the first-generation M4 series tanks, but incorporated into some second-generation M4 series tanks armed with either the 75mm or 76mm main gun, was a hydraulic traverse control joystick mounted on a bracket on the inner turret wall next to the tank commander's position. By repositioning a control lever, the tank commander could automatically disable the gunner's turret traverse control handle, allowing the tank commander to point the turret in the direction he chose with his joystick.

WET AMMUNITION STORAGE DETAILS

With the change to liquid-protected main gun ammunition stowage on second-generation M4 series tanks, a partial turret basket, which occupied only about one-third of the space, replaced the full turret basket. The partial basket permitted access to the main gun ammunition stowed under the sub-floor on the loader's side of the tank.

In second-generation M4 series tanks equipped with the 76mm main gun and wet storage, the majority of the 71 main gun rounds were stored on either side of the drive shaft enclosure under or behind armored hatches. Thirty-five of the main gun rounds sat in vertical racks on one side, and 30 sat in horizontal racks on the other side of the interior behind the bow gunner's seat. The remaining six main gun rounds, referred to as "ready rounds," resided in an armored box upon which the gunner's seat perched.

With the second-generation M4A3(75)W, the ammunition storage arrangement was somewhat similar to that on the 76mm main-gun-equipped tanks, except for an additional 29 rounds stored within the vehicle, for a total of 104 main gun rounds. Like the 76mm main-gun-equipped tanks, the M4A3(75)W had an armored box on the turret floor with a number of ready rounds.

The second-generation M4A3(75)W main gun ammunition storage arrangement retained the full turret basket that equipped the older first-generation M4 series tanks. Under the turret basket (in the sub-floor) were ten vertically oriented ammunition storage racks, each containing ten main gun rounds. There were another four ready rounds in an armored box on the turret basket floor.

The only access for the loader to the 100 main gun rounds stored in the sub-floor of the M4A3(75)W came from opening two armored hatches in the bottom of the turret basket floor (located directly under his feet). In prolonged firing of the main gun, this storage arrangement called for turning the tank's turret every so often to allow the loader to access all the main gun round ammunition storage racks.

World War II Tank Production (acceptances) M4 Series Tanks

Manufacturer	1941	1942	1943	1944	1945	Total
American Locomotive			126	2,174		2,300
Baldwin Locomotive		12	1,190	43		1,245
Detroit Arsenal Tank Plant		2,432	6,611	5,338	3,369	17,750

Manufacturer	1941	1942	1943	1944	1945	Total
Federal Machine (M4A2)		21	519			540
Fisher Body		1,540	2,240	5,627	2,148	11,555
Ford (M4A3)		514	1,176			1,690
Lima Locomotive (M4A1)		820	835			1,655
Montreal Locomotive	31	1,113				1,144
Pacific Car & Foundry (M4A1)			266	660		926
Pressed Steel		1,174	3,526	2,171	1,276	8,147
Pullman Standard		1,112	2,314			3,426
Totals	31	9,130	21,245	13,179	6,793	50,378

A SOLUTION TO A PROBLEM

One of the biggest problems with the 76mm tank gun fitted to the second-generation M4 series tanks was the weapon's muzzle blast and the resulting target obscuration from smoke and dust. This is seen in a postwar U.S. Army staff memorandum titled "Characteristics for Tank Guns": "Also the muzzle blast from the 76mm was increased so that it was practically impossible to observe fire from the tank. The net result was that when the 76mm gun fired at an enemy tank the crew was temporarily blinded. They could not observe the strike, and if they had missed, were subject to being destroyed before they could try another shot."

This issue was of such serious concern to then Major-General Alvan Cullom Gillem, Jr. (commander of XIII Corps of the Ninth Army in the ETO) that he felt it was important to retain the 75mm main gun on the M4 series tanks as it presented a far less serious problem with muzzle blast.

To correct the problem of target obscuration, the Ordnance Department came up with a two-part solution: first, a longer primer in the 76mm cartridge case that improved the burning rate of the propellant, and second, a redesigned combination muzzle brake and blast deflector to be fitted to the 76mm gun M1A1.

Those 76mm tank guns equipped with the threaded barrels for the fitting of a muzzle brake were designated the 76mm Gun M1A1C, while a later model having rifling with a tighter twist became the 76mm Gun M1A2. The tighter rifling improved projectile stability in flight, which in turn produced a slight increase in penetration at longer ranges. When no muzzle brake was available for fitting on threaded 76mm guns, a protective ring was installed to protect the threads from damage.

Major Paul A. Bane, Jr., of the 2nd Armored Division, discussed tank muzzle brakes in *United States vs. German Equipment*, a March 1945 U.S. Army report: "Recently we have received a few M4A3E8 tanks equipped with muzzle brakes. Test firing and combat operations have proven the muzzle brake to be a great help. We consider muzzle brakes an essential part of the tank gun."

DRUMMING UP INTEREST IN THE 76MM TANK GUN

Prior to the invasion of France in June 1944, a number of demonstrations of the new M4A1(76)W tanks occurred in Great Britain for senior U.S. Army officers. Everyone who saw the firepower demonstrations came away impressed with the 76mm cannon. Yet, most high-level armor commanders decided that they would rather stick with the familiar 75mm main gun on their existing first-generation M4 series tanks than retrain their troops to use a new tank main gun. They were also concerned with introducing a new gun and ammunition into the already complex logistical problem of keeping thousands of tanks up and running.

In the April 1944 memo "Conference Notes: Distribution of medium tank M4 Series (76mm. gun)," from Brigadier-General Joseph A. Holly (in charge of the Armored Fighting Vehicle and Weapons Section of the ETO) to the G-3 (in charge of operations) of the ETO, the feeling of senior officers who initially rejected the perceived troublesome 76mm gun on the M4 series was summed up: "an excessive price for the additional inch of armor penetration obtained."

This lack of interest in a bigger, and supposedly better, gun for the M4 series tanks seems strange for those who look back in hindsight. At the time, though, the Army Ground Forces (AGF), under the command of Lieutenant-General Lesley J. McNair,

The 75mm-gunned M4A2 series medium tanks were also built with second-generation or large hatch hulls. However, these tanks retained the earlier ammunition stowage arrangement, with the ammo bins positioned in the sponsons. Ordnance retrofitted the designation "dry" to any current or earlier production M4 series medium tank made without wet stowage. Thus, the large hatch vehicle shown was designated M4A2(75)Dry. (TACOM Historical Office)

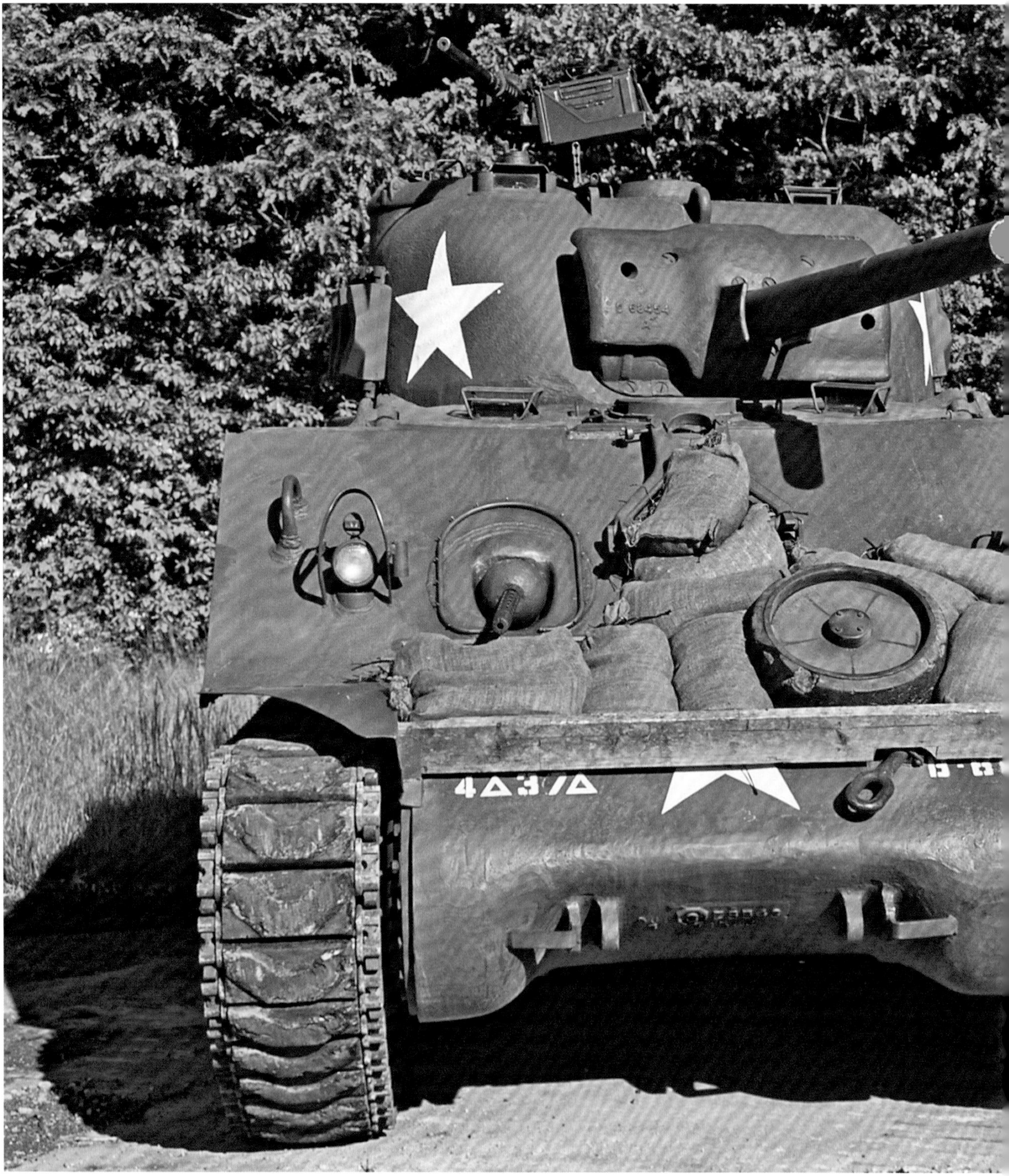

still had complete faith in the 75mm gun-armed M4 series tanks. McNair was still the champion of the tank destroyer concept, and felt that improvements in the armor-defeating capabilities of tank guns would encourage unit leaders to plan on tankers engaging German tanks instead of creating tactical strategies leaving them to his tank destroyers. In addition, the HE round of the 75mm gun was better than that of the 76mm, which further suited those officers who believed the primary role of U.S. Army tanks was infantry support.

Even then, Lieutenant-General Patton, who witnessed a demonstration of the M4A1(76)W six days after the invasion of France took place, decided he would take the new gun only if first placed into separate tank battalions for a combat test. Patton changed his tune after the Battle of the Bulge (December 1944 through January 1945), when his Third Army made plans to make up for the shortage of second-generation 76mm gun-armed M4 series tanks by mounting 76mm main guns in first-generation M4 series tanks. Pictorial evidence indicates some of these vehicles saw combat at the end of the war in Europe with Third Army.

76MM MAIN GUN ROUNDS

The standard complete APC-T round for the 76mm main gun was the approximately 25-pound M62. The projectile portion of the round weighed about 15 pounds and attained a muzzle velocity of 2,600ft/sec, which could penetrate armor about 1 inch (25mm) thicker than the M61 APC-T projectile fired from the 75mm main gun, which had a muzzle velocity of only 2,030ft/sec.

At a range of 500 yards, the 76mm tank gun firing the M62 APC-T round was supposed to be able to penetrate 3.7 inches (93mm) of armor, and at 2,000 yards 3 inches (75mm) of armor. Late-production examples of the M62 APC-T round came with an HE filler and a base-detonating fuze, earning it the designation M62 APC-HE-T.

In the field, the M62 APC-T round's performance was a big disappointment to most. Second Lieutenant Frank Seydel, Jr., of the 2nd Armored Division, described a combat action in the March 1945 U.S.

The US Army wanted to discontinue the use of the 75mm gun on the M4 series by the end of 1943. However, the British Army and the U.S. Marine Corps stated that they did not want the 76mm gun-armed M4 series tanks, and would require 75mm gun-armed versions into 1945. As a consequence, it was decided to continue production of a single model. This picture shows an example of the only second-generation M4 series tank built that was armed with a 75mm gun and designated M4A3(75)W. Units of this type built in 1944 featured a VVSS suspension system, while those produced in 1945 were equipped with HVSS. (Michael Green)

At pointblank range

An encounter between a second-generation M4 series tank armed with a 76mm main gun and German antitank gunners is described by First Lieutenant Chauncey C. Lester in a book published in Germany shortly after the war in Europe ended titled *Dare Devil Tankers: The Story of the 740th Tank Battalion*:

An 88 opened up at point blank range. I could see the powder blast blow its way through the bushes when he fired but couldn't see the gun. I judged him to be about 100 feet to my left front, so swung the gun there and at the same time told my driver to back up and give it hell. He did just that, and my gunner was doing the same with the 76. I counted five shots from the 88, but all were high. I could almost see them going by in front. They seemed close enough to touch, and it took all the guts I could muster to keep my head outside the tank. The Jerry was a very poor shot, thank God, because we backed up 150 feet before he hit us. I don't think he would have hit us then if I hadn't hit a building that I was trying to back behind. As we hit the building he hit us, but the angle was such that the round bounced off. The instant we were hit I told the boys to bail out, and I dropped inside to let my gunner out. It took them about one second to vacate, and then I followed. I went back down the street and found that my platoon sergeant had stopped an AP with the final drive of his tank. Neither of the crews were hurt, except for their feelings, so we felt pretty lucky. After the infantry had cleared the trouble, we found two 88s, the crew of one having been blown to bits. Think maybe I got them.

Army report *United States vs. German Equipment*, in which the M62-T APC round saw use:

On March 3rd at Bosinghoven, I took under fire two German Mark V [Panther] tanks at a range of 600 yards. At this time, I was using a 76mm gun, using APC for my first round. I saw this round make a direct hit on a vehicle and ricochet into the air. I fired again at a range of 500 yards and again observed a direct hit, after which I threw about ten rounds of mixed APC and HE, leaving the German tank burning.

There was also another AP round for the 76mm gun designated the M79 Shot AP-T. The complete round weighed 24 pounds. While the 15-pound projectile featured the same muzzle velocity of the M61 APC-T, the lack of a ballistic cap affected its armor penetration performance at longer ranges. At 2,000 yards it could only penetrate 2.5 inches (64mm) of armor. At 500 yards it should have been able to penetrate 3.7 inches (93mm) of armor.

Besides firing a smoke round designated the M88, the up-gunned second-generation M4 series tanks with the 76mm main gun fired a 22-pound HE round, designated the M42A1. The projectile portion of the round weighed approximately 12 pounds and left the muzzle of the gun tube at 2,700ft/sec.

The use of HE was much more common than APC-T during World War II for American tankers in all theaters of operation. As an example, the U.S. Army's 13th Tank Battalion, which formed part of the 1st Armored Division in Italy between August 3, 1944 and December 31, 1944, expended 19,634 rounds of HE and only 55 rounds of M62 APC-T. Typical targets for American tankers late in the war in the ETO were enemy antitank guns and enemy infantry armed with hand-held antitank weapons firing shaped-charge warheads.

IMPROVED TANK AMMUNITION

To increase the penetrating ability of the 76mm gun, the Ordnance Department developed a new main gun round, designated the M93 hypervelocity armor-piercing tracer (HVAP-T) solid shot. It was nicknamed "Hyper-Shot" by the tank crews. While the complete HVAP-T round weighed 18.9 pounds, the lightweight 9.4-pound projectile consisted of a dense core of tungsten carbide with an aluminum outer body, nose, and windshield.

With a muzzle velocity of 3,400ft/sec, the projectile portion of the M93 HVAP-T could penetrate 6.2 inches (157mm) of armor at 500 yards (almost double that of the M62 APC-T main gun round). It was supposed to penetrate up to 5.3 inches (135mm) of armor at 1,000 yards. At 1,500 yards, penetration was 4.6 inches (116mm) of armor, and at 2,000 yards, 3.9 inches (98mm) of armor.

As soon as the first M93 HVAP-T rounds came off the production line, they were transported by air to Northwest Europe around August 1944 and then to the tankers in the field. Each 76mm gun-armed M4 series tank was supposed to have a small supply of this special ammunition because priority for this tank-killing round went to McNair's beloved tank destroyers.

Major Paul A. Bane, Jr., of the 2nd Armored Division, commented in the March 1945 U.S. Army report *United States vs. German Equipment*: "Our tank crews have had some success with the HVAP 76mm ammunition. However, at no time have we been able to secure more than five rounds per tank and in recent actions this has been reduced to a maximum of two rounds, and in many tanks all this type has been expended without being replaced."

A POSSIBLE M4 SERIES TANK REPLACEMENT

In spring 1942, an Ordnance Department developmental program began that was intended to come up with a possible replacement for the M4 series tanks (that were just entering into production) that would address the continued improvements in firepower and armored protection seen on German tanks during the fighting in North Africa and expected in the future. It was desired that the possible successor to the M4 series tanks would incorporate both combat experience gained during the fighting in North Africa, as well as any new technical advances conceived since the M4 series tanks were first designed.

The point men in this developmental project were Major-General Gladeon Marcus Barnes, from the Chief of the Ordnance Department Research and Development division, and Colonel Joseph Colby of the Tank-Automotive Center. They proposed a state-of-the-art design that kept the five-man layout of the M4 series tanks, but did away with the upper hull and the sponsons that projected out over the vehicle's tracks. In their place would be a simple box-like lower hull upon which the tank's turret was mounted. All of the nonessential equipment stored within the hull of the M4 series tanks was to be kept in storage bins located on top of the fenders that extended out from the hull over the tracks on either side of the vehicle.

A wooden mockup of the proposed new medium tank design was presented in May 1942 and was a great hit among all the senior officers attending. It offered a better armed tank, possibly mounted a more powerful 76mm gun M1 in place of the 75mm gun M3 in the first-generation M4 series tanks, as well as 0.5 inches (13mm) more armor and a possible automatic transmission. This was all accomplished without exceeding the weight of the first-generation M4 series tanks. The vehicle was quickly approved and was designated the medium tank T20 by Action of the Ordnance Committee that same month.

Reflecting the fact that many different components were being considered for inclusion in the final design of the T20, a number of designations were added to distinguish between the various versions of the vehicle. Two pilot vehicles to be fitted with a modified version of the synchromesh transmission in the first-generation M4 series tanks were designated the medium tank T22, and two pilot vehicles to be fitted with an electric transmission and steering system developed by General Electric (GE) were designated the medium tank T23. These vehicles would also be further subdivided into various models by the armament fitted.

The T20 series of medium tanks was developed to replace the M4 series medium tanks. Pictured is the T20E3 pilot vehicle in July 1943. It was armed with a 76.2mm main gun and rode on a torsion bar suspension system. (TACOM Historical Office)

THE TESTING PROCESS

The first T20 pilot appeared in May 1943 with an early type of HVSS system. Another T20 pilot equipped with a torsion bar suspension system was designated the T20E3. Due to problems with the T20's transmission, the T20 and the T20E3 were canceled in December 1944. The engine, Torqmatic transmission, and controlled differential steering units on the T20 and T20E3 were all located in the rear hull of the vehicle and could be removed as a single unit.

Coming off the factory floor in June 1943, the T22 pilots also demonstrated serious power train problems. The T22E1 boasted a turret armed with a 75mm gun M3 hooked up to an automatic loader. Although this gun featured a phenomenal 20 rounds per minute rate of fire, the technology was not mature enough to work properly and the gun itself was considered inadequate by that time. The project was suspended and then canceled in February 1944.

The first T23 pilot was completed in January 1943, and appeared with the GE electric drive that worked with the liquid-cooled, gasoline-powered Ford V-8 Model GAA engine. The vehicle rode on a VVSS. In contrast to the other pilot vehicles in the T20 series that featured cast-armor turrets, the first T23 pilot had a turret constructed of welded armor. An improved cast-armor turret was mounted on the second T23 pilot that featured the 76mm gun M1A1.

Early testing of the T23 pilots was so successful that 250 series production units were eventually ordered. However, testing of the first ten series production units led the Army Ground Forces (AGF) to conclude that the complexity of the GE electric transmission and steering system was beyond the skill set of the typical U.S. Army mechanic and the vehicle was therefore unsuitable for use in the field. The vehicle was also offered informally to the ETO in February 1945, but the offer was rejected as it would have involved retraining maintenance personnel and acquiring a new inventory of spare parts.

Three of the T23 pilots were fitted with the late-model HVSS as seen on second-generation M4 series tanks and designated as the T23E4. There was also another version of the T23 that rode on a torsion bar suspension system and was designated the T23E3. Like the T23, the T23E3 and T23E4 would never enter into field use. The one claim to fame of the T23 series was that its cast-armor turret armed with the 76mm gun M1A1 formed the main feature of the majority of second-generation M4 series tanks.

There was another version of the T23 built that was fitted with the 90mm gun M1 and designated the T25. A more heavily armored version of the T23 armed with a 90mm gun M1 was designated the T26. As events unfolded, the T25 would never be built, as U.S. Army interest in the summer of 1944 turned to more thickly armored tanks. A version of the T26 that rode on a torsion bar suspension system was designated the T26E3 and would eventually become the heavy tank M26 that saw service in World War II and the Korean War.

Pictured is the T22E1 pilot vehicle which generally followed the same basic hull design of the T20 pilot vehicle. The modified 75mm first-generation M4 series medium tank turret included an automatic loader. The vehicle rode on an early, narrow track version of the HVSS system. (TACOM Historical Office)

Pictured is the T23 pilot number 2 vehicle. A dramatic departure within the design of the various versions of the T20 series was the use of an electric drive transmission. Despite the advantages promised, the concept was dropped as it was found to be overly complex and difficult to maintain. (TACOM Historical Office)

M4 SERIES TANK VARIANTS

In March 1943 the number of M3 series medium tanks available for conversion into the tank recovery vehicle M31 was dwindling. It was decided to convert surplus early-production first-generation M4 series tanks into tank recovery vehicles. This program called for the fielding of the tank recovery vehicles M32, M32B1, M32B2, M32B3, and the M34B4, based respectively on the M4, M4A1, M4A2, M4A3, and M4A4. All versions of the M32 series entered production except for the M32B4, which was based on the M4A4 tank.

The M32 series of tank recovery vehicles all came equipped with the same 30-ton Gar Wood winch that was driven by a power takeoff from the vehicle's propeller shaft. The winch was located inside the front hull just behind the driver's seat on all versions of the series. The winch cable was fed out through a tow winch dragline roller located on the lower front upper hull of the vehicles. All versions of the M32 series also featured a crane type boom that could be used for lifting or towing purposes and a fixed superstructure on top of the upper hull. In total, 1,582 units of the M32 series would be built or converted between June 1943 and May 1945.

Not everybody was thrilled with the M32 series as is seen in this extract from a postwar research report, *Armor in Mountain Warfare*, prepared at The Armored School in Fort Knox, Kentucky: "The M32 series tank retriever has such obvious limitations as a recovery vehicle that it seems unnecessary to mention more than two of the most serious limitations: (1) the narrow tracks prohibit its use in soft terrain: and (2) the open turret prohibits the use of the vehicle under fire."

In British Army service the M32 series was referred to as the "ARV III." The British also converted a number of standard first-generation M4A2s into ARVs by removing their gun-armed turrets and providing them with a 3.7-ton jib crane for removing engines or power trains from tanks and drawbars for towing disabled tanks if the need arose. These makeshift ARVs were referred to in the British Army as the "Sherman ARV I."

The Sherman ARV II was a more advanced ARV and was fitted with both a jib crane and a winch. To fool the enemy, the vehicle's box-like superstructure, fitted on the top of the upper hull, sported a dummy 75mm main gun.

The British Army converted 52 of its ARV Is into Beach Armored Recovery Vehicles (BARVs) prior to the Allied invasion of France on June 6, 1944. Their intended role was to rescue any tracked or wheeled vehicles that became stuck in the surf during the beach landings. With a waterproofed hull and superstructure, the vehicle could wade through 9 feet of water to push or pull stranded vehicles out of the surf onto a beach.

Reflecting the eventual switch to second-generation M4 series tanks, the Ordnance Committee decided to field new tank recovery vehicles riding on the HVSS system. There were a number of modifications made to the second-generation M4 series tank recovery vehicles. These included adjustable rod-type stabilizers employed to lock out the springs on the front and rear bogies when lifting operations were being conducted.

The initial batch of 80 units of the M32 riding on the HVSS system was constructed by the Baldwin Locomotive Company and International Harvester and rolled off the factory floor by the end of 1945. They were originally designated as the T14E1. Reflecting the various changes to the second generation tank recovery vehicles riding on the HVSS system, the M32, M32B1, M32B2, and M32B3 became, respectively, the M32A1, M32A1B1, M32A1B2, and M32A1B3.

An M32B1 tank recovery vehicle on display at the annual "War and Peace Show" held in England every July. The 81mm mortar on the vehicle's glacis would have been employed in wartime to lay smoke to mask its location while performing recovery operations. (Christophe Vallier)

For the invasion of France, the British Army decided there was a requirement for a full-tracked recovery vehicle able to assist others in transiting from their landing craft onto the beaches of Normandy. In response, a number of M4A2 medium tanks were converted to Beach Armored Recovery Vehicles, as seen here. (Christophe Vallier)

FLAME-THROWER TANKS

The U.S. Army Chemical Warfare Section (CWS) had been studying the installation of flame-thrower guns in medium tanks prior to America's official entry into World War II. The initial project began when the CWS installed an experimental E2 flame-thrower gun in the turret of a medium tank M2. This was followed by a shortened version of the E2 flame-thrower gun mounted in the turret of the medium tank M3 in August 1942. Maximum range of the flame gun was only about 35 yards. User interest was nil at that time, as it was hard to conceive of any practical purpose for this short-range weapon in the wide open desert wastes of North Africa.

There was a dramatic peak in interest in flame-thrower gun equipped tanks after fighting against the Japanese infantry in the jungles of Guadalcanal in fall 1942 demonstrated that such a vehicle could be extremely useful. Initially, light tanks would be converted, but their thin armor protection made them vulnerable to enemy return fire as they attempted to close to within the short effective range of their flame-thrower guns. To rectify this problem, it was decided to modify first-generation M4 series tanks to serve as carriers for the new E4-5 auxiliary flame gun.

The new E4-5 auxiliary flame gun was not mounted in the turret of the first-generation M4 series tank, but mounted in place of the front hull-mounted .30 caliber machine gun. In this arrangement, the weapon had a range of about 60 yards. A half-dozen E4-5 auxiliary flame guns were installed on M4A2s and saw service with the Marine Corps during the fighting for the island of Guam between July and August 1944.

A small number of M4s equipped with the bow-mounted E4-5 flame gun would see limited use in Northwest Europe during the latter stages of World War II. It was not a popular vehicle with users for a number of reasons. Crews felt it made their M4 series tanks even more susceptible to vehicle fires, the range of the flame thrower was shorter than that of the German hand-held rocket launchers, and it was not a mechanically reliable device. Furthermore, the arrangement of the device inside the M4 series tanks blocked the escape hatch located under the bow gunner's seat, making it more difficult for him to leave the vehicle. The E4-5 flame gun was standardized by the US Army as the M-3-4-3 in 1945.

Also developed during that same time period were a couple of other auxiliary flame guns designated the E6R3 and E12R3. They were also referred to as periscope-type flame guns because they were configured to fit alongside the bow gunner's overhead hatch M6 periscope. These were employed during the fighting for the islands of Iwo Jima and Okinawa that took place between February and June 1945. Like the E4-5, the E6R3 and E12R3 were not that highly valued due to their short range and limited fuel capacity of only 24 gallons, which provided them with 20 to 30 seconds of flame time before the fuel tanks were empty.

What the Marine Corps and U.S. Army really wanted was a turret-mounted flame-thrower gun arrangement with a larger onboard fuel capacity. A composite group from the U.S. Army CWS, Marine Corps, and U.S. Navy set about fulfilling that requirement as is described in an extract from an article by Captain John W. Mountcastle in the March–April 1976 issue of *Armor* magazine, titled "Inferno: A History of American Flamethrowers":

> Working around the clock in late 1944, this group succeeded in fabricating a very effective flame-thrower which was mounted in the gun tube of the standard M4 medium tank. Tanks of this type participated in the Iwo Jima invasion… and were so successful that many more were requested by the Marine Corps for the upcoming Okinawa campaign. The Army too wanted flame tanks. A [U.S. Army] medium tank battalion [54 tanks] was converted completely to flame and performed magnificently on Okinawa. The 713th Armored Flamethrower Tank Battalion was in combat for 75 days and was credited with killing 4,788 Japanese. Approaching enemy positions in caves and bunkers, the flame tanks threw long (100 yards) jets of flaming napalm at the Japanese. Those who escaped death by burning or suffocation were machine gunned as they ran screaming from their holes.

The M4 series tanks modified to accommodate the turret-mounted flame guns were designated as the POA-CWS-H1. The troops nicknamed them "Zippos," the brand name of a popular cigarette lighter renowned for its reliability. The POA-CWS-H1 carried 290 gallons of fuel in four tanks located below the vehicle's turret.

On display at the U.S. Army Chemical Corps Museum at Fort Leonard Wood, Missouri, is this POA-CWS-H5 flame-thrower tank, based on a first-generation medium tank M4A1. This concept allowed for the use of the 75mm main gun, as well as the flame gun contained within the upper barrel. (Lorén Hannah)

The first batch of POA-CWS-H1 tanks consisted of eight first-generation Marine Corps M4A3s. Four were issued to the 4th Tank Battalion, 4th Marine Division, and the other four to the 5th Tank Battalion, 5th Marine Division. All eight would see action during the conquest of Iwo Jima between February and March 1945. Results were mixed, according to a Marine Corps report titled *Armored Operations on Iwo Jima*, dated March 16, 1945:

> This weapon gave excellent results when it worked and could reach the target. In rubble brush and defiladed positions it caused casualties where no other weapon could reach the target… A combination of mechanical trouble and poor fuel reduced the efficiency of these weapons about 75 percent. The average length of flame produced was about 30 yards while the potential range of the same gun is over 100 yards farther than this… Unfavorable winds completely rendered the weapon useless on several occasions.

The 54 POA-CWS-H1s employed by the U.S. Army during the fighting for Okinawa between April and June 1945 were based on late-model M4 tanks referred to as composite hulls.

An extract from First Lieutenant Patrick J. Donahoe's article "Flamethrower Tanks on Okinawa," in the January–February 1994 issue of *Armor* magazine, sums up the effectiveness of the POA-CWS-H1s:

> A key to the flame tank's success was the mounting of the weapon in the sturdy M4 medium tank, which was survivable enough to take the fight directly to the enemy's position. Also, the inclusion of the armored flamethrower as an intrinsic part of the tank-infantry combined arms team was instrumental to it use [sic]. Had the main armament flamethrower been perfected earlier in the war, American casualties could have been reduced and the tank infantry team could have been more effective in the Pacific War.

Some within the U.S. Army and Marine Corps felt that the POA-CWS-H1 would be better off if the 75mm main gun was retained for self-protection. In response to this desire, starting in late 1944 all newly converted M4 series tanks would have their flame guns mounted alongside the vehicle's main armament, be it a 75mm main gun or a 105mm main gun. In this configuration the tanks were designated the POA-CWS-H5. Due to the space taken up by the fuel tanks for the flame gun inside the hull, their main ammunition storage was greatly reduced.

The Marine Corps wanted 72 units of the POA-CWS-H5 for the planned invasion of Japan, which fortunately never happened. The U.S. Army had hoped to employ ten of the POA-CWS-H5s during the fighting for Okinawa, but they arrived after the fighting concluded. A number of other projects to mount flame guns on the M4 series tanks were in the pipeline for use by the American military. However, none would reach the field before World War II ended and they would all be eventually canceled.

There was some early interest before the invasion of France by the U.S. Army in modifying the M4 series tanks to accommodate the British-developed and built Crocodile flame-thrower system, which included a two-wheel towed trailer that held the fuel for the vehicle's front hull-mounted flame gun. In the end, only four would be built and go on to see action with the U.S. Ninth Army in the ETO. They proved very useful, and a request for additional units was made; but it was rejected as it was felt that American-designed and built medium tank-mounted flame-thrower guns would soon be available.

ACCESSORIES

There would be a number of accessories designed for use by the M4 series tank during World War II to allow it to perform roles for which it was not originally intended. The largest number of these accessories proved to be for clearing enemy minefields and could be broken down into two major types of mine exploders: pressure-type mine exploders and concussion-type mine exploders. Pressure-type mine exploders activated mine fuzes with pressure applied by a mechanical device such as a flail, roller, disc, or plunger. The second type of mine exploder activated mines fuzes by sympathetically detonating mine charges with the force of a concussion or blast from a nearby explosive charge.

A pressure-type mine exploder developed and fielded by the British Army was a rotary flail system nicknamed the "Crab." It consisted of a cylindrical rotor with 43 large chains attached to the front of a first-generation M4 series tank that beat the ground in the hope of detonating any mines in its path during the breaching process. The cylindrical rotor was turned by a power takeoff from the vehicle's main engine. Due to the amount of dust generated by the Crab when in operation, visibility proved to be a serious issue. To assist the crews of the Crabs in maintaining the proper direction when breaching minefields, they were fitted with both a directional gyro and a magnetic compass.

There were two versions of the Crab fielded by the British Army during World War II, the "Crab I" and "Crab II." With the Crab I, a hydraulic ram on either side of the vehicle's front hull was employed to raise and lower the rotor assembly. Once locked into position it could not be adjusted for terrain deviations. On the Crab II,

A medium tank M4A4 is seen here in postwar Israel Defense Forces service mounting a British-designed flail-type mine exploder, referred to as the "Crab II." When flailing, the Crab could only be operated at a maximum speed of 1.25mph. (Patton Museum)

the rotor assembly was balanced by an adjustable weight mounted on one side of the vehicle's front hull that maintained a constant height of 4 feet 3 inches in spite of terrain variations.

On a number of occasions, small numbers of Crabs were transferred to the U.S. Army by the British Army for use in the ETO during World War II. The Crabs made a favorable impression on the U.S. Army units that employed them in combat, who liked the fact that the vehicle could be employed as a standard tank when not involved in breaching minefields and that they provided the quickest breaching method available. Several disadvantages of the Crab identified by the U.S. Army included being vulnerable to off-set mines, delay action mine fuzes, and large explosive charges.

AMERICAN MINE-CLEARANCE DEVICES

The simplest mine-clearing device fielded by the American military during World War II was designated the mine excavator T5E3, of which 100 were constructed by La Plante-Choate between March and May 1945. Designed to be mounted on the front of first-generation M4 series tanks, it consisted of a V-shaped plow arrangement with scarifying teeth to extract mines without detonating them and, using a moldboard, to cast them to either side of the vehicle's tracks. To prevent the extracted mines from running up over the top of the moldboard it had a curved upper lip. The T5E3 weighed approximately 4.5 tons.

The medium tank M4 pictured is equipped with the T1E3 mine exploder. The drive chains attached to the tank's drive sprockets drove the roller assemblies. The large plate at the rear of the tank served as a bumper to allow the vehicle to be pushed by another tank if the need arose. (Patton Museum)

The Ordnance Department had also developed a rotary flail system, designated the mine exploder T3, and nicknamed the "Scorpion." Unlike the Crab's cylindrical rotor, which was powered by a takeoff from the vehicle's engine, the Scorpion's cylindrical rotor was turned by an auxiliary engine. Forty-one of the T3 devices were adapted to fit on the front of first-generation M4 series tanks by the Pressed Steel Car Company between April and July 1943. Once completed, the modified tanks were rushed to the Italian Theater of Operations. They did not impress the users due to their weight and lack of off-road mobility and were quickly removed from service.

There were also two improved rotary flail systems developed by the Ordnance Department following the Scorpion; however, neither would see field service. The first was referred to as the T3E1, and the second as the T3E2, which was also nicknamed the "Rotoflail." There were also some improvised rotary flail systems developed in the field for fitting onto the front of U.S. Army and Marine Corps first-generation M4 series tanks.

The U.S. Army Ordnance Department favored pressure-type mine exploders consisting of large steel rollers/discs in various configurations that mounted on the front of the tank recovery vehicle T32 or first-generation M4 series tanks. Those eventually fielded in the ETO in very small numbers included the original T1E1 and improved versions. The T1E1 consisted of three sets of steel rollers mounted on the front of the recovery vehicle T32 that had a combined weight of 18 tons. Gar Wood Industries would complete 75 units of the T1E1 in April 1944, which were nicknamed "Earthworm."

The improved T1E3 was intended to be mounted on the front of first-generation M4 series tanks, with the first four pilot examples of the T1E3 being shipped overseas in May 1944, two going to England and two to Italy. The T1E3 consisted of two large sets of steel rollers with a combined weight of 29.5 tons. The Whiting Corporation built 200 units of the T1E3 between March and December 1944.

The Ordnance Department never stopped tinkering with the design of its pressure-type mine exploders, and came up with another variation, designated the T1E4. It consisted of a single roller fitted with 16 serrated steel discs that fitted to the front of first-generation M4 series tanks and weighed 24 tons. The T1E4 was followed that same year by the T1E6, which was slightly wider with 19 serrated steel discs and was designed to be mounted on the front of second-generation M4 series tanks.

All the Ordnance Department-developed pressure-type exploders that employed roller/discs rendered their carrying vehicle generally road-bound. They also proved vulnerable to relatively large explosive charges as well as off-set mines and delayed-action mine fuzes. The large and bulky roller/disc assemblies also proved extremely difficult to move from one location to another. This meant they had to be constantly assembled and disassembled – a very time-consuming process.

A description of the effectiveness of German minefields appears in this passage from a late-1944 report by Major-General E. N. Harmon titled *Notes on Combat Experience during the Tunisian and African Campaigns*:

> The extensive use of mines, both antitank and antipersonnel, by the Germans is one of the greatest menaces of the present war. No area, either forward or back, is safe from the mine. The most effective enemy mining was the sporadic mining of long stretches of road, road shoulders, craters, and areas upon withdrawal. Heavily mined fords strewn with metal fragments to render detectors useless were also effective delays. In general, the enemy's mine technique and mine equipment were superior to our own.

CONCUSSION-TYPE MINE EXPLODERS

Besides the pressure-type exploders developed by the Ordnance Department, a number of concussion-type mine exploders were developed by the U.S. Army Corps of Engineers. Most would not see field service during World War II. One of those that did see use in combat with mixed results was the 400-foot-long demolition snake M2. Based on a Canadian design, it was built by the Armco International Corporation and pushed into firing position by M4 series tanks.

The M2 weighed 6.25 tons and consisted of a double row of explosive cartridges fitted inside corrugated steel plates. It was followed in production by the lighter weight M2A1 and M3 fitted inside lighter aluminum plates. Once emplaced in their firing position, the various types of snakes were detonated by machine-gun fire aimed at an impact-sensitive fuze or, in an emergency, by a direct hit from a tank's main gun.

A description of the single operational use of the demolition snake in Northwest Europe appears in a postwar U.S. Army report titled *Armored Special Equipment*, prepared for the General Board U.S. Forces, European Theater:

The only tactical employment of snakes on record was in the Third U.S. Army assault of Fort Driant of the Metz defense works on 31 October 1944, and this attack was unsuccessful. One dummy snake was assembled on 1 October and used for training, and

The T1E3 mine exploder was not a full-width device and had very poor off-road capabilities due to its size and weight, and proved extremely unpopular with the user community. Work continued on refining the concept, with the next step being the smaller and lighter full-width T1E4 mine exploder, seen here mounted on an M4A3(76)W with VVSS. (TACOM Historical Office)

four "live" snakes were ready the night of 2 October for movement to the fort. The snakes had to be moved approximately one mile to the fort by tanks of the 735th Tank Battalion. One tank was to tow each snake to the approximate site of detonation, and another tank was to follow, maneuver the snake into the exact location desired, and detonate it.

Unfortunately, the plans using the M2s for the assault on the German-held Fort Driant quickly fell apart as two of them broke up when the tanks towing them into position had to make sharp turns on the approach to the enemy fort. Another M2 was lost a half mile from the fort when the towing tank drove over some logs. The remaining M2 was dropped just short of the enemy fort by the towing tank, but it buckled and twisted when another tank attempted to push it into its firing position. The accompanying U.S. Army assault on the enemy fort was beaten back by the German defenders.

A description of the effectiveness of the M2 during the breakout at the Anzio beachhead in Italy in May 1944 appears in this passage from Harmon's *Notes on Combat Experience during the Tunisian and African Campaigns*:

> The "snake" was successfully used in the breakthrough at Anzio [May 1944] and has possibilities where the mine field can be located ahead of time and where conditions are favorable for approaching with a tank to push a "snake" through the field. Several lanes were blown through minefields with a great demoralizing effect on the enemy. However, the danger with the "snake" is its susceptibility to being set off by artillery and causing heavy casualties by our troops in the immediate vicinity.

Another pressure-type exploder developed by the Corps of Engineers was the projected-line charge M1. It consisted of up to four plywood pallets towed behind an M4 series tank. Each pallet had a 300-foot-long charge containing plastic explosives that were launched from the edge of a minefield by a rocket propulsion unit mounted on the same pallet. Upon crossing a minefield the plastic explosives carried within the line charges were detonated in the hope of successfully breaching a path wide enough for vehicles and personnel to cross safely.

The Corps of Engineers also experimented with using liquid explosives for breaching minefields. It was intended that three 150-foot-long hoses, 3 inches in diameter and filled with liquid explosives (desensitized nitroglycerine), be launched from a modified M4 series tank and propelled over a minefield by a turret-mounted rocket launcher. Once emplaced, the three liquid explosive-filled hoses would be detonated to breach a path through a minefield for friendly forces to pass through.

Using the turret-less chassis of an M4 series tank, the Corps of Engineers developed the mine exploder T12. In place of the vehicle's turret was a platform level with the tank's upper hull that was fitted with 25 spigot mortars capable of launching 115-pound T13 HE shells in a long narrow impact pattern intended to breach a suitably wide enough path through a minefield. Three pilot examples of the T12 were built for testing. However, superior results obtained with rocket-propelled charges for breaching minefields led to the termination of the T12 project.

ENGINEER-ARMORED VEHICLES

It did not take long for the American military to conclude that unarmored bulldozers were not suitable for use in combat areas. As a short-term fix, a number of improvised armor kits were constructed by units in the field to offer the vehicle operators a measure of protection from enemy fire. The Corps of Engineers determined in January 1942 that the optimum solution would involve the fielding of a bulldozer kit that could be attached to the front of first-generation M4 series tanks. The end result of this line of development appeared during the second half of 1943 and was designated the bulldozer, tank mounting, M1. It would be widely distributed wherever U.S. Army or Marine Corps armored units served.

The straight toothless blade on the M1 was 10 feet 4 inches long and mounted on side arms that pivoted from the center bogie assemblies of first-generation M4 series tanks. A redesigned version designated the M1A1 had an 11 foot 6 inch wide straight toothless blade and could be adapted to mount on the front of both first-generation M4 series tanks and second-generation M4 series tanks riding on the HVSS system. It could also be adapted to mount on the front of first-generation M4 series tanks featuring the E9 spaced-out suspension system. By early 1945, a new bulldozer kit appeared that could be attached to the front hull of any M4 series tank. Known as the bulldozer, tank mounting, M2, it did away with the side bracket arms seen with the M1 and M1A1.

The U.S. Army in the ETO evaluated the various tank-mounted bulldozer kits for the M4 series tanks on the conclusion of the war in Europe. They proved to be a valuable addition to tank battalions as they could be readily employed to reduce obstacles in their path or prepare defensive positions if the need arose. It was also favored because it was a simple device that was reliable, with few maintenance needs. The big disadvantage of the various kits

One of the most useful M4 series medium tank accessories was the tank-mounted bulldozer M1, seen here mounted on the front of a first-generation medium tank M4A1. Field expedient dozers had been used in 1943, but the M1 was first employed in Italy in spring 1944. (Patton Museum)

was that they overloaded the front of the M4 series tanks and caused suspension system problems.

A description of the effectiveness of the tank-mounted bulldozer appears in this extract from Harmon's *Notes on Combat Experience during the Tunisian and African Campaigns*: "One of the most important items of equipment is the armored bulldozer. It played an important part in the pursuit of the Germans in Italy. Without it we couldn't have advanced as all bridges and culverts were blown, and houses were often blown into the streets. A greater proportion of armored bulldozers should be set up."

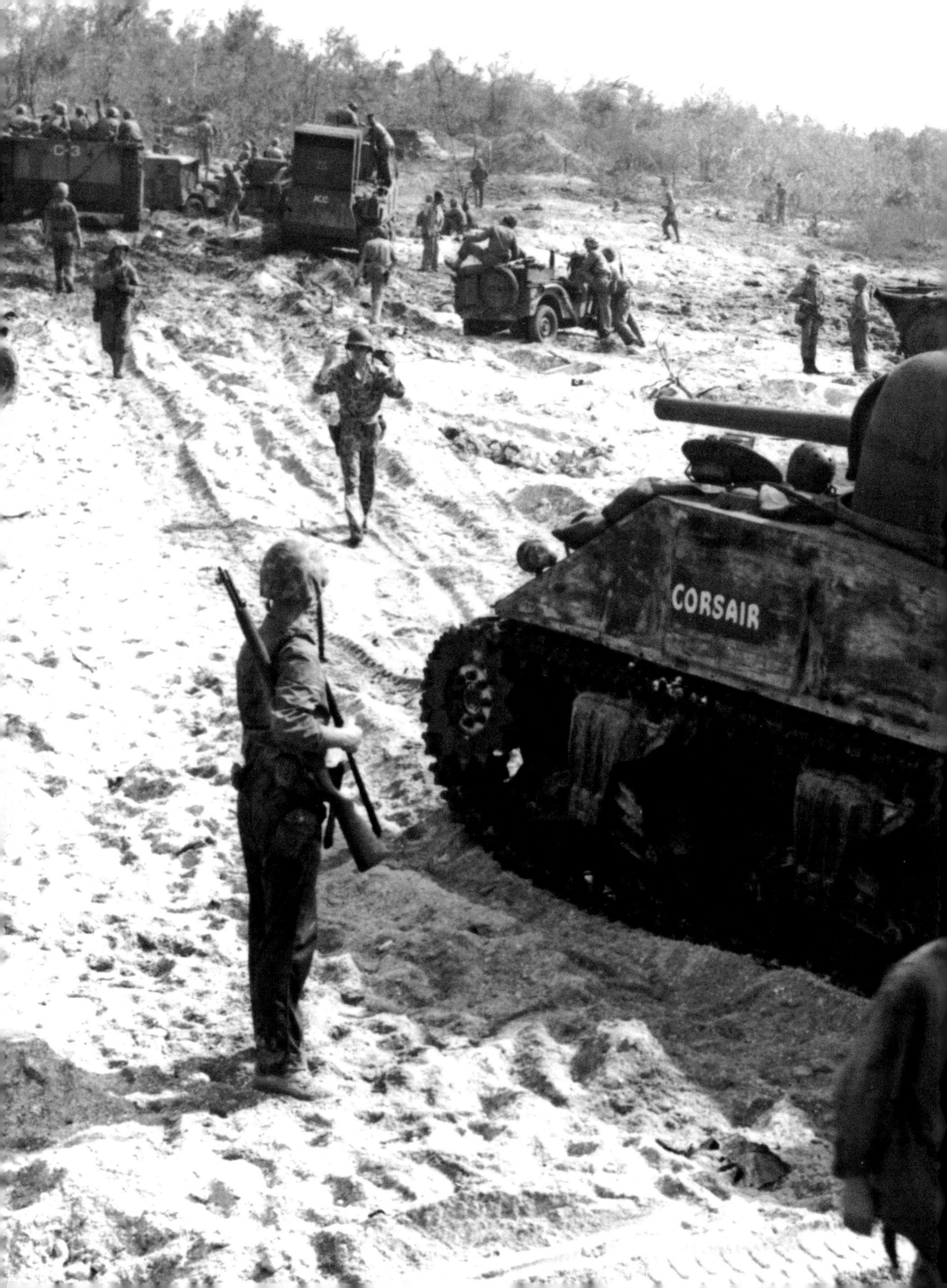
CORSAIR

A U.S. Marine Corps medium tank M4A2 fitted with the sheet metal deep water fording kit is being directed to its beach staging area. In combat conditions, the top section of the wading trunks could be dropped immediately upon reaching shore. However, de-waterproofing stations were directed to be set up as soon as possible, in order to remove sealants and tapes that might interfere with the proper operation of the vehicle. (Patton Museum)

FLOTATION DEVICES AND FORDING EQUIPMENT

The planned Allied invasion of France in summer 1944 called for a method to assist tanks in getting ashore during the initial assault wave to provide much-needed fire support to the infantry coming in on naval landing craft. What was eventually acquired by the U.S. Army and the British Army for use on their tanks was a flotation device developed by Nicholas Straussler, a prolific inventor of Hungarian descent who became a British citizen in 1933. He came up with an inflatable, rubberized, waterproof canvas screen that could be erected around the upper hull and turret of first-generation M4 series tanks.

In the water, the vehicles were powered by two propellers temporarily mounted to the rear hull and connected to the tank's tracks. Straussler's flotation device was referred to as "Duplex Drive" or "DD tanks."

Once on land, the inflatable canvas screen on the DD tanks could be lowered, allowing for use of the tank's turret-mounted weapons. At the same time, the rear hull-mounted propellers were raised so as not to interfere with the vehicle's mobility on land. The debut of the DD tanks took place on June 6, 1944, (D-Day) with the Allied invasion of France. Three U.S. Army tank battalions (the 70th, 741st, and 743rd) had been trained in their use along with three British and two Canadian armor battalions.

Of the two U.S. Army tank battalions assigned to Omaha Beach (the 741st and 743rd), the 741st lost 27 of their 29 DD tanks when swimming ashore due to rough water swamping the vehicles. The 743rd had their DD tanks delivered to the landing site by naval landing craft for fear that the same fate that befell the 741st might happen to them. Losses among the other DD tank-equipped units on D-Day were not nearly as high, as the seas were generally calmer at their landing locations.

DD tanks would continue to see employment by the U.S. and British Armies during the fighting in the ETO. Forty-eight of them would take part in Operation *Anvil*, the U.S. Army invasion of Southern France, in August 1944. With calm seas and little enemy resistance, 20 of the DD tanks

safely transitioned from sea to shore, with the remainder delivered to shore by naval landing craft.

U.S. Army and British Army DD tanks would next see use in the Allied crossing of the Rhine River in March 1945, codenamed Operation *Plunder*. As the U.S. Army had a shortage of DD tanks in early 1945, they borrowed a number from the British Army inventory. All the U.S. Army DD tanks that attempted to cross the Rhine did it successfully. A British Army unit used DD tanks to cross the Po River in Italy on August 24, 1944 and the Adige River in Italy on August 28, 1944. The last use of DD tanks in the ETO occurred on April 29, 1945, when a British Army unit used them to cross the Elbe River in Germany.

Developed for the invasion of France was a British-designed swimming device referred to as "Duplex Drive" or "DD." It consisted of a canvas screen with a supporting structure that could be erected around the hull for flotation. Power takeoffs attached to the tank's idler wheels rotated a pair of propellers for propulsion. (Patton Museum)

DD tanks never saw service in the PTO during World War II. The only flotation device employed by the American military in that part of the world was originally referred to as the T6, and was later standardized as the device M19.

The M19 consisted of large compartmented steel floats attached to the front, rear, and sides of M4 series tanks. The steel floats were filled with plastic foam encased in waterproof cellophane to lessen their vulnerability to damage from enemy fire and natural obstacles. Propulsion in the water for the M19 was by the turning of the tank's tracks. The biggest drawback of the M19 was its size, which was 11 feet by 47 feet 8 inches. In spite of this drawback, the U.S. Army employed a few of them in the invasion of Okinawa in April 1945.

Another flotation device tested by the Ordnance Department was the T12, which consisted of two 15-ton engineer pontoons strapped on either side of an M4 series tank. Propulsion was provided by outboard motors fitted to each pontoon. The T12 would not see use in World War II and was considered only as a field expedient method of moving a tank across a water barrier.

Pictured is a late-production M4A3(75)W swimming with the T12 flotation device. This consisted of a pair of 15-ton engineer pontoons attached to the sides of the hull. Propulsion in the water was provided by an outboard motor attached to the rear of each pontoon. (Patton Museum)

The most widely employed piece of equipment used to assist M4 series tanks in moving from ship to shore through a surf line came from an idea generated by the Fifth Army Invasion Training Center (5AITC) located in Algeria in early 1943. They conceived the concept of waterproofing first-generation M4 series tanks, and then two tall sheet metal trunks were fitted on top of the rear hull engine compartment – the forward trunk to provide air for the engine or engines, and the rear trunk to vent the engine exhaust. This deep water fording arrangement would allow a suitably equipped tank to cross a surf line 6 feet deep.

Improvised deep water fording kits were first employed during Operation *Husky*, the invasion of Sicily in July 1943. By the time the invasion of France occurred in June 1944 standardized deep water fording kits had been developed. The fording kits would also be used during the U.S. Army invasion of Southern France, in August 1944, codenamed Operation *Anvil*. Both improvised deep water fording kits and factory-produced examples would see use in the PTO with the U.S. Army and Marine Corps.

CHAPTER THREE

LIGHT TANKS

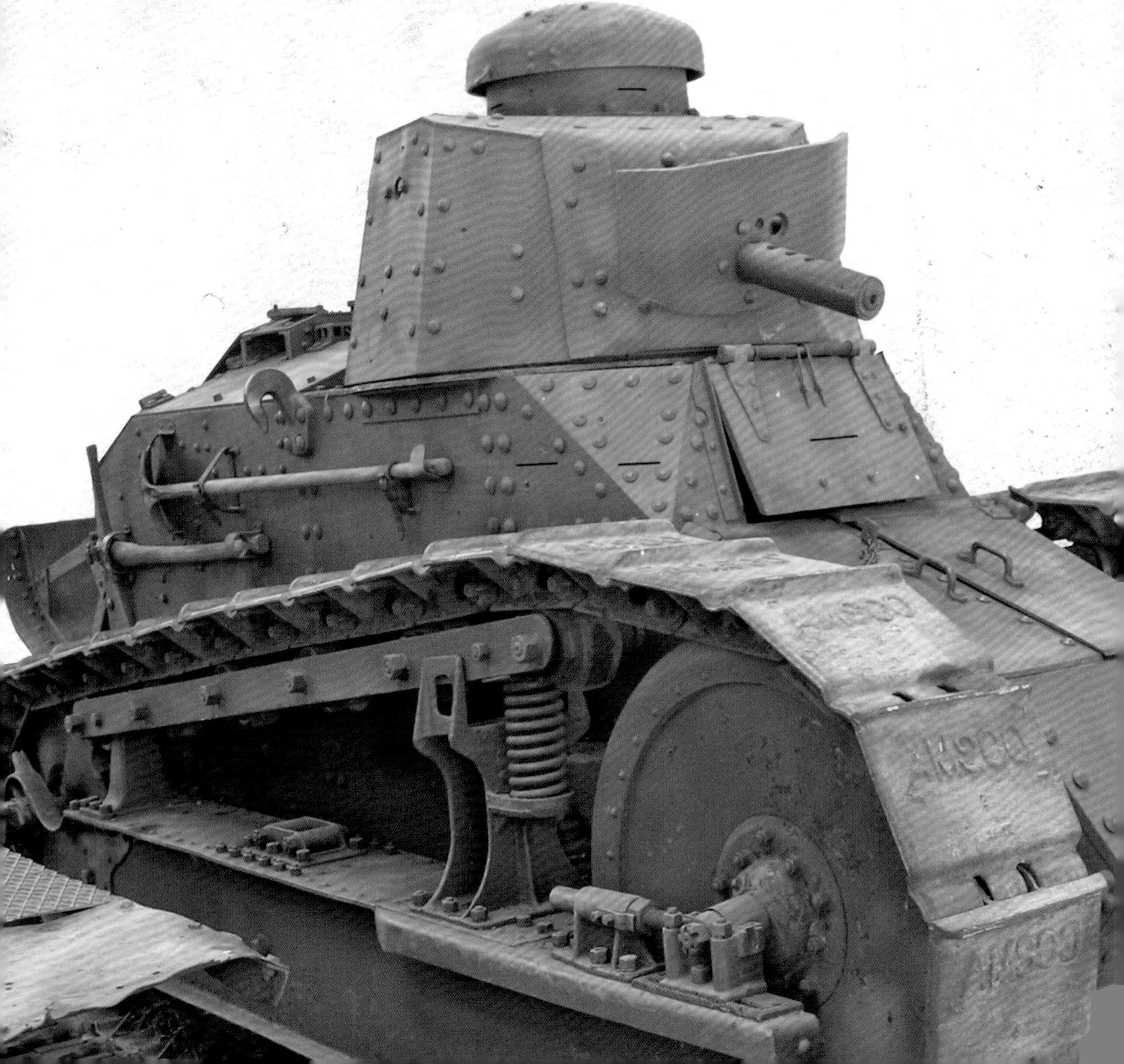

In May 1922, the Ordnance Department inquired of the Infantry Branch what they wanted to see in a new light tank. The Infantry Branch responded that any new light tank design had to be armed with a 37mm main gun and a coaxial .30 caliber machine gun. Frontal armor protection had to defeat AP .30 caliber machine gunfire. It also could not weigh more than 5 tons and had to be transportable by heavy truck as was the 6-ton Tank M1917, the American-made, near-identical copy of the French Renault FT-17 light tank employed by the U.S. Army during World War I.

As medium tank designs had the highest priority in the early and mid-1920s, little developmental work was completed on a new light tank design until March 1927, when the James Cunningham, Son and Company was awarded a contract by the Ordnance Department to build a pilot of a new front-engine light tank that would be designated the T1. It was powered by an eight-cylinder, gasoline-powered, liquid-cooled engine that produced 110hp at 2,700rpm.

Originally outfitted with a dummy turret and main gun, the T1 pilot tank was demonstrated to interested parties in September 1927. Testing of the vehicle showed that it was far superior to the 6-ton Tank M1917. By January 1928, the Ordnance Committee approved the standardization of an improved version, designated the T1E1, which would become the light tank M1. Much to the chagrin of the Ordnance Committee, the standardization of the vehicle was ordered withdrawn by the Secretary of War in March 1928.

Despite the light tank M1 being canceled, developmental work continued with the T1E1, which resulted in improved versions, subsequently designated the T1E2 and T1E3. By the time the Ordnance Department had gotten to the T1E4, the engine had been relocated from the front hull to the rear hull, and the final drives and track sprocket moved from the rear hull to the front hull.

The U.S. Army was keen on fielding its own light tanks for use in World War I. This was not to happen and in the sake of expediency it was decided to build a copy of the Renault FT-17 light tank referred to as the 6-ton Tank M1917. Here we see an M1917 armed with a machine gun that belonged to the former Military Vehicle Technology Foundation (MVTF). (Michael Green)

The first light tank in U.S. Army service was the French-designed and built Renault FT-17. Pictured is an unrestored and battle-damaged example on display in France. The two-man tank entered American military service in 1918 and would see combat in World War I. (Pierre-Olivier Buan)

CHRISTIE'S LIGHT TANK EFFORTS

When the Cavalry Branch of the U.S. Army in the early 1930s decided to start considering using the internal combustion engine in place of its beloved horses, J. Walter Christie saw a business opportunity and submitted his 11-ton M1931 tank for consideration. Seven of his M1931s had been ordered by the Ordnance Department in June 1931 and designated the convertible medium tank T3.

Of the seven convertible medium tanks ordered by the Ordnance Department, three went to the Infantry Branch and were armed with a turret-mounted 37mm main gun. The other four vehicles went to the Cavalry Branch and were re-designated as the combat car T1. In lieu of the 37mm main gun on the Infantry Branch version of the vehicle, the Cavalry Branch combat car T1 was armed with a turret-mounted .50 caliber machine gun. The Cavalry Branch was looking for a light tank that could perform its traditional roles of long-range reconnaissance and act as a covering force for the main body of the Army when in the field.

The first post-World War I light tank to be tested for consideration by the U.S. Army was the T1 pilot, in September 1927. It had a front hull-mounted engine and was armed with a turret-mounted 37mm main gun and a coaxial .30 caliber machine gun. Another pilot vehicle, seen here, had thicker armor and was designated the T1E2. (Ordnance Museum)

The Christie M1931 vehicle weighed too much to be carried by a heavy truck and was therefore viewed by the Infantry Branch as a medium tank, even though it only had a two-man crew. The Cavalry Branch considered its version of the M1931, the combat car T1, a light tank even though they were forced to call it a combat car, since only the Infantry Branch could have tanks at this time. This exercise in semantics with the different versions of Christie's M1931 tank was in the end all for naught as the U.S. Army found dealing with Christie too difficult and decided to pursue the concept of convertible light tanks on its own.

The front hull-mounted engine arrangement of the first few T1 pilot vehicles proved unpopular. It was therefore decided to move the engine to the rear of the vehicle on one of the T1 pilot tanks, creating the T1E4, and later the T1E6, seen here. The T1E6 featured a more powerful engine than the T1E4. (Ordnance Museum)

The most contentious individual in the early development of American light and medium tanks proved to be the civilian inventor J. Walter Christie. One of his innovative tank designs that was tested by the U.S. Army in the early 1930s for consideration as a light tank, but ultimately rejected, was the combat car T1 seen here. (Patton Museum)

In August 1933 approval was given to construct a pilot light tank, designated the combat car T5, pictured here. The new four-man tank had two turrets and was armed only with machine guns. It had a controlled differential steering system and rode on an Ordnance Department-developed volute spring suspension system. (Patton Museum)

Development of the combat car T5 continued and resulted in a single-turret version, standardized as the combat car M1, first coming off the assembly line in 1935. The machine-gun-armed vehicle had no cupola, only a roof consisting of a triple-sectioned hinged cover that could be folded together for protection for the turret crewmen. (Patton Museum)

THE NEXT STEP

With Christie out of the picture, the Ordnance Department designed its own convertible light tank that was built at Rock Island Arsenal in 1931. The strange-looking vehicle was originally referred to as the convertible armored car T5 and later designated the combat car T2. This interesting three-man vehicle was armed with three machine guns: a .50 caliber machine gun and a .30 caliber machine gun in the turret, and a .30 caliber machine gun that could be mounted in the front hull or mounted on the turret as an antiaircraft gun.

Power for the T2 came from a Continental Motors A70 gasoline-powered, air-cooled radial engine that produced 156hp at 2,000rpm. Results from testing the vehicle were unsatisfactory and the vehicle was modified at Rock Island Arsenal in 1932. Additional testing of the T2 showed that the modifications to the vehicle had failed to result in any marked improvement and the project was terminated.

The T2 was followed by the four-man T4 in 1934, which was also armed with an assortment of machine guns. As with the T2, the T4 was powered by a Continental Motors gasoline-powered, air-cooled radial engine and could run on its tracks or large road wheels. It weighed about 9 tons and could attain a maximum road speed of 40mph when operating on its road wheels and 22mph on its tracks. Despite the Cavalry Board coming out in favor of an improved T4 designated the T4E1, they were overruled and the vehicle's design did not progress any further.

In the early part of 1933, the Secretary of War ordered that in the future, any light tanks or combat cars being developed be restricted to a maximum weight of 7.5 tons. This would result in the construction at the Rock Island Arsenal of the four-man combat car T5. Like the combat cars that came before it, armament consisted of a number of machine guns and it was powered by a Continental Motors gasoline-powered, air-cooled radial engine. In this case the engine was designated the R-670 and produced 235 net horsepower at 2,400rpm.

The most important feature of the T5 proved to be the adoption of the new VVSS developed by the Ordnance Department. Another important feature found on the T5 was its use of the Ordnance Department's new T16 track system made up of smooth rubber blocks (also known as pads) vulcanized around steel links held together by rubber-bushed track pins.

The 6-ton T5 was tested in April 1934 at APG with the Infantry Branch's near-identical light tank T2. Test results with the T5 were excellent, and the vehicle was eventually standardized as the combat car M1. The now 9-ton vehicle featured two small turrets, one armed with a .50 caliber machine gun and the other a .30 caliber machine gun.

Improvements to the combat car M1 resulted in the combat car M2 seen here. It differed externally from its predecessor due to its trailing idler, which increased the ground contact length of the track, and in turn reduced ground pressure. (Patton Museum)

Rock Island Arsenal built 89 units of the combat car M1 between 1935 and 1937. An improved version of the vehicle with a reworked rear hull engine compartment and a single machine-gun-armed turret was designated the M1A1, and 24 were built in 1938. Seven of these vehicles were fitted with a Guiberson diesel engine as an experiment.

Not resting on its laurels, the Ordnance Department continued to improve the combat car M1 series using the original T5 pilot vehicle as a test bed. In 1937, they came up with a trailing idler that extended the ground combat length of the track, reducing the vehicle's ground pressure. This improvement and others were added onto the production line of the combat car M1A1 and resulted in the new vehicle being redesigned as the combat car M2 in 1940 with 34 built by Rock Island Arsenal.

With the outbreak of World War II in September 1939 and the growing threat from abroad, the Cavalry Branch was authorized 292 additional combat car M2s. Also authorized was the modernization of 88 combat car M1s. The original plans called for all of the M2s to be fitted with the T-1020 Guiberson diesel engine when they became available in sufficient numbers, something that would not come to pass.

With the formation of the Armored Force in July 1940 and tank development being taken away from the Infantry and Cavalry Branches, the need to maintain the silly pretense of a separate class of "combat cars" disappeared and they soon became known as light tanks. The combat car M2 became the light tank M1A1, and the modernized combat car M1A1 became the light tank M1A2 in August 1940. Neither vehicle would ever see combat, and they would be retained in the United States as training vehicles during World War II.

The end result of the Cavalry Branch's interest in convertible light tanks was the T7, which was constructed by the Rock Island Arsenal and sent to Aberdeen Proving Ground (APG) in August 1938. There was no T6. The T7 ran on six very large steel road wheels (three on either side of the lower hull) rimmed with pneumatic tires fitted with bullet-resistant tubes. Testing of the T7 was judged a failure and in October 1939 the Cavalry Board recommended that all further development be canceled.

In a diversion of effort, the Infantry Branch worked on the development of the light tank T2 at the same time the Cavalry Branch was developing the combat car T5. The T2 seen here rode on a strengthened version of the Ordnance Department-developed volute spring suspension system placed on the T5. (TACOM Historical Office)

DEVELOPMENTAL WORK CONTINUES

The four-man light tank T2 demonstrated at APG in April 1934 alongside the combat car T5 was also designed by the Ordnance Department and constructed by the Rock Island Arsenal. Both vehicles used the same power train. However, the T2 had a British-inspired suspension system with steel tracks and a single turret.

Testing at APG showed the British-inspired suspension system of the original version of the light tank T2 to be inferior to the Ordnance Department's VVSS. The T2 was therefore converted to the VVSS as fitted to the T5. It was also fitted with the Ordnance Department's new T16 track system. After these changes, the vehicle became the light tank T2E1. Tests of the T2E1 at APG in early 1935 were positive, and the vehicle soon found itself standardized as the light tank M2A1.

Fully loaded, the M2A1 weighed 9 tons and could attain a top road speed of 45mph. It was protected by FHA 0.625 inches (16mm) at its thickest portion. Armament consisted of a turret-mounted .50 caliber machine gun and a coaxial .30 caliber machine gun. There was also another .30 caliber machine gun in the front hull and provisions for mounting another one on the turret for antiaircraft use. Nine examples of the M2A1 were built at Rock Island Arsenal in 1935.

With the addition of the Ordnance Department-developed volute spring suspension system and a new Ordnance Department-developed double pin, rubber block track, the light tank T2 became the light tank T2E1 seen here. After successful testing, the vehicle was standardized as the light tank M2A1 in 1935. (TACOM Historical Office)

At the same time the light tank M2A1 was under development, the Infantry Branch had decided to field a twin-turreted version of the vehicle, seen here and designated the light tank T2E2. When standardized the vehicle became the light tank M2A2. (TACOM Historical Office)

A light tank M2A2 is shown climbing onto a railroad flatcar during a training exercise. One of the vehicle's two turrets was armed with a .50 caliber machine gun and the other, a .30 caliber machine gun. Maximum armor thickness on the vehicle was 0.625 inches. (Patton Museum)

Based on tests conducted with the twin-turret arrangement on the T5, it was decided to come up with a twin-turret scheme for the M2A1. Vehicles so modified were designated the M2A2. Rock Island Arsenal would construct 237 units of the M2A2 in 1937. The one-man turret on the left side of the upper hull was armed with a .50 caliber machine gun, while the other one-man turret mounted a .30 caliber machine gun. There was some variation in the turrets and hulls of the M2A2 during its production run. The M2A2 was affectionately known as the "Mae West" for its dual-turret design.

The M2A2 was continuously modified to improve its performance during its production run. These improvements included spacing out the suspension system to increase the ground contact length of the tracks in order to lower the vehicle's ground pressure. This modification and others resulted in the designation M2A3 and the tank's weight rising to 10 tons. Rock Island Arsenal constructed 73 units of the M2A3 in 1938.

Like the other combat cars and light tanks built before it, the M2A3 was powered by a Continental Motors gasoline-powered, air-cooled radial engine. In this case, the engine was designated the W-670 series 9 engine and produced 235 net horsepower at 2,400rpm. With a governor fitted the maximum speed of the vehicle was 30mph.

In a continuing effort by the Ordnance Department to find ways to improve the performance of its M2 light tank series, the vehicle was modified in a number of ways and was designated the light tank M2A3. The most noticeable difference between the M2A3 and earlier models was the increase in spacing between the bogie wheel assemblies, as seen in this photograph. (TACOM Historical Office)

ROCK ISLAND ARSENAL

812-42594 Sept. 22, 1938
Tank, Light, M2A3, with
Diesel Engine. Actual Weight
less fuel 19,500 pounds.

THE INFLUENCES OF WAR INTRUDE

Observations on the employment of light tanks in the Spanish Civil War (July 1936 until April 1939) convinced some of the U.S. Army's senior leadership that its existing inventory of light tanks was both under-armored and under-gunned. This realization pushed forward the development of what would eventually become the four-man light tank M2A4 that was standardized in December 1938. The first example of the M2A4 was converted from the last M2A3 built.

Unlike the M2A3, which had two machine-gun-armed turrets, the M2A4 had a single turret armed with a 37mm gun M3A1 and a coaxial .30 caliber machine gun in a combination gun mount designated the M20. The 37mm gun M3A1 was a 5-inch shorter version of the standard Infantry Branch towed weapon, designated the 37mm antitank gun M3. Ordnance Committee action re-designated the 37mm gun M3A1 as the 37mm gun M5 in October 1939.

On display at the former Patton Museum of Armor and Cavalry is this weaponless light tank M2A3. It is the only surviving example of 73 built in 1938. The M2A3 had slightly thicker armor than its predecessors and there was a bit more spacing between the two turrets. (Chun-Lun Hsu)

Reflecting the desire of the Infantry Branch to field a better-protected and better-armed light tank, it was directed in December 1938 that a single light tank M2A3 be taken off the production line and be both up-armored and up-gunned. The resulting vehicle, seen here coming off a naval landing craft, was designated the light tank M2A4. (Patton Museum)

Maximum armor thickness on the four-man light tank M2A4 was 1 inch. Its slab-sided turret was armed with a 37mm main gun and a coaxial .30 caliber machine gun. Most production M2A4s were powered by a gasoline engine, a few, like the one pictured, were powered by diesel engines. (Patton Museum)

The 37mm gun M5 in the M2A4 turret had a manually operated breechblock operated by the vehicle commander who doubled as the loader. The vehicle commander/loader stood on the right side of the weapon's breech ring, while the gunner stood on the left side of the breech ring. There was no turret basket in the M2A4.

It was up to the vehicle commander/loader to turn the turret of the M2A4 in the direction of the chosen target with his manual turret traverse control. Once the gunner identified the target by looking through his M5A1 telescopic sight, he would use a shoulder rest attached to the main gun breech ring to fine tune his aim. This was possible due to the 20 degrees of free play built into the M20 combination gun mount, which was accomplished with a rack and pinion traverse gear. Depression of the main gun was also done manually by the gunner.

Besides the coaxially mounted .30 caliber machine gun there were three front hull-mounted .30 caliber machine guns fitted to the M2A4. One was in ball mount directly in front of the bow gunner, with the other two mounted in a fixed forward-firing position on either side of the upper hull. These machine guns were fired by the driver with triggers on his two steering laterals. There was provision to mount another .30 caliber machine gun on the outside of the turret for antiaircraft protection.

The thickest FHA on the light tank M2A2 was 0.625 inches (16mm) and that on the light tank M2A3 0.875 inches (22mm). The thickest FHA on the M2A4 was 1 inch (25mm). This extra armor and the addition of the 37mm gun M5 on the vehicle pushed up its combat weight to 12 tons.

Power for the M2A4 came from a gasoline-powered, air-cooled radial engine designated the Continental W-670-9A, which gave the vehicle a maximum road speed of 36mph and a cruising range of approximately 70 miles. Production of the M2A4 began in May 1940 and continued until April 1942 with 375 units completed. Reflecting the need for ever greater numbers of light tanks and the limited production capacity of the Rock Island Arsenal, the contract to build the M2A4 was awarded to the American Car and Foundry Company.

In U.S. Army service from 1940 to 1942, the M2A4 served only as a training vehicle in the United States; however, with the U.S. Marine Corps it saw combat use during the fighting on the island of Guadalcanal between August 1942 and February 1943. The Marine Corps had originally requested 36 light tanks in July 1940 of a newer U.S. Army model, but made do with the M2A4 as that was what was available at the time. The Marine Corps would dispose of its M2A4s when newer light and medium tanks became available. The British Army received 36 M2A4s under Lend-Lease but never issued them to field units.

U.S. MARINE CORPS LIGHT TANK DETOUR

Before asking the U.S. Army for some of its Ordnance Department-designed light tanks, the U.S. Marine Corps had been involved with the civilian firm of Marmon-Herrington in an effort to have a light tank designed and built to meet what it perceived as its own unique requirements. That light tank turned out to be an approximately 5-ton, turret-less vehicle referred to by the company as the CTL-3. It was armed with three machine guns placed in ball mounts on the front hull. Power came from a Lincoln V-12 gasoline-powered, liquid-cooled engine that produced 110hp and could propel the vehicle to a maximum road speed of 33mph.

During U.S. Navy Fleet Exercise Number 4 conducted in the Caribbean in 1937, the first five series production units of the CTL-3 suitably impressed those who saw them in action during landing operations. Less impressed was the Marine Corps lieutenant in charge of the platoon of five light tanks, who saw them as both mechanically unreliable and underpowered. He also felt that asking the two-man crew to operate the vehicle's three machine guns in combat and deal with their other crew duties was unrealistic. The lieutenant's company commander couldn't figure out why the Marine Corps wanted such a lightweight tank when the U.S. Navy lifting booms could handle up to 21 tons.

Despite the misgivings some had about the capabilities of the CTL-3, the Marine Corps senior leadership wanted to continue with the acquisition of the vehicle. At the time, they were strong believers in the theory that the more tanks that could be placed ashore on foreign beaches the better, regardless of actual capability. In an effort to please some of the critics of the CTL-3, a slightly heavier and improved version equipped with an upgraded suspension and a more powerful Hercules gasoline-powered, liquid-cooled engine that produced 124hp was constructed. It was assigned the company designation CTL-3A, and five more were ordered by the Marine Corps.

OVERLEAF
Taken during a training exercise, this picture shows several light tank M2A4s. The vehicle would not see combat duty with the U.S. Army. However, the M2A4 went on to see fighting on the island of Guadalcanal with the U.S. Marine Corps, beginning in August 1942. (Patton Museum)

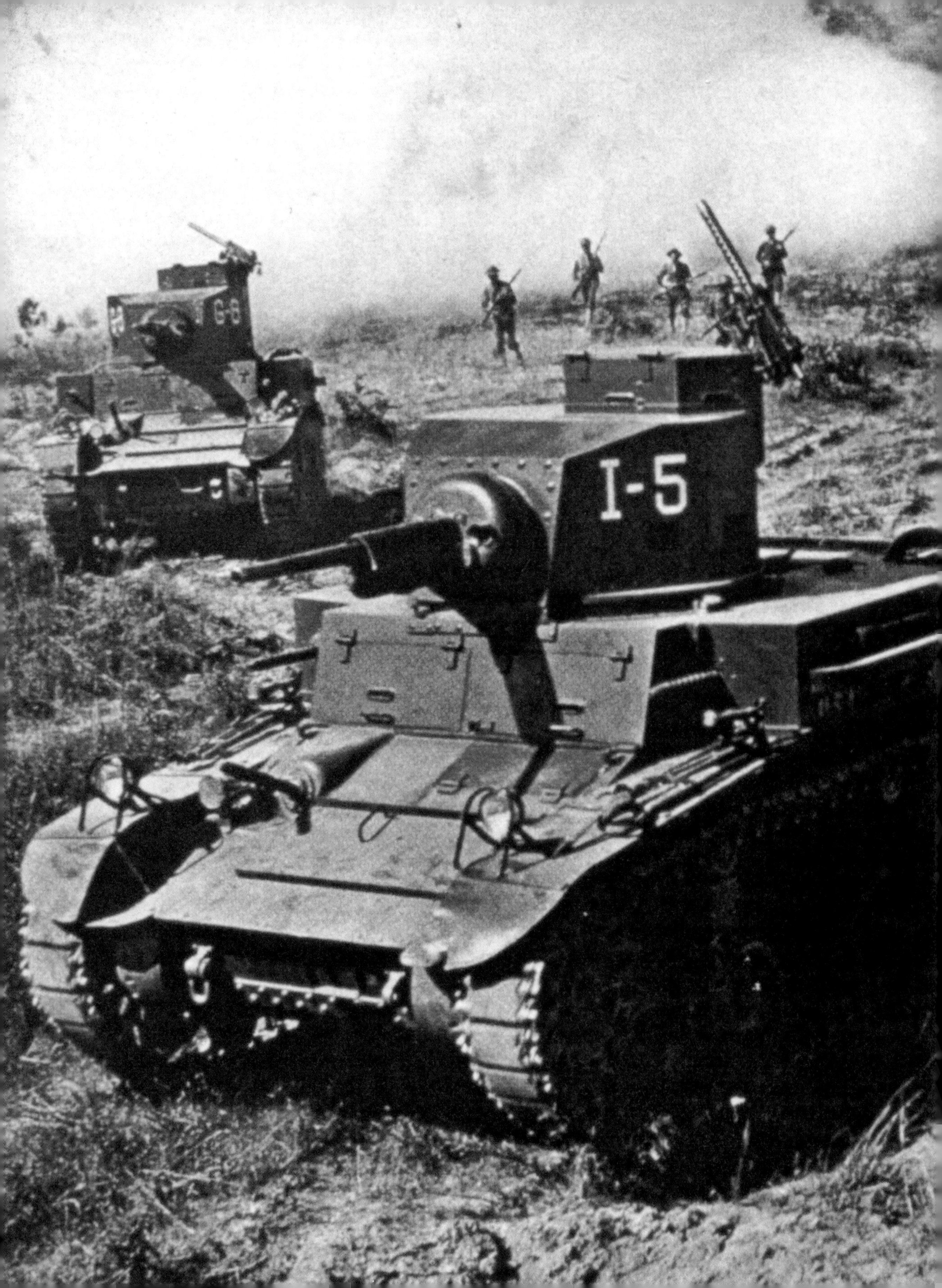
I-5

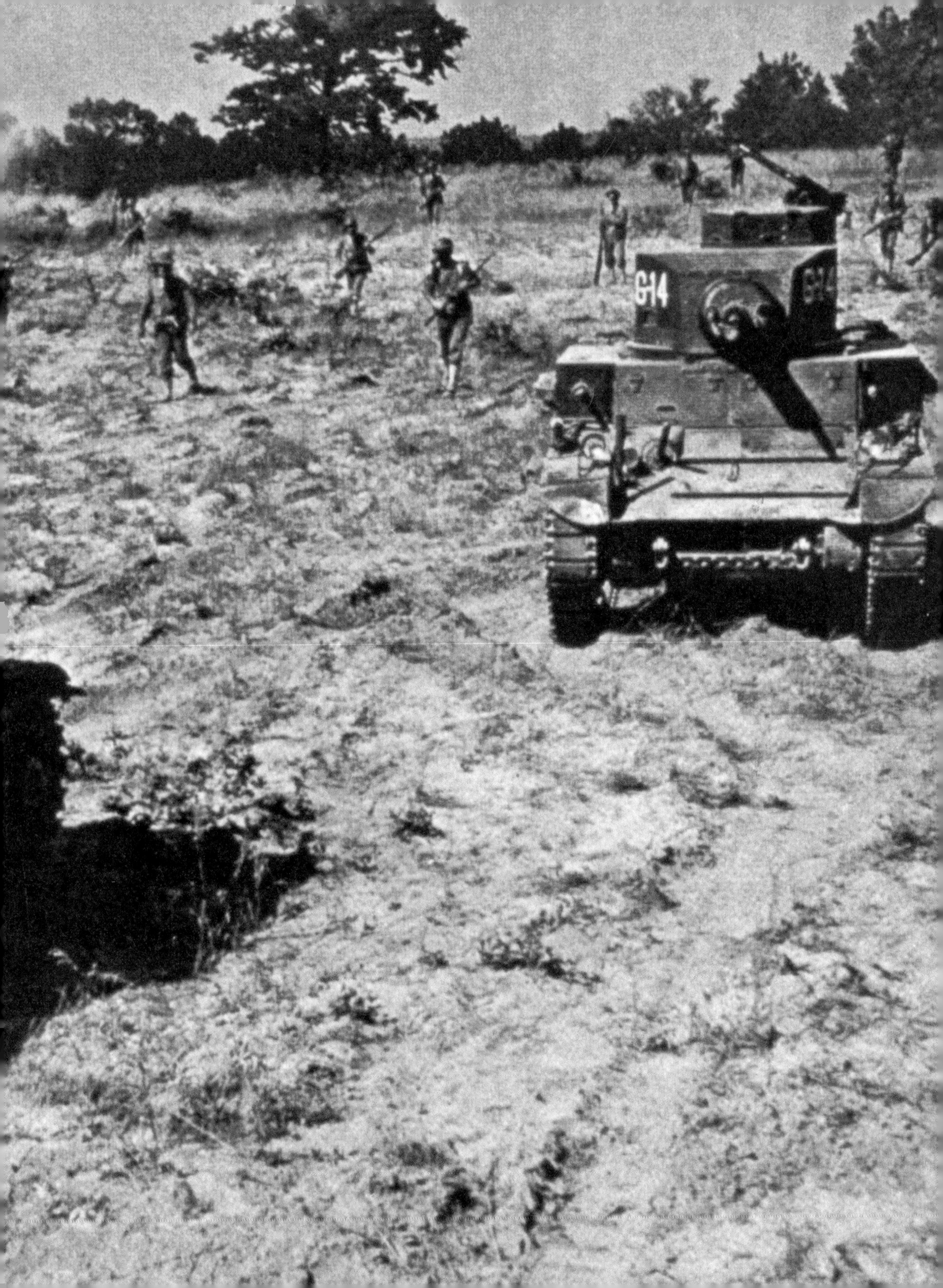
614

Before America's official entry into World War II the Marine Corps had gone its own way on tank development. An example of what the Corps was initially interested in is this turret-less light tank, referred to as the CTL-3 and designed and built by the firm of Marmon-Herrington. (Marine Corps Historical Center)

The Marine Corps liaison officer assigned to the Marmon-Herrington factory where the improved CTL-3A was being built was so disappointed in the pilot vehicle's performance that he recommended the entire project be terminated. He was overruled by the Marine Corps Commandant who received a personal appeal from the company not to halt production of the vehicle. The five CTL-3A tanks eventually entered into Marine Corps service in June 1939 and would go on to disappoint all who came into contact with them due to their lack of sufficient armor protection and general mechanical unreliability.

The CTL-3A was tested alongside a U.S. Army pilot light tank M2A4 and the combat car M1 in April 1939. The two U.S. Army vehicles outperformed the Marine Corps light tank in every trial. However, the Marine Corps senior leadership remained committed to the troublesome CTL-3 as they felt the U.S. Army light tanks were too heavy and that Marmon-Herrington would improve their vehicles. Production delays in building the CTL-3 by Marmon-Herrington eventually forced the Marine Corps senior leadership to adopt U.S. Army light tanks for service – hence the request for 36 light tanks from the U.S. Army inventory in July 1940, all of them the M2A4.

It didn't take long after America's entry into World War II for Marine Corps tankers to be faced with going into battle against the Empire of Japan. In contrast to their earlier enthusiasm, U.S. Marine Corps leaders decided to leave their existing Marmon-Herrington light tanks behind in areas where there was no threat of enemy action and take the much more capable U.S. Army light tanks into battle.

In addition to the CTL-3A, the Marine Corps took into service in 1942 20 units of the Marmon-Herrington CTL-6 and five units of the Marmon-Herrington CTM-3TBD. The CTL-6 was an improved CTL-3 with a VVSS and only two front hull-mounted machine guns. The three-man CTM-3TBD boasted a turret armed with two .50 caliber machine guns and three .30 caliber machine guns in the front hull. These vehicles would all be disposed of by the end of World War II.

One of the last Marmon-Herrington-designed and built tanks acquired by the Marine Corps was the turreted vehicle, designated the CTM-3TBD. Like all the Marmon-Herrington tanks taken into service by the Corps, it proved mechanically unreliable and, compared with its U.S. Army counterparts, under-armed, being equipped only with machine guns. (Marine Corps Historical Center)

M2A4 FOLLOW-ON

The Ordnance Department was well aware that the M2A4 was only a stopgap vehicle and that testing had uncovered a number of shortcomings. In June 1940, the Ordnance Committee recommended that any follow-on light tanks built in 1941 should have frontal armor 1.5 inches (38mm) thick. To maintain the lowest ground pressure possible and improve the weight distribution of the next-generation light tank, the addition of a trailing idler (first seen on the combat car M2) was deemed crucial.

The recommendations by the Ordnance Committee were approved in July 1940, and the new vehicle was designated the light tank M3. Production of the M3 was awarded to the American Car and Foundry Company, the builders of the M2A4, who began construction of the M3 as soon as the last M2A4 came off their assembly line in March 1941. Production of the M3 would continue until October 1942, with 4,526 units completed. Due to the thicker frontal FHA on the M3, the vehicle's combat weight rose to 14 tons, compared with the 12-ton combat-loaded weight of the M2A4.

Design shortcomings in the light tank M2A4 were addressed in the light tank M3 pictured here. This particular example is a late-production unit and has a welded-armor turret and hull. Earlier production vehicles had a partly welded turret and riveted-armor hull. As with the light tank M2A3, the M3 had a trailing idler, also visible in this picture. (Ordnance Museum)

55497   1-9-42   ABERDEEN PROVING GROUND   ORDNANCE DEP'T.

Project No. 2-7-2-1. Light Tank M3, #3213, (A. C. F. Welded Hull). Right side view.

BRITISH ARMY USE

British industry was unable to meet the production numbers required to replace the large number of tanks lost early in World War II. The British government turned to the United States in May 1941 to see what could be had quickly and in sufficient numbers under Lend-Lease to aid in the re-equipping of their tank units fighting in North Africa. What was available at the time was the relatively new light tank M3, which the British Army officially called the "Stuart," after the famous American Civil War General J. E. B. Stuart.

The first batch of 84 M3s arrived in the Middle East in July 1941. By October 1941 almost 300 were on hand. Suitably modified by their shop personnel to meet the needs of British Army tankers serving in North Africa, which included the addition

The first to use the American-designed and built a M3 series of light tanks in combat was the British Army in North Africa. The M3 pictured is from the 8th King's Royal Irish Hussars, 4th Armoured Brigade, 7th Armoured Division, in November 1941. Over the base olive drab, the British have applied a multicolored paint scheme using three different colors. These included the new base color of BSC No. 64 Portland Stone, with overlays of BSC No. 28 Silver Grey, and BSC No. 34 Slate. Khaki Green 3 was also later authorized as an alternative to the slate color. The vehicle's American registration number is in white on the original base olive drab. On the turret and hull of the tank is a national symbol consisting of a white background with a vertical red bar bisecting a yellow rectangle. In British Army service the M3 was referred to as the "Stuart I." (© Osprey Publishing Ltd.)

of sand shields and smoke grenade launchers, these M3s entered combat for the first time on November 19, 1941. Things did not go well for the British Army-manned M3s and the British Army as a whole against their German opponents. Of the 165 M3s committed to battle that month, there were only 35 left at the conclusion of the series of engagements referred to by the British Army as Operation *Crusader*, although the losses were more reflective of poor British performance on the battlefield rather than deficiencies of the light tank M3.

Combat experience showed that the light tank M3 was both under-gunned and under-armored compared with the German medium tanks it was meeting in battle. As the first of the American-supplied M3 series medium tanks began arriving in the Middle East in spring 1942, and the M4 medium tanks in summer 1942, the M3 light tanks were quickly reassigned to other duties as reconnaissance vehicles.

Early-production M3s went into battle with the same 37mm gun M5 that was mounted in the M2A4. Late-production units of the M3 were armed with an improved version of the 37mm gun M5 that featured a semi-automatic breech and was designated the 37mm antitank gun M6. The vehicle commander doubled as the loader on the M3, as he had on the M2A4.

The M3 inherited from the M2A4 the fixed forward-firing .30 caliber machine guns mounted on either side of the vehicle's upper hull. They could only be aimed by turning the tank in the direction of the intended target or targets. British Army tankers in North Africa who manned the M3 quickly deduced the uselessness of the two fixed forward-firing machine guns and removed them to provide additional storage within the vehicle.

M3 DETAILS

There were a couple of external features on the M3 that distinguished it from its earlier brethren besides the rear trailing idler. These included an improved cast-armor gun shield, designated the combination gun mount M22. Unlike the cast-armor combination gun mount M20 on the M2A4, the cast-armor gun mount on the M3 had a redesigned and shortened recoil mechanism that did not extend out from the bottom of the gun shield as it did so prominently on the M2A4 with its armor protected sleeve. Also, instead of the seven pistol ports on the turret of the M2A4, the M3 had only three.

Power for the four-man M3 came from a seven-cylinder Continental W-670 gasoline-powered, air-cooled radial engine that produced 250 net horsepower at 2,400rpm. It was this radial engine that most impressed the British tankers about their M3s due to its outstanding reliability compared with the engines that powered British tanks. So enamored were the British tank crews with the radial engines in their M3s that they began referring to the American-supplied tanks by the unofficial nickname "Honey."

The light tank M3 retained both the vehicle commander's cupola and sponson-mounted .30 caliber machine guns from the light tank M2A4 as seen here. There were also three pistol ports in the M3 turret, one of which is seen here, and six vision slots in the vehicle commander's cupola. (Patton Museum)

ENGINE ISSUES

Due to a constant shortage of gasoline-powered, air-cooled radial engines, the American Car and Foundry Company built 1,285 units of the M3 powered by the Guiberson T-1020 diesel engine between June 1941 and January 1943. With the diesel engine fitted, the M3 was designated the light tank M3 (diesel). The gasoline-powered M3s employed by the British Army became the "Stuart I," and the diesel-powered version the "Stuart II." The maximum speed of the vehicle on a level road with a governed gasoline or diesel engine was 36mph.

With the gasoline-powered engine, the M3 had a maximum cruising range on roads of 70 miles; with the diesel-powered engine that went up to 90 miles. To increase the limited operational range of the M3, the Ordnance Committee directed that jettisonable fuel tanks be developed for the M3. What sprang forth from this project was an arrangement of two 25-gallon drums that sat on top of the tank's upper hull and bracketed the rear portion of the vehicle's turret.

The British Army also sought out ways to improve the operational range of the M3s that arrived in North Africa under Lend-Lease. Rather than have two external fuel tanks mounted on the top of the upper hull in a vulnerable location and preventing the turret from being rotated 360 degrees, the British Army attached a

Obviously somebody told the Marine Corps crew to look busy for the guy who took this picture of a mid-production light tank M3 with a welded-armor turret and riveted hull. Like the early M4 series medium tanks, the loader had to enter and exit the vehicle through the vehicle commander's cupola. (Marine Corps Historical Center)

single external fuel tank at the rear of the tank's upper hull on the armor fixture that slanted back over the rear hull engine access doors.

The only way to distinguish between a gasoline-powered M3 and a diesel-powered M3 is the arrangement of the flexible intake pipes from the two external air cleaners located on either side of the vehicle's rear upper hull. On the gasoline-powered M3, the flexible intake pipes curve directly from the air cleaners into the top of the rear-hull engine deck. With the diesel-powered M3, the flexible intake pipes extend out of the air cleaners across the top of the rear engine deck and enter the engine compartment through an overhead metal screen.

The replacement on the production line for the light tank M3 was the improved light tank M3A1 seen here on display at the Tank Museum, Bovington, in the markings of the Brazilian Army. This vehicle has the original riveted-armor hull. A spotting feature for the M3A1 was the lack of a vehicle commander's cupola. (Tank Museum)

OTHER THEATERS OF OPERATIONS

British Army M3s also saw action in the Far East when an armored brigade, withdrawn from the fighting in North Africa, was hastily dispatched in January 1942 to Burma (now known as Myanmar) to stem the Japanese military invasion of that country. Outnumbered and outgunned, the British Army M3s fought a series of costly rearguard actions that cost them all their tanks except one as they were forced to cross back into India in May 1942.

The first American-designed and built tanks the Australian Army received under Lend-Lease were M3s in fall 1941. These vehicles were used successfully by the Australian Army during their campaign to rid the island of New Guinea of Japanese forces between September 1942 and January 1943. As the Japanese Army was short of dedicated antitank weapons, they often relied on close-in attacks by Japanese infantrymen using everything from iron bars to jam the turret ring of tanks to magnetic mines. For protection from the former threat, the Australian Army fitted a fairly wide piece of circular armor around the bottom of the M3's turret, which is noticeable in pictures.

According to Paul Handel of the Australian Army Tank Museum, the Australian Army placed an extra crewman in the very small and cramped turret of their M3s to relieve the vehicle commander from the job of loading the 37mm main gun, allowing him to concentrate on managing the vehicle in action.

AMERICAN M3S SEE ACTION

The first American-crewed M3s to see combat occurred during the defense of the Philippines. Prior to the Japanese attack on Pearl Harbor, the senior military leadership of the U.S. Army had sought to bolster the defenses of the Philippines under the command of General Douglas MacArthur. Among the reinforcements sent in September 1941 were two U.S. Army National Guard tank battalions with a combined total of 108 M3s. The Japanese Army's initial landing in the Philippines took place on December 8, 1941, the same day as their attack on Pearl Harbor. Due to the Philippines sitting on the other side of the International Date Line, December 8 in the Philippines was December 7 in Hawaii. Both attacks took place on the same day, about ten hours apart.

Before the Japanese attack on Pearl Harbor, the U.S. Army decided to strengthen its forces stationed in the Philippines. Among those sent were two federalized U.S. Army National Guard tank battalions, the 192nd and the 194th. Prior to being shipped overseas the tank battalions were equipped with brand new M3 light tanks, armed with a 37mm main gun. The vehicle pictured is a representative of one of those vehicles, belonging to Company C, 192nd Tank Battalion. It is in its base olive drab with the registration numbers in blue drab. There is no national symbol applied to the vehicle, only the nickname "Helen" done in yellow. (© Osprey Publishing Ltd.)

The main Japanese Army landing took place on Luzon, the main island of the Philippines, on December 22, 1941. That same day, a platoon of five American M3s was ambushed by a Japanese Army unit equipped with three-man Type 9 light tanks called the "Ha Go" that were armed with a 37mm main gun and three machine guns. The Japanese tanks prevailed in the engagement and successfully knocked out one M3 and captured the crew. The remaining four M3 tanks were also struck by Japanese gunfire but managed to withdraw. Short of ammunition and fuel, the ever thinning ranks of M3s would fight a continuing series of rearguard actions until American forces in the Philippines were forced to surrender in April 1942.

Marine Corps M3s, alongside M2A4s, would see combat for the first time during the August 1942 invasion of Guadalcanal. They continued supporting Marine Corps and U.S. Army units on the island until the Japanese were finally vanquished on

During the Guadalcanal campaign, which took place between August 1942 and February 1943, the U.S. Marine Corps employed the M2A4 light tank, the forerunner of the M3 light tank. Pictured is an M2A4 belonging to Company A, 1st Marine Tank Battalion, on Guadalcanal in September 1942, in its base olive drab, with the various tactical markings in white, as well as a hand-painted national symbol. Black and white photographic evidence shows that the tanks of the battalion had a hand-painted bar, in various colors, around their turrets as an identifying feature to prevent their fellow Marines from mistaking them for their opponent's light tanks. (© Osprey Publishing Ltd.)

February 9, 1943. Captain Robert L. Denig, Jr. (postwar promoted to colonel), who commanded Company B, 2nd Tank Battalion of the U.S. Marine Corps on Guadalcanal, equipped with the M3, sums up his feeling on the effectiveness of the vehicle:

> During the Guadalcanal Operation, it became apparent that light tanks, with their 37mm guns and .30 caliber machine guns, were of little value in the jungle, because they were not heavy enough to push their way through the heavy underbrush. In addition, the 37mm gun did not have enough explosive power to destroy machine gun emplacements or concrete bunkers. Because of these defects, it was decided that medium tanks with 75mm guns would be the main battle tank for the U.S. Marine Corps.

The final nail in the coffin for the continued use of light tanks by the Marine Corps was the Marine Corps' capture of the small island of Betio, which formed part of the Tarawa Atoll, between November 20 and November 23, 1943. During that hard-fought struggle, the Marine Corps M3A1s were not up against the jungle, but hundreds of Japanese bunkers made of thick logs or concrete. Their 37mm main guns lacked the firepower to penetrate the walls of the enemy pillboxes, forcing the tankers to drive right up to their embrasures to fire within them. The Marine Corps M3A1s on Betio proved more useful in employing their machine guns and 37mm main guns to fire canister rounds at the Japanese infantry attempting to climb onto the Marine Corps M4A2s also operating on the island.

ARMOR PROTECTION UPGRADES

Even as the gasoline and diesel-powered M3 tanks were rolling off the factory floor, they were the subject of continual improvements, the most important of these being the upgrading of the tank's armor protection. The very early-production units of the M3 had a turret made up of FHA plates riveted and bolted together. The front of the turret was 1.5 inches (38mm) thick, not counting the 1.5-inch-thick (38mm) cast-armor gun shield. The sides and rear of the turret were 1 inch (25mm) thick.

Ballistic tests quickly demonstrated that a turret riveted and bolted together was seriously deficient. Impact strikes from 37mm AP projectiles that did not penetrate deformed and loosened the various FHA armor plates that made up the turret and created a great deal of spalling. The term "spalling" is when a projectile impacts with the exterior portion of an armor plate and causes fragments or splinters from the rear of an armor plate to break off or "spall" into the interior of an armored vehicle.

Impact strikes from .50 caliber AP projectiles drove the rivets and bolt heads located on the exterior of the M3 tank turret into the interior of the turret. To correct the problem posed by the original riveted and bolted turret of the M3, the Ordnance Committee recommended in December 1940 that all future turrets for the M3 be fabricated by welding. This occurred on the 279th example of the M3 built. However, the factory initially retained a bolted-on welded turret front plate for the M3 that was later replaced by a welded-armor turret front plate.

In March 1941, the Ordnance Committee recommended that the FHA plates that made up the sides and rear of the M3 turret be replaced with a large, one-piece horseshoe-shaped welded-armor plate to simplify production and improve ballistic protection. It was bolted or welded to a flat turret front plate that was also made out of welded armor. Attached to the very bottom of the new horseshoe-shaped armor plate was a forged steel ring to protect the tank's turret ring. The vehicle's hull continued to be constructed of FHA.

A turret ring is located in the hull roof of a tank and is the circular opening in which the turret ball bearing race fits. The turret of a tank rotates on the ball bearing race. The M3 had a turret ring opening of 46.75 inches; for comparison, first-generation M4 series tanks had a turret ring diameter of 69 inches.

To improve the light tank M3 prior to the building of the light tank M3A1, some of the components intended for the M3A1 were shoehorned into late-production M3 tanks. They can be identified by the lack of the vehicle commander's cupola and the retention of the front hull sponson-mounted .30 caliber machine guns, as seen on these Marine Corps late-production M3s. (Patton Museum)

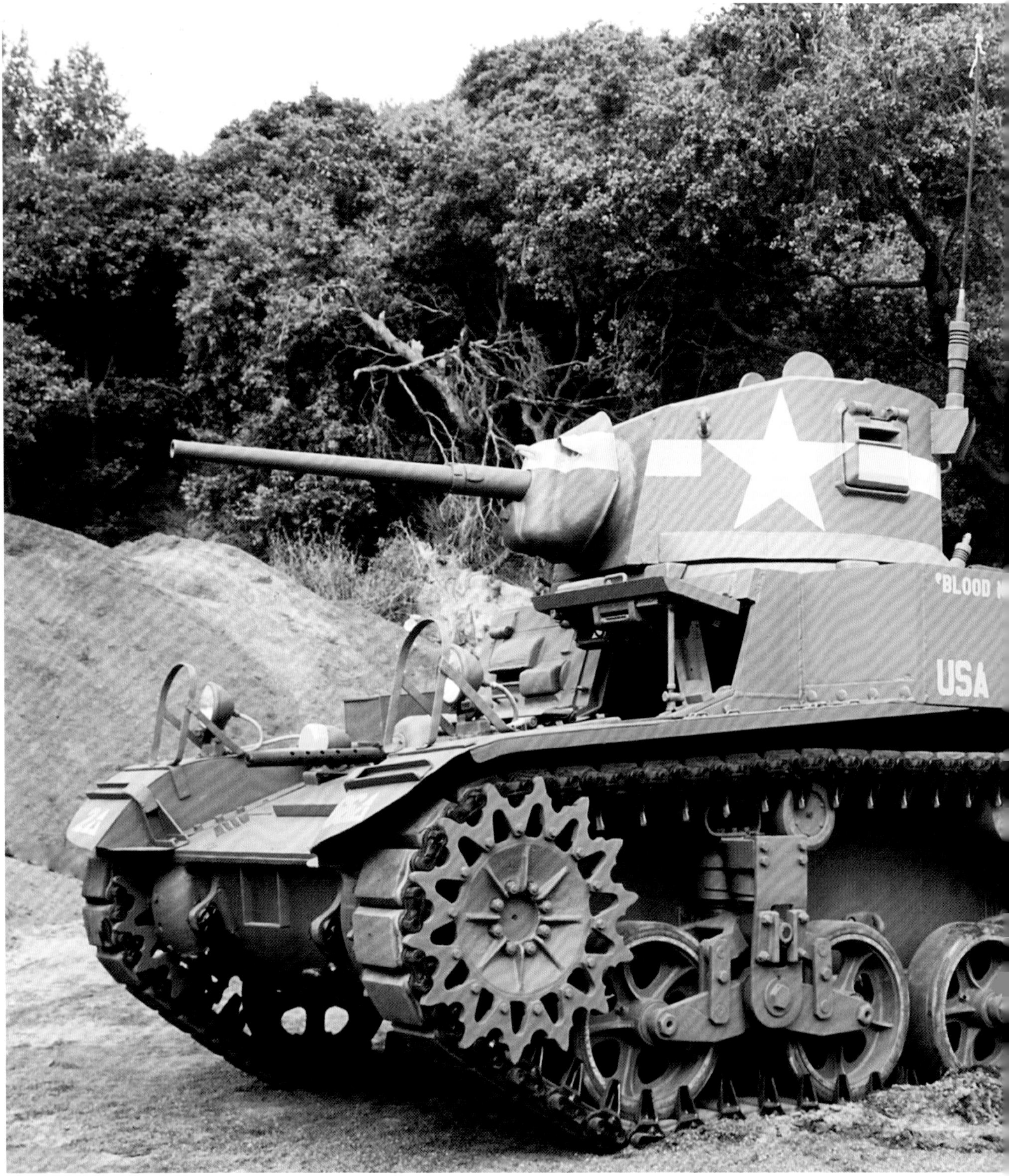
'BLOOD
USA

The welded-armor front plate of the new M3 turret remained 1.5 inches (38mm) thick; however, the cast-armor gun shield was now 2 inches (51mm) thick. The curved sides and rear of the new welded-armor turret were boosted up to 1.25 inches (32mm) thick. A total of 1,946 units of the M3 were built with the new welded-armor turret. The three pistol ports on the new welded armor turret were replaced with protectoscopes. Protectoscopes were also provided for the driver and bow gunner on the M3 in addition to their existing vision slots with moveable armor covers.

The original multi-sided riveted welded-armor vehicle commander's cupola on the M3 was replaced with a one-piece, curved, welded-armor cupola on the new welded armor turret. This new cupola lacked the protected vision slots seen on the original version. Eventually, user feedback resulted in late-production M3s with the new welded-armor vehicle commander's cupola having their protected vision slots reinstalled.

According to Charles Lemon, the retired curator of the Patton Museum of Armor and Cavalry, the vision slots on all American prewar and wartime light tanks were protected by replaceable 0.375-inch-thick (9.5mm) laminated glass blocks, about 6 inches long and 2 inches high and edged with steel. They were held in place by simple spring clamps behind the vision slots.

ANOTHER VERSION OF THE M3 APPEARS

User feedback on the M3 in combat seemed to stress the importance of an integrated fighting compartment for the tank. This meant that the vehicle's two-man turret required a turret basket underneath (with seats) to allow the vehicle commander/loader and gunner to rotate with the turret as it was turned. With the turret basket would come an Oilgear turret power traverse system.

This private collector light tank M3A1 is a late-production unit with an all welded-armor hull. The M3A1 was never fitted with the front-hull sponson-mounted .30 caliber machine guns seen on the M2A4 and M3 light tanks. Early-production units had the gun ports plugged, whereas here they have been done away with completely. (Michael Green)

It was also decided to delete the welded-armor vehicle commander's cupola on the latest production model of the M3. In its place, the vehicle commander was provided with a 360-degree rotating M6 periscope directly in front of his overhead hatch. M3s missing the vehicle commander's cupola in Army service were unofficially referred to as "low profile" or "streamlined," while in the Marine Corps they were referred to as "low tops." Those M3s that retained the vehicle commander's cupola became known as "high tops." The gunner was also provided with an overhead hatch.

Another planned improvement to the newest version of the M3 was the addition of a Westinghouse elevation gyrostabilizer, which would greatly improve accuracy while firing on the move. All these improvements were incorporated into an M3 tested at APG in May 1941. Everybody who observed the various tests with the upgraded M3 liked what they saw and the vehicle was quickly standardized as the light tank M3A1.

The M3A1 differed from the M3 in that it was provided with a new cast-armor gun shield, designated the combination gun mount M23. However, the tank retained the 37mm gun M6 that first showed up on the M3. While the M3A1 retained the gunner's M5A1 telescopic sight, as was mounted in the M2A4 and M3, it was sometimes not fitted as gunners could not use it at high angles of elevation. To supplement the gunner's M5A1 telescopic sight, the gunner on the M3A1 was provided with an M4 periscope sight.

With the introduction of the turret power traverse system on the M3A1, the gunner now took on the task of turning the main gun in the direction of a chosen target, freeing the vehicle commander/loader from that job. There was no more horizontal free play with the main gun, and the rack and pinion traverse gear and shoulder rest were dispensed with as the main gun was now locked in traverse with the M23 gun mount.

To assist the tank crew in communicating with each other, the M3A1 had an interphone system. Before the M3A1, vehicle commanders on the various light tanks and combat cars used their feet to tap on their driver's upper torso and head a series of pre-arranged signals that included right turn, left turn, speed up, slow down, etc. This physical interaction between the vehicle commander and the driver on the M3A1 was no longer possible with the vehicle's turret basket being in the way.

The M3A1 lacked the fixed forward-firing .30 caliber machine guns mounted on either side of the tank's upper hull as seen on the M2A4 and M3. Testing had shown this arrangement to be fairly useless, and the openings were plated over on the M3A1 until eventually they were done away with completely. The hull of the M3A1 was constructed of both FHA and welded armor.

Production of the M3A1 began in May 1942 and continued until February 1943, with 4,621 units built by the American Car and Foundry Company. Out of the total number of M3A1s built, the Continental W-670 gasoline-powered, air-cooled radial engine powered 4,410 of them, with the rest receiving power from a Guiberson T-1020 diesel engine and designated the M3A1 (diesel).

As with the M3, a great many M3A1s (gasoline and diesel powered) went overseas under Lend-Lease, with the British Army receiving 1,590 units and the Red Army picking up 340 units. In British Army service the gasoline-powered M3A1 was referred to as the "Stuart III," while the M3A1 diesel-powered version was known as the "Stuart IV."

INTERIM M3

As the M3A1 was being built concurrently with late-production M3s, there was some merging of the two vehicles' components. At some point on the production line they started mounting the M3A1 turret, minus the power traverse system and the turret basket, on the M3 hull. Some writers have labeled this as an "interim M3." However, this was never a factory or official U.S. Army designation. The exact number of interim M3s built is not known as the American Car and Foundry Company did not distinguish between them and the other M3s it assembled. Externally, there was no way to tell the difference between an interim M3 without the vehicle commander's cupola and the standard M3A1.

Testing of this interim version of the M3 by the U.S. Army led to the conclusion that it was unsatisfactory, especially the internal turret arrangement, because without the power traverse system it fell back upon the vehicle commander/loader to turn the turret with his manual turret traverse control. However, the gunner could no longer fine tune his aim by manually traversing the main gun with the aid of a shoulder rest as the main gun was locked in place with the M23 combination gun mount that came with the M3A1 turret. Despite the recommendation that the interim M3 should not be taken into service by the U.S. Army, some were employed for training duties in the United States. A few of the interim M3s also made it into Marine Corps service during World War II.

The British Army disliked the turret basket on the M3A1 since it blocked the escape route of the driver and bow gunner if not turned in a certain direction when utmost speed was required to disembark from the vehicle. At least one unit of the British Army equipped with the M3A1 took out the turret baskets on their vehicles. To satisfy a British Army preference at the time for the M3 over the M3A1, the production line for the M3 was kept running for an additional three months; this meant they began receiving the late-production or interim M3s. These vehicles were referred to as the "Stuart Hybrid" by the British War Office.

Due to the poorly designed internal layout of the Stuart Hybrid, the British War Office, like the U.S. Army, initially declared them unacceptable for use in combat. Despite their initial misgivings, they eventually modified them enough to make them acceptable for field use by moving the manual turret traverse control from the vehicle commander/loader's side of the turret to the gunner's side of the turret.

The hybrid M3 would go on to see service in North Africa with the British Army. Some of the British Army's Commonwealth allies, including the armies of Australia and New Zealand, would also take the Stuart Hybrid into service during World War II. The New Zealand Army received 292 units of the Stuart Hybrid before the M3A1 started showing up at the docks and decided not to use them in combat and restrict them instead to a training role only.

The Australian Army received a total of 370 M3 and M3A1s under Lend-Lease, including gasoline- and diesel-powered versions. Paul Handel of the Australian Army Tank Museum and author of the book *Dust, Sand, and Jungle: A History of Australian Armor 1927–1948* states that they also received the Stuart Hybrid, both the gasoline and diesel-powered versions. However, the Australian Army reserved them for training duties only. The gasoline-powered version was referred to as the "Stuart Hybrid I" and the diesel-powered version as the "Stuart Hybrid II."

DEVELOPMENT CONTINUES

As with its medium tanks, the U.S. Army was confronted in early 1941 with the fact that there would not be enough gasoline-powered, air-cooled radial engines available to power all the light tanks in the pipeline. To overcome this anticipated bottleneck, the Ordnance Department began exploring additional power-plant possibilities. To speed up this process, they initially concentrated their search for engines that might be suitable for powering light tanks to those already in production.

As an experiment, the Ordnance Committee approved the mounting of two eight-cylinder, gasoline-powered, liquid-cooled, Cadillac car engines mated together and combined with a Hydramatic automatic transmission in an M3 in June 1941. The test vehicle was designated the M3E2 and had the late-production M3A1 welded-armor horseshoe-shaped turret without the vehicle commander's cupola.

The follow-on to the M3 light tank was the M3A1 light tank, pictured. It is marked as belonging to the 4th Marines on Emirau Island, Bismarck Archipelago, in March 1944. It is in the factory-applied base olive drab. The tactical designation on its turret is yellow. The diamond outline surrounding the geometric symbol on the vehicle's turret identifies the company the tank is assigned to. U.S. Marine Corps tanks in the Pacific Theater of Operations seldom featured the national symbol, which was more common on U.S. Army tanks. Some U.S. Army tankers eventually came to the conclusion that the national symbols on the hulls and turrets of their vehicles were being used as aiming points by their opponents and quickly covered them over with whatever was at hand, be it paint or mud. (© Osprey Publishing Ltd.)

The M3E2 was also fitted with an automatic auxiliary transmission, referred to as a "transfer unit" or "transfer case." Testing of the vehicle went extremely well and quickly proved the reliability of this new engine arrangement. The positive test results with the M3E2 resulted in the Ordnance Committee designating it as the light tank M4 on November 13, 1941.

At the same time the Ordnance Committee decided to meld the M4 with another train of development that incorporated the ballistic advantages of using welded armor plates for the M3A1's lower and upper hull. A single example of the M3A1 was chosen for this experiment and designated the M3A1E1. The designation light tank M3A2 was

The next step in the evolution of the M3 series of light tanks was the light tank M3A3, seen here at a military museum in France. The obvious difference between the two light tanks was the addition of sloped armor on the M3A3 hull, a feature first applied to the glacis of the light tank M5. (Pierre-Olivier Buan)

reserved for the production version of this vehicle. As events transpired, the M3A2 would never be built, as the M3A1E1 was combined with the M4 to create the M3E3.

The M3E3 appeared with a welded armor glacis plate that was 1.125 inches (29mm) thick and sloped at 48 degrees. By moving the glacis plate forward there was now room to provide the driver and bow gunner in the front hull with overhead hatches, each fitted with a 360-degree rotating M6 periscope. The driver and bow gunner no longer had to enter and leave the vehicle via the turret overhead hatches.

As there were dual driving controls in the front hull of the M3E3, the bow gunner was also referred to as the assistant driver. Provisions were made to allow the driver and assistant driver/bow gunner to raise their respective seats vertically so their heads could project out over the top of the glacis plate for greater visibility in non-combat situations.

As a backup to their M6 overhead periscopes, both the driver and assistant driver/bow gunner on the M3E3 were provided with small, 2-inch-diameter holes directly in front of their seats to look out of if all else failed. When not in use, these small peep holes were closed with steel plugs attached to chains within the front hull.

Other distinguishing external features visible on the M3E3, besides the new sloped glacis, included the addition of a fixed forward-firing .30 caliber machine gun in the glacis operated by the driver. This design feature was later deleted from series production vehicles. To make room for the engine cooling system in the M3E3, the engine compartment roof was raise.

Owing to the lower crankshaft height of the V-8 engines compared with the radial engines they replaced, the drive shaft tunnel that connected the rear hull-mounted engine with the front hull-mounted transmission was much lower in the M3E3 than that found with earlier models. This made it possible to move the turret power traverse components originally fitted on the turret basket floor of the M3A1 turret to new mountings beneath the turret basket floor, increasing the working space for the turret crew in what was always a very cramped environment.

The light tank M3A3 pictured is in Free French Army service. There were a couple of other new features on the vehicle besides the sloped-armor hull. One was the bustle on the rear of the tank's turret for a radio, and second, the overhead hatches for the driver and bow gunner combined with seats that could be raised or lowered. (Patton Museum)

With the commencement of production of the medium tank M4 in February 1942, it was decided to designate the light tank M4 as the light tank M5 to avoid confusion with the medium tank M4. Series production of the M5 began in April 1942 at the Cadillac Division of the General Motors Corporation with other companies joining in the production run. By the time series production of the M5 concluded in December 1942, a total of 2,074 units had been built. The vehicle weighed 16 tons combat loaded.

There had been some concerns in the U.S. Army that adding another light tank to the existing inventory would overburden the logistical system and the complexity of the various automotive components in the M5 would prove more than the tankers or support personnel could manage. These concerns did not become an issue, and the U.S. Army would soon come to the conclusion that the M5 was superior to any of the light tanks that had preceded it into service. The vehicle was eventually reclassified as limited standard as newer vehicles replaced it in service.

IMPROVED LIGHT TANK M5

Following on the heels of the M5 in September 1942 was an improved version, designated the light tank M5A1. It is readily identified by its different turret design from that of the M5. The new turret design came from another light tank, designated the M3A3, and featured a bustle at the rear of the turret for the mounting of a radio. There was also a removable plate at the rear of the new turret bustle on the M5A1 that would allow for the extraction of the M6 37mm gun without removing the turret if the need arose.

Early-production M5A1s were fitted with pistol ports on either side of the turret in place of the protectoscopes mounted on the M5. As the U.S. Army no longer saw a need for them on the M5A1 as they were a ballistic weak spot, they were eventually welded shut and then done away with altogether on later production units. Late-production units of the M5A1 also saw the addition of a large stowage box on the rear engine deck of the vehicle. Not normally visible on the M5A1 was the addition of an escape hatch at the bottom of the hull floor just behind the assistant driver/bow gunner's seat.

Late-production M5A1 turrets can be identified by a curved armored storage box on their right side to hold the .30 caliber machine gun that was sometimes mounted on the top of the turret for antiaircraft purposes. On the M5, the .30 caliber machine gun was mounted at the rear of the turret, which meant that the vehicle commander/loader had to climb out of the turret to employ the weapon. On the M5A1, the mount for the antiaircraft machine gun was relocated to the right side of the turret, adjacent to the vehicle commander's overhead hatch, making it much easier for him to engage targets to his front or right side without exposing his entire body to enemy fire.

Series production of the M5A1 began in November 1942 with a number of companies eventually taking part. Final production ended at the Massey Harris Company in June 1944. Total production of the M5A1 reached 6,810 units. It was the most common light tank to see service in Northwest Europe from June 1944 until the German surrender in May 1945. The vehicle was reclassified as substitute standard in July 1944.

Under Lend-Lease the British Army received 1,131 units of the M5A1 and referred to it in service as the "Stuart VI." The Free French Army received 226 units of the M5A1 and the Red Army only got five. Combat loaded, the vehicle weighed in at 16 tons.

The light tank M5 was 14 feet 6 inches long and had a width of 7 feet 4 inches. It was 7 feet 6 inches tall and weighed 16 tons. The maximum speed on level roads was 36mph, with a cruising range on roads of approximately 100 miles. It carried 123 rounds of main gun ammunition. (TACOM Historical Office)

COMBAT RESULTS

United States Army M3, M3A1, M5, and M5A1 light tanks would see combat during the invasion of North Africa on November 8, 1942, referred to as Operation *Torch*. Not having a sufficient number of landing craft able to transport medium tanks from ship to shore meant that the light tanks would be in the vanguard of the invasion force, until such time that the medium tanks could be unloaded. There would be some minor combat encounters with the light tanks of the French Vichy Army. With their heart not in the fight, the Vichy French Army military forces surrendered on November 10, 1942.

A redesigned light tank M3A1 powered by twin Cadillac automobile engines and featuring a glacis sloped at 48 degrees that was 1.125 inches (29mm) thick became the light tank M5. It retained the existing light tank M3A1 turret. The M5 shown is on display at the First Division Museum in Wheaton, Illinois. (Paul Hannah)

The true test of the combat worth of the U.S. Army's inventory of light tanks didn't occur until they met the combat-proven German Army during the fighting for French Tunisia that lasted from November 12, 1942 through May 13, 1943. The results were extremely disappointing. When forced to assume a main combat role, the U.S. Army's light tanks were under-gunned and under-armored compared with the German medium tanks they encountered in battle.

At the end of the campaign in French Tunisia, some senior American military officers that served in North Africa recommended what the British Army had concluded the year before: the only remaining duties that light tanks were suitable for were reconnaissance and, with the U.S. Army, flank security.

The Armored Force leadership back in the United States was already coming to the same conclusion as the senior officers that served in North Africa. They were planning on changing the table of organization and equipment (TO&E) for the existing armored divisions that would drastically reduce the number of light tanks in service. This action took place in September 1943 and would eventually involve 14 of the U.S. Army's 16 armored divisions. The new armored division TO&E, referred to as "light," had only 77 light tanks, whereas the former armored division TO&E, now referred to as "heavy," had 158 light tanks.

The decision to retain light tanks in the U.S. Army was a mixed bag. Many saw little or no use for the vehicle, as seen in a passage from a March 1945 U.S. Army report titled *United States Armor vs. German Armor* by Corporal Clarence E. Land:

It was soon decided that the light tank M5 should have the same enlarged turret with a radio mounted in its rear bustle as developed for the light tank M3A3. With the new turret, the M5 became the light tank M5A1. An early-production unit is seen here during a World War II reenactment event. (David Marian)

I have been in the M5A1 tank for about sixteen months and all the value that they have in my view point is against infantry. They might have more maneuverability than other tanks, but that isn't any good unless you have more firepower than a 37mm gun… The M5A1 doesn't have a large enough fighting compartment. The 37mm gun is too small. It is just a waste of ammunition to use such a small gun on a tank, when a larger one could be mounted on it.

U.S. Army tank battalions typically had three medium tank companies labeled A, B, and C of 17 tanks each and a light tank company of 17 tanks labeled D. The lack of value assigned to the M5/M5A1 by U.S. Army tank battalions fighting in the ETO can be surmised by the fact that some of them would strip men from their light tank company, or even go as far as to stand the company down, to replace heavy losses among the crews of their medium tank companies. By the end of the war in Europe, some tank battalions had replaced all their light tanks with medium tanks. However, there were also some all light tank battalions that remained in the field until the war in Europe concluded.

The AAR of the 759th Light Tank Battalion describes how their vehicles were employed with infantry units in the ETO:

> We found that the best success was obtained by attaching not less than a company of light tanks to an infantry Bn [Battalion] and then employing the tanks in sections, with an infantry team, made up of a tank director and a squad of two men, who actually guided the tanks by finding targets, routes of advance etc., for them and then covering the vulnerable flanks and rear of the tanks with small arms fire as they took the advance to keep-off enemy bazooka and grenade men.

Joseph J. Graham, who served in the U.S. Army during World War II in Northwest Europe, recounts one of the positive features about the M5A1 tanks he commanded:

Later production units of the light tank M5A1 had the .30 caliber antiaircraft machine gun mount moved slightly rearward on the right side of the turret. Another change was the addition of a shielded compartment affixed to the right side of the turret as seen here to store the antiaircraft machine gun when not in use. (Michael Green)

> The same Ford 500 horsepower aluminum engine we had tested at Fort Knox in the early days of our battalion powered all the Shermans [M4A3s] of the other tank companies. They were extremely loud. You could hear them long before you ever saw them. It was impossible to sneak up on the Germans in a Sherman. The German Tiger tanks were the same, so I guess we came out even on that score.
>
> The Stuart in contrast came equipped with two relatively quiet Cadillac gasoline engines of 110 horsepower each. Because of this fact, it was often possible for us to maneuver some of our tanks to sites from which, with a squad or two of the 100th infantry on our back decks, we could dash into German positions. The infantry would grab a couple of Krauts, throw them on the back of our tanks, and rush back to our own lines with the prisoners who could have valuable information for the intelligence officers.

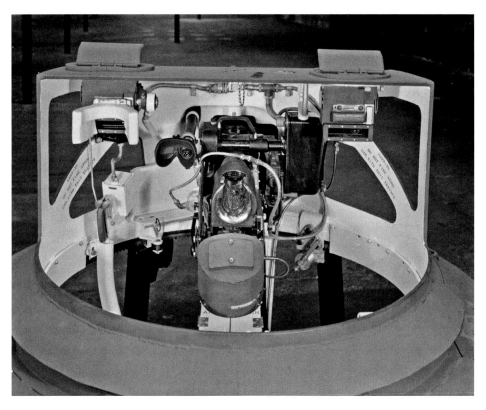

Shown is the cutaway turret trainer for the light tank M5A1. The gunner sat on the left side of the 37mm main gun and the vehicle commander on the right side. Visible in this picture are the gunner's overhead periscope sight and direct telescopic sight. The gunner's turret traversing control is also seen in the photograph. (Patton Museum)

Operating around the grounds of the Tank Museum, Bovington, during their annual "Tankfest" event is a light tank M5A1 in British markings. As portrayed here, it was not unheard of for the crews of these tanks in the ETO to add sandbags and logs to the glacis of their vehicles for added stand-off protection from shaped-charge warheads. (Christophe Vallier)

THE END OF THE ROAD

The last version of the M3 series was designated the light tank M3A3. It came about because the Armored Force requested that the M3A1 be modified to have a hull similar in configuration to that of the M5, especially the sloped welded armor front glacis. This sloped glacis meant that the driver and assistant driver/bow gunner would be provided their own separate overhead hatches, each fitted with a 360-degree rotating M6 periscope.

Where the M3A3 differed in hull configuration from the M5 was the 20-degree sloping welded armor sides of its upper hull, which was in sharp contrast to the vertical sides seen on all the light tanks and combat cars that came before it. Since the M3A3 retained the gasoline-powered, air-cooled radial engine found in the M3A1, there was no need for the pronounced raised rear engine deck seen on the M5.

Like the M5, the enlarged hull of the M3A3 meant that the air cleaners could be moved inside the rear engine compartment, instead of the vulnerable location on top of the rear engine deck. A new turret with larger overhead hatches for the vehicle commander/loader and gunner was designed and built for the M3A3. The new turret also featured a rear bustle in which a radio could be mounted. The radio antenna was moved from the top of the rear engine deck as seen on the M3A1 and placed at the rear of the new turret bustle on the M3A3.

The new turret on the M3A3 came with an improved combination gun mount, designated the M44, which featured the M54 telescopic sight on the left side of the mount. Unlike the telescopic sight M5A1 fitted in the M3A1 that was mounted too

The replacement in U.S. Army service for the M3 series of light tanks was the M5 light tank, shown. The vehicle appears in the base olive drab and is marked as belonging to Company C, 70th Tank Battalion (Light) in Oran, Morocco, in January 1943. The extremely large national insignia painted on the front hull of the tank, along with the American flag painted on the side of the upper hull, with a smaller national insignia, were an attempt to convince the Vichy French forces not to fire upon them. The large white national insignia on the front hull of the tank shown was also authorized to be done in yellow until later rescinded. The vehicle's registration numbers are in blue drab. The white star within a blue circle on the tank's rear engine deck was authorized for Operation *Torch* to prevent attack by friendly aircraft. (© Osprey Publishing Ltd.)

low to use when the main gun was firing at its highest elevations, the M54 telescopic sight was mounted higher on the mount and could be employed at any elevation. In contrast with the M5, the M3A3 did not have a periscope sight for the gunner.

Production of the 16-ton M3A3 began in September 1942 at the American Car and Foundry Company and did not end until September 1943 with a total of 3,427 units completed. Of that number, the majority went overseas under Lend-Lease, as the U.S. Army decided to replace its existing inventory of M3 series tanks with the M5 and M5A1. The British Army received 2,045 units of the M3A3 and referred to them as the "Stuart V." The Free French Army picked up 277 units of the M3A3, with another 100 units being sent off to serve with the Chinese Nationalist Army.

The final version of the M5 series of light tank to enter service with the U.S. Army and the U.S. Marine Corps was designated the M5A1. The vehicle pictured belonged to the 601st Tank Destroyer Battalion and is located at the Volturno River, Italy, in October 1943. Painted in the base olive drab, the vehicle has a national symbol in white surrounded by a white circle. This version of the national symbol was approved for the invasion of Sicily, which took place between July and August 1943. The unit's tactical markings are red on a yellow background. There are no registration numbers on the vehicle's hull. (© Osprey Publishing Ltd.)

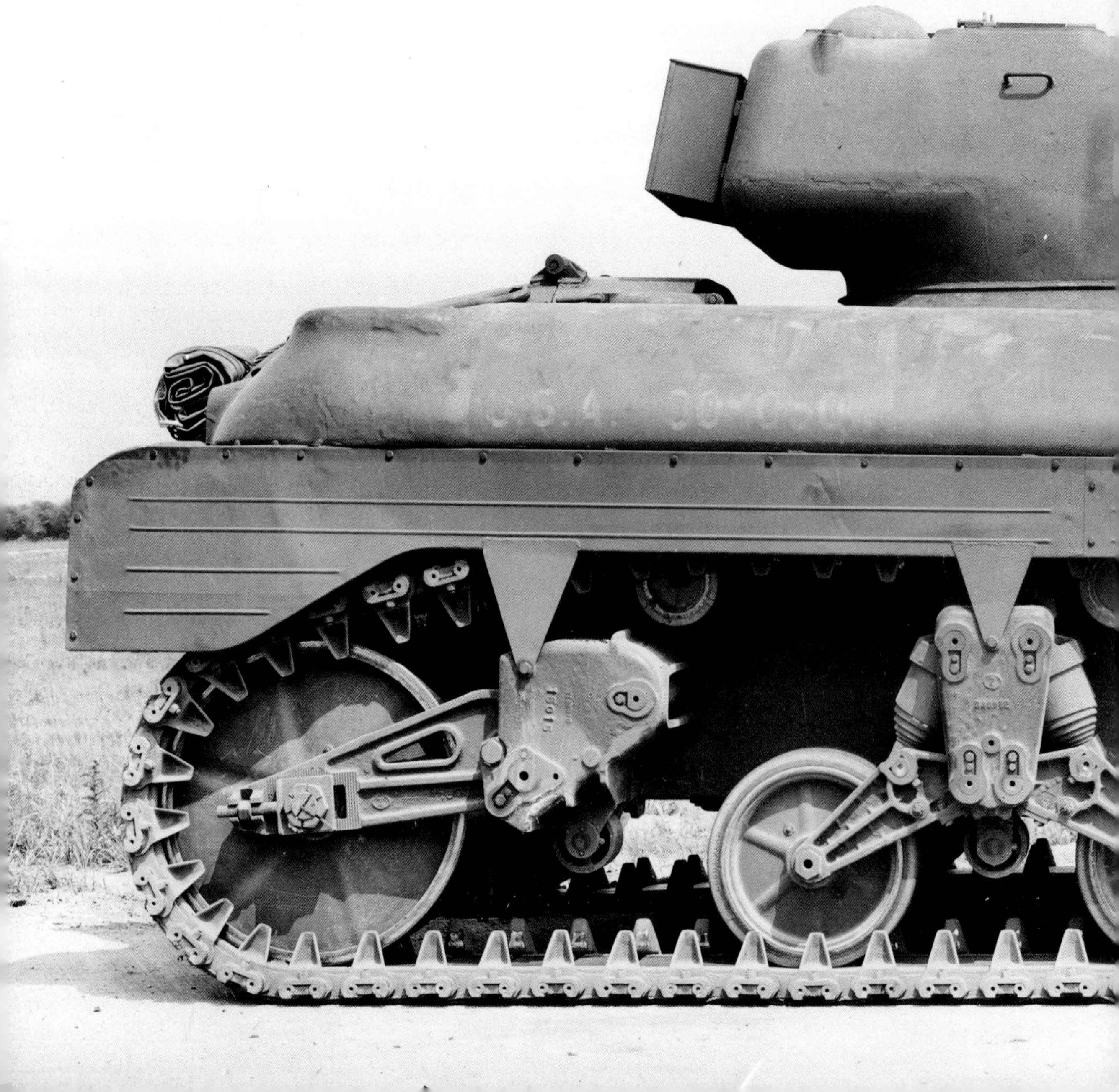

Based on its experience gained with the development and fielding of the light tank M2A4, the U.S. Army generated a list of desirable characteristics. The end result was the construction of a number of pilot vehicles, designated the T7 light tank series, one of which is seen here. Due to issues with weight, the project was later canceled. (Patton Museum)

A DESIGN DEAD END

Not everything the Ordnance Department came up with in terms of light tank development would go into service. An example of that would be the Light Tank T7 series. This line of development was pursued in January 1941 when the Armored Force came up with a list of requirements they wanted to see in a new light tank design based on their service experiences with the light tank M2A4.

The requirements for a new five-man light tank design included a weight limit of 14 tons and a maximum armor thickness of 1.5 inches (38mm). The new light tank was to be armed with a turret-mounted 37mm gun M6 and a coaxial .30 caliber machine gun. There were three .30 caliber machine guns mounted in the front hull. The Ordnance Committee assigned the designation light tank T7 to this proposed vehicle.

Two pilot tanks were authorized for construction in February 1941. As they had different design features they were designated the T7 and T7E1. A wooden mockup of the T7 was demonstrated to interested parties in July 1941. By August 1941 it was decided to have three more pilot vehicles built highlighting different design features, including a combination of cast- and welded-armor turrets and hulls. Without even the completion of a single pilot tank it was estimated that the T7 would be 2 tons heavier than originally requested.

The first soft-steel pilot of the T7 was finished at Rock Island Arsenal in January 1942. It featured a soft-steel cast-armor hull and turret armed with the 37mm gun M6 and was powered by a gasoline-powered, air-cooled radial engine connected to an automatic transmission. However, at this time, the Armored Force had already decided they wanted a lower and wider version of the vehicle.

In response to the Armored Force's preference, Rock Island Arsenal came up with another pilot, designated the T7E2. At the instructions of Major General Gladeon M. Barnes, from the Office of the Chief of Ordnance, a British-designed and built 6-pounder (57mm) main gun was installed in the vehicle. The T7E2 showed up for testing at APG in May 1942. This version of the T7 weighed in at almost 26 tons.

The Armored Force rejected the 57mm main gun on the T7E2 and insisted on its replacement with a 75mm gun M3. This re-arming of the T7E2 took place in August 1942 and resulted in the vehicle being designated the T7E5. The mounting of a larger, heavier main gun caused a 1-ton weight gain to the vehicle. Reflecting the 27-ton weight of this latest version of the T7, the Ordnance Committee decided to standardize the light tank T7 as the medium tank M7. In the end, only seven units of the medium tank M7 were built before being canceled in favor of the more practical medium tank M4A3.

IMPRESSED LIGHT TANKS

Following the Japanese attack at Pearl Harbor, the U.S. Army quickly seized all the military equipment being built for friendly overseas users. Among the grab bag of weapons taken into service was a fleet of 600 two-man light tanks that were being built by the Marmon-Herrington Company as a private venture for the Chinese Nationalist Army. These vehicles were equipped with three .30 caliber machine guns, one in the manually operated turret and two in ball mounts in the front hull.

These two-man light tanks weighed in at about 8 tons and had two different factory designations. The CTLS-4TAC had the driver sitting on the left side of the front hull, while the CTLS-4TAY had the driver sitting on the right side of the front hull. After shipping off the bulk of their CTLS-4s to friendly foreign armies, the U.S. Army had 250 units left over and decided to take them into service. The CTLS-4TAC was designated the light tank T16, and the CTLS-4TAY as the light tank T14.

Reflecting the dire straits in which America found itself in early 1942, and with an unfounded belief that a Japanese military invasion was possible at any moment, the CTLS-4s were deployed along the western borders of the United States from the Aleutian Islands all the way down to southern California. A handful also went to other locations, such as Newfoundland in Canada and even Bermuda in the Caribbean. None of the CTLS-4 light tanks ever saw combat and all were ordered disposed of in November 1942.

The U.S. Army would also take control of a number of Marmon-Herrington light tanks being built for the Royal Netherlands East Indies Army. Their job was to protect the Dutch East Indies from possible Japanese invasion, but the German invasion of the Dutch homeland had taken the Dutch out of the war in both Europe and Asia. These impressed light tanks included a model with the factory designation CTMS-1TB1, and another with the factory designation MTLS-1G14.

The three-man CTMS-1TB1 was armed with a turret-mounted 37mm main gun built by the American Armament Corporation and a coaxial Colt .30 caliber machine gun. There were provisions for mounting three additional Colt .30 caliber machine guns in the vehicle's front hull. Weighing about 13 tons, the vehicle had a maximum armor thickness of 0.5 inches (12.7mm) of FHA and rode on a VVSS.

The four-man MTLS-1G14 was very similar to the CTMS-1TB1 in overall

Pictured is the light tank referred to as the CTMS-1TB1 belonging to the former Military Vehicle Technology Foundation. It was designed and built by the Marmon-Herrington Company for a foreign buyer but was impressed into U.S. Army service for a short time after the Japanese attack on the naval base at Pearl Harbor, Hawaii. Testing showed the vehicle was unfit for service. (Michael Green)

Pictured on the Marmon-Herrington Company production line are a number of machine-gun-armed light tanks, referred to by the firm as the CTLS-4TAY. Built for a foreign buyer, the vehicles were impressed into U.S. Army service as the light tank T14 following America's official entry into World War II. (Patton Museum)

appearance. The biggest difference between the two vehicles was that the MTLS-1G14 had thicker FHA and was therefore heavier at 21 tons. As with its lighter counterpart, the vehicle rode on a VVSS.

Instead of a single 37mm main gun, the MTLS-1G14 sported two turret-mounted side-by-side 37mm main guns in the same gun mount. Besides a Colt .30 caliber machine gun fitted in the turret in a ball mount to the right of the twin 37mm main guns, there was also a mount for another Colt .30 caliber machine gun on the top of the turret for antiaircraft protection. The vehicle had provisions for mounting another three Colt .30 caliber machine guns in the front hull with ball mounts.

For consideration of possible use by the U.S. Army, a single example of the CTMS-1TB1 was tested at APG in February 1943 alongside two examples of the MTLS-1G14 in April 1943. The results were not very impressive. According to a report issued after the testing, the Marmon-Herrington light tanks "were unreliable, mechanically and structurally unsound, under-powered and equipped with unsatisfactory armament."

AIRBORNE LIGHT TANKS

Impressed with the German military employment of paratroopers and glider-delivered infantry during the early part of World War II, the U.S. Army began to consider forming its own airborne units. The U.S. Army was also aware of the Red Army's experiments in the 1930s with attaching light tanks under the fuselage of modified four-engine bombers that could then be flown into airfields captured by their paratroopers. The Red Army light tanks employed in these demonstrations were armed only with machine guns.

The Ordnance Department called a group of interested parties together for a meeting on February 27, 1941, which included the Armored Force and Army Air Corps, to consider the development of an airborne light tank. The British Purchasing Commission also expressed their interest in acquiring such a tank if built for use by the British Army. It was decided at the meeting that the Ordnance Department would come up with a suitable light tank and the Army Air Corps an aircraft that could transport it into battle.

In May 1941, the Ordnance Committee came up with the desired characteristics for the proposed new airborne vehicle, now designated as the light tank T9. Based on the carrying capacity of anticipated future transport aircraft, it could not weigh more than 7.5 tons in a stripped-down configuration with no crew or onboard stowage. The vehicle's approximate dimensions were to be 11.5 feet long, 7 feet wide and 5.5 feet tall. Armament was to be either a 37mm or 57mm main gun with a coaxial .30 caliber machine gun in a power-operated turret having an elevation gyrostabilizer. Additional machine guns were to be fitted if possible. The crew complement for this proposed airborne light tank was to be either two or three men.

At the invitation of the U.S. Army, a bid to build the proposed new airborne light tank was solicited from J. Walter Christie, who had pushed the concept of such vehicles as far back as the 1920s. However, the designs put forward by Christie did not meet the Ordnance Committee requirements and were rejected. The Pontiac Division of General Motors also submitted a proposal to build a T9 pilot tank but were beaten out by the Marmon-Herrington Company bid which cost less.

By November 1941 Marmon-Herrington had completed a partial wooden mockup of the upper hull of their T9 pilot tank. The mockup was shipped to the Douglas Aircraft Company to see if it could be successfully attached to the bottom fuselage of their prototype C-54 "Skymaster" four-engine transport aircraft that would enter service the following year. It was intended that the vehicle's hull be raised and lowered from the plane's lower fuselage with a hydraulic hoisting mechanism. There was no need for a T9 turret mockup at this time. Because the C-54 ground clearance was too low to carry a completely assembled tank, plans called for only the vehicle's hull to be carried under the aircraft's fuselage. The turret was to be carried within the plane's fuselage and reattached to the hull upon landing at a captured enemy airfield.

Marmon-Herrington had a mild-steel pilot model of the T9 ready in April 1942. It had a three-man crew with the driver in the front hull and the vehicle commander/loader and the gunner located in the turret armed with a 37mm gun M6 and a coaxial .30 caliber machine gun. The maximum armor thickness on the T9 was 1 inch (25mm). The turret was constructed from cast armor and the hull of welded armor.

Located in the front hull of the T9 were two fixed, forward-firing .30 caliber machine guns to be operated by the driver. Power came from a six-cylinder gasoline-powered, liquid-cooled, Lycoming 0-435 engine that produced 162 net horsepower at 2,800rpm. Maximum speed on level roads was 35mph. The completed pilot vehicle was shipped to the Douglas Aircraft Company to confirm it could still be carried by the C-54, which it could.

Subsequent testing of the pilot T9 at various locations revealed that it had grown in weight and now exceeded the original 7.5-ton limit. The vehicle's mobility had been compromised by the weight gain. As both the AAF and the British indicated that the maximum weight of the vehicle could not be above 7.9 tons, it was decided that the T9 had to go on a diet and a number of features were deleted. These included the turret power traverse system, the elevation gyrostabilizer, and the two front-hull-fixed machine guns.

Reflecting the various improvements made in tank designs and components since the pilot T9 was first constructed by Marmon-Herrington, it was also decided that an improved version would be built incorporating as many new features as possible. Two improved pilot vehicles were completed by Marmon-Herrington in November 1942 and designated the light tank T9E1. One went to APG and the other was shipped off to England for evaluation by the British Army.

Testing of the two pilot T9E1s must have gone well, as series production was authorized even before the light tank was standardized as the M22. Marmon-Herrington would go on to build 830 units of the vehicle between April 1943 and February 1944. As plans for the airborne element of the invasion of France in summer 1944 did not call for the seizure of any enemy airfields, the M22 would not take part in this one great opportunity to prove its worth. The vehicle was reclassified as limited standard in September 1944.

The British Army was allocated 260 units of the M22 under Lend-Lease. In British

The Marmon-Herrington Company's track record with tanks before and during World War II was not a sterling one. Added to their list of design failures was the light tank M22 seen here. It was intended for use by the U.S. Army's airborne forces and would have been landed by large, multi-engine cargo aircraft when enemy airfields were seized. (Christophe Vallier)

Army service the light airborne tank was assigned the terrifying nickname "Locust." The British Army modified their M22s with the addition of a smoke grenade launcher. They also developed a tapered bore adaptor, referred to as the Littlejohn Muzzle Adaptor, originally developed for their 2-pounder (40mm) gun, for use on the vehicle's American 37mm main gun. The Littlejohn fired a 1.46-inch-diameter projectile which was swaged down to 1.195 inches as it traveled through the tapered section. It was intended to fire a special AP round with a tungsten core that had superior armor-penetration abilities compared with the standard AP projectile. It is unknown how many M22s were modified to mount this device, but in any event the adaptor tended to fail after only a few rounds, and was not fielded in American military use.

The only time the M22 would see combat in World War II was when nine of them took part in the airborne portion of the British Army's crossing of the Rhine River on March 24, 1945. Rather than being flown in by the AAF C-54 Skymaster, the M22s were carried in by large British Army Hamilcar gliders. According to author George

Forty in his book *United States Tanks of World War II*, over half of the gliders carrying the M22s were lost to enemy fire upon landing. However, of the remaining vehicles, one that was immobilized just outside the landing site zone was reported to have killed over 100 German soldiers.

LESSONS LEARNED

In August 1942 the Ordnance Department met with the Armored Force leadership to talk about the next generation of light tanks that would replace the existing M3 series of light tanks, which were already acknowledged as being under-armored and under-gunned. The Ordnance Department informed the Armored Force leadership that they would prepare design studies for a new light tank that would reflect the characteristics they wanted to see.

The Ordnance Department decided to base their new light tank design on the T20 medium tank, but with thinner armor. Designated the light tank T21, the vehicle was to have a five-man crew and feature a 76mm main gun. The final layout drawings for the vehicle were completed at the Fisher Body Division of General Motors in March 1943. At this stage of development, the weight of the proposed T21 was estimated to be at least 25 tons, 5 tons over the maximum weight ceiling mandated by the Armored Force for consideration as a light tank. Having learned a painful lesson from the light tank T7 fiasco – that the series production vehicle would grow even more in weight from the pilot vehicle – the Armored Force canceled the T21 project in July 1943.

Having successfully mounted and fired a 75mm gun M3 from the open-topped turret of the 75mm howitzer motor carrier M8, which was based on the chassis of the light tank M5, the Ordnance Department now knew that the chassis could absorb the recoil of such a large weapon. The problem remained that the turret on the M5A1 was too small to house the weapon. To overcome this problem, the Ordnance Committee decided in March 1943 to use the power train of the M5A1 and build a new light tank around it, which would be designated the light tank T24.

The Ordnance Department originally believed that by taking the existing 75mm gun M3 and fitting it with the lighter and more compact recoil system developed for the light tank T7, a suitable turret for the T24 could be designed to accommodate it. However, it was soon clear to all that the 75mm gun M3 and its associated gun mount imposed such a

The last light tank to enter U.S. Army service during World War II was the M24, pictured here in Italy in spring 1945. With the main gun forward the vehicle was 18 feet 3 inches long. It had a width of 9 feet 10 inches and was 9 feet 1 inch tall with the antiaircraft gun mounted on the top of the turret. (Patton Museum)

weight burden that it would be impossible to mount it in a vehicle weighing 20 tons or less. An answer to the Ordnance Department's dilemma came about when it was suggested that the lightweight 75mm gun T13E1 developed at the behest of the Army Air Forces for mounting in the nose of the B-25H Mitchell medium bomber might be a suitable candidate.

The 75mm gun T13E1 proved to be the weapon the Ordnance Department was looking for to mount in the turret of its next-generation light tank. The weapon was later designated the 75mm gun M6. In place of the long recoil system mounted in the B-25H bomber, the Ordnance Department came up with a new concentric short recoil system designated the T33. The 75mm gun M6 with the T33 recoil system was fitted into a new gun mount that was standardized as combination gun mount M64, which included a coaxial .30 caliber machine gun.

The M64 gun mount with its cast-armor gun shield was fitted inside a newly designed welded-armor turret that had the loader on the right side of the main gun and the vehicle commander and gunner on the left side of the main gun breech ring. This was the opposite turret crew arrangement seen on the first- and second-generation M4 series medium tanks. To accommodate the three-man turret crew, the larger 75mm gun M6, and its gun mount, the turret ring on the T24 was 60 inches wide. This in turn drove the width of the vehicle's chassis to 118 inches, with sand shields fitted – nearly 10 feet.

The turret of the T24 sat on top of a brand new, welded-armor chassis that rode on an Ordnance Department-developed torsion bar suspension system in place of the VVSS seen on all previous light tanks and combat cars going back to the 1930s. Torsion bars became the suspension system of choice for late-war American tanks and

Belonging to the Virginia Museum of Military Vehicles is this M24. The vehicle was powered by the same power train as the light tanks M5 and M5A1. As there had been problems with the automatic transfer unit in the light tanks M5 and M5A1, the M24 was provided with a manually operated transfer unit. (Michael Green)

tank destroyers because of their simplicity, combined with their ability to absorb more energy in relation to their weight than any other spring-types of suspension mechanism.

The torsion bar suspension system of the T24 rode on T72 cast-steel tracks that were 16 inches wide. The twin Cadillac engines in the vehicle provided a maximum speed on level roads of 35 miles per hour. The M24 also got a new transfer unit, combined with a Hydramatic transmission that provided the driver and assistant driver/bow gunner with eight speeds forward and four in reverse.

The driver and assistant driver of the M24 were both located in the vehicle's front hull – the driver on the left side and the assistant driver on the right. Both were provided with overhead hatches fitted with 360-degree rotating M6 periscopes. There was an emergency escape hatch in the bottom of the vehicle's hull located just behind the assistant driver's seat. The assistant driver had a .30 caliber machine gun installed in a ball mount directly in front of his seat. The crew of the M24 were provided with an interphone system to communicate with each other inside the vehicle.

Pleased with the T24's performance, the Ordnance Department quickly authorized production and the first vehicles came off the assembly line of Cadillac Motors in April 1944, followed by the Massey Harris Company in July 1944. It was that same month that the vehicle was standardized as the light tank M24. It would first arrive in small numbers in Northwest Europe beginning in December 1944, just in time for the Battle of the Bulge.

The M24 could not arrive fast enough for many tankers fighting in Northwest Europe, as is reflected in this extract from *Armored Section Report, First U.S. Army, Report of Operations, 23 Feb. till 8 May 1945*: "The inadequacy of the M5 series light tank made it highly desirable that they be replaced as quickly as possible by the M24 series. However, it is apparent that shipping conditions would prohibit complete replacement for a very long time."

Fortunately for all concerned, production of the M24 was expedited to such an extent that the numbers of vehicles joining units in the field in early 1945 dramatically increased. Before the end of the war in Europe, almost all of the U.S. Army's mechanized cavalry reconnaissance squadrons were equipped with the M24, and the various armored divisions were also receiving them in great numbers.

The M24s that showed up in Northwest Europe in late 1944 and into 1945 were a bit different from the pilot versions. The external changes included a raised vehicle commander's cupola, the same that appeared on second-generation M4 series tanks. There was a repositioned turret ventilator and the replacement of a lightweight machine gun mount located at the rear of the turret roof with a shorter and beefier mount at the same location. Both mounts were intended for a .50 caliber machine gun.

Even with the 75mm main gun firing AP, the M24 was never intended to fight it out with late-war German tanks or heavily armored self-propelled (SP) guns as is seen in this quote from a March 1945 U.S. Army report titled *United States vs. German Equipment* by First Lieutenant Harold Shields:

On 2 March 1945, the 2d Battalion, 66th Armored Regiment, was making a drive to the Rhine River. Upon taking the battalion objective at Fichlen, Germany, three of our medium tanks were knocked out by a German SP gun (a long barreled 75 on a Mark IV chassis). I took this SP gun under fire with my platoon of M24 light tanks at 800 yards. The platoon fired a total of twenty-five rounds, the majority of which were AP. None of the AP pierced the front slope plate of the SP, but ricocheted off.

The main role of the M24s arriving in Northwest Europe was to replace the M5 and M5A1 in escorting the mechanized infantry and in reconnaissance and flank security duties. Armor thickness on the M24 topped out at 1 inch (25mm) except for the cast-armor gun shield that was 1.5 inches (38mm) thick. The larger 75mm HE round fired by the new light tank was the most welcomed feature, as it was far more destructive to enemy personnel, bunkers, buildings, trucks, etc., than the puny HE round the 37mm gun in the M5A1 fired.

A feature that received positive reviews from the tankers that operated the M24 in the ETO was the vehicle's ruggedness, as described in the U.S. Army *Report on the Light Tank M24*, by the headquarters of the 744th Light Tank Battalion:

> The tank has demonstrated the quality of ruggedness time and again. It has been able to remain in the fight with minor maintenance difficulties and even when hit by antitank weapons. In one instance a tank received three direct hits from an antitank gun. The right front and left rear bogie wheels were knocked off, but the tank was able to proceed under its own power to a place

The final light tank to enter service with the U.S. Army in World War II was the M24. It appears here in the base olive drab, with all the markings in white. On the rear engine deck is an air identification panel to prevent attack by friendly aircraft. This unhappy situation occurred so often during the fighting in Western Europe that some U.S. Army tank units became loath to radio for air support lest they be attacked by their own planes. The M24 first began appearing in Western European field units in summer 1944. Those units that received them found them a big improvement over the previous light tanks in service, being both better armed and with far superior off-road mobility. (© Osprey Publishing Ltd.)

The 75mm main gun on the light tank M24 was ballistically the same as the 75mm main gun on the first generation of M4 series medium tanks but the gun tube was thinner, meaning it heated up faster and had a shorter life span. Pictured is a restored M24 in France. (Christophe Vallier)

Perceiving a need for a dedicated flame-thrower vehicle the U.S. Marine Corps modified 24 M3A1 light tanks in 1944 to feature a turret-mounted flame gun, in place of the standard 37mm main gun. Earlier experiments with the flame gun mounted in the front hull of light tanks proved unsatisfactory. Pictured is one of the 24 M3A1s converted into flame-thrower tanks in May 1944. Their official nickname was the "Satan." The vehicle is painted in the base olive drab, with all the tactical markings applied in yellow. (© Osprey Publishing Ltd.)

where it was repaired and put back in action in less than 12 hours. In another action a tank received two direct hits in the suspension system, but was not put out of action.

Production of the M24 would continue to the end of World War II, with 4,731 units completed by the two firms. Two hundred and eighty-nine units of the M24 went off to the British Army under Lend-Lease; however, none would see combat before the end of the war in Europe. Two examples of the M24 went to the Red Army. The British Army named the M24 the "Chaffee" after U.S. Army Major-General Adna R. Chaffee, Jr., the first commander of the Armored Force, who died of cancer in August 1941.

FLAME-THROWER TANKS

Fighting by the Marine Corps in the jungles of the South Pacific in 1942 and 1943 convinced them of the desirability of a dedicated flame-thrower vehicle. Early improvised mounting of the infantry's portable M1A1 flame thrower in the bow gunner's position of the light tank M3A1 did not yield the results hoped for and work began on a more suitable arrangement. At first, the Marine Corps looked for a medium tank chassis to convert into a flame-thrower vehicle, but none were available. Instead, they decided to use the M3A1, as that vehicle was then being replaced in Marine Corps service by the M5.

What eventually evolved was the removal of the 37mm main gun from the M3A1 and its replacement by a British-designed flame gun named the "Ronson." The first trial of this new configuration took place on April 15, 1944 and must have pleased all those present, as the Marine Corps requested that 24 of these vehicles be constructed and assigned the name "Satan." They were all completed by May 17, 1944 and placed on ships heading for the Marianas.

The Satan lacked the turret basket and the power traverse system seen on the

The fighting in the Pacific quickly convinced both the Marine Corps and the U.S. Army that flame-thrower tanks might prove very useful when tackling heavily fortified enemy bunkers. As a short-term fix, a number of improvised arrangements were made. Pictured is a U.S. Army light tank M3A1 with a flame-thrower gun in place of the bow machine gun. (Patton Museum)

standard M3A1. These were done away with to make room for fuel storage tanks, which had a capacity of 170 gallons of fuel for the flame gun. The maximum effective range of the American copy of the British-designed flame gun named the Ronson was 60 to 80 yards. The two-man-crewed Satan would take part in the Marine Corps' invasions of the islands of Saipan in June 1944 and Tinian in July 1944.

Eventually, an American-designed and built flame gun that was designated the E7, offering an increase in range over the British Ronson, was proposed for mounting in the M5A1 by the U.S. Army. Test results were positive in November 1943, and the E7 flame gun, along with the fuel system E7, was installed in the M5A1 and the vehicle was designated the E7-7.

Unlike the conventional gun-armed M5A1 that had a crew of four men, the E7-7 had three crewmen, consisting of the driver, assistant driver/bow gunner, and a gunner in the turret.

Only four examples of the E7-7 were completed before interest shifted to medium tank-based flame-thrower vehicles. The four E7-7s served with the 13th Armored Group, which assisted the U.S. Army's 25th Infantry Division on the fighting for the Philippine main island of Luzon in April 1945.

ANTIAIRCRAFT VEHICLE

Of the many antiaircraft gun arrangements proposed for mounting on the chassis of U.S. Army light tanks during World War II, only one made it into series production at the closing stages of the conflict. That vehicle proved to be the 40mm gun motor carriage M19, which was based on the heavily modified chassis of the M24. The vehicle was standardized in June 1944 and production began at the Cadillac Motor Car Division of the General Motors Corporation in April 1945. The program was canceled in August 1945, when the Empire of Japan surrendered. As a result, the production run of the vehicle ran to only 300 vehicles, none of which would see combat use.

The M19 consisted of the lengthened chassis of an M24 with its power plant moved from the rear of the hull to a position just behind the driver's compartment. This left the rear of the vehicle's hull open to install the power-operated, twin 40mm automatic gun mount M2, which had 360 degrees of traverse and sat on top of the lower hull stowage compartment. Besides the driver and assistant driver in the front hull, the twin 40mm guns were serviced by a crew of four men: two loaders and two gunners. The maximum firing rate for the dual gun arrangement was 240 rounds per minute (120 per tube). The vehicle carried 352 rounds onboard of 40mm ammunition and could tow a wheeled trailer with storage for another 320 rounds of 40mm ammunition.

One of the few variants of the light tank M24 to enter into production during World War II was designated the 40mm gun motor carriage M19. The example seen here belongs to the Virginia Museum of Military Vehicles. No M19s were fielded before the end of World War II. (Michael Green)

CHAPTER FOUR
HEAVYTANKS

T he successful German invasion of Poland in 1939 prompted some within the U.S. Army to look at the development of a suitable heavy tank to counter future German armored threats. The preliminary studies for this project took place at APG that same month. By May 1940, the Chief of the Infantry Branch recommended that a requirement be established for a heavy tank and that designs be undertaken for a vehicle that could weigh between 50 and 80 tons. To give you an idea of how monstrous this vehicle was for its time, today's M1A1 Abrams tank is approximately 70 tons. In 1940, any tank over 30 tons was regarded as a heavy tank by the U.S. Army. This concept was later altered and tanks weighing up to 35 tons were classified as medium tanks by the U.S. Army.

The first iteration of this new heavy tank concept anticipated a multi-turret vehicle as seen with some of the Red Army's late 1930s heavy tanks, such as the T35 series. The planned new American heavy tank was envisioned as having four rotating turrets along with multiple machine guns. The two primary turrets, each armed with low-velocity T6 75mm guns, would be limited to 250 degrees of traverse, while the two smaller secondary turrets – one armed with a 37mm gun and the other with a 20mm gun – would have 360 degrees of traverse. This concept found enough support to be approved in July 1940, and the proposed vehicle was to be designated the heavy tank T1.

As quickly as the multi-turret T1 project was approved, it was dropped and a full-scale wooden mockup appeared that proposed a single, large, electric power-operated turret mounting a modified version of the 3-inch T9 antiaircraft gun, later standardized as the M7. Instead of a coaxial .30 caliber machine gun, there was to be a coaxial 37mm gun M5E1. This revised arrangement was approved in November 1940, and in the same month the Ordnance Department awarded a contract to the Baldwin Locomotive Works to build a single pilot heavy tank to be followed by 50 series production units.

The German military invasion of Poland in September 1939 forced a rethink by the U.S. Army regarding the need for heavy tanks in its inventory. The result was the consideration of a multi-turret heavy tank weighing up to 80 tons. Cooler heads prevailed and what was developed proved to be the M6 heavy tank, pictured here. (TACOM Historical Office)

The only heavy tank developed for the U.S. Army during World War I, but fielded too late to see combat in that conflict, was the Mark VIII, pictured here at Aberdeen Proving Ground. It was 34 feet 2.5 inches long, 12 feet wide, and 10 feet 3 inches tall. (Charles Kliment)

The single-turret T1 wooden mockup proposed that, in addition to the cannon armament, a large number of machine guns be fitted to the vehicle. The vehicle commander, located on the left side of the main gun, would be provided with a 360-degree rotating overhead cupola borrowed from the medium tank M3, armed with a .30 caliber machine gun. At the right rear of the large turret would be a rotor-mounted .50 caliber machine gun intended to be operated by the loader, who, like the gunner, was located on the right side of the main gun. The gunner was to be provided with both a direct vision sight as well as a periscope sight.

Mounted on either side of the front hull of the proposed T1 would be two fixed, forward-firing .30 caliber machine guns, operated by the driver electrically. The bow gunner position on the new heavy tank was to feature two .50 caliber machine guns in a rotor mount. Both the driver and bow gunner would have overhead hatches. There would be an escape hatch at the bottom of the hull floor directly behind the bow gunner seat. Also situated in the front hull of the proposed T1 would be a sixth crewman, referred to as the ammunition passer.

Even at the wooden mockup stage of the T1, it was estimated that the vehicle would come in at around 50 tons when built. To assist them in deciding on a suitable engine and power train for the vehicle, the Ordnance Department sought out the assistance of the Society of Automotive Engineers (SAE). Eventually the Ordnance Department decided to use a nine-cylinder, gasoline-powered, air-cooled radial engine in the T1. The official designation for the engine was the Wright G-200 Model 781C9GC1, and it produced 700 net horsepower at 1,950rpm. The vehicle would ride on an HVSS system.

Taking part in a mobility demonstration is a U.S. Army Mark VIII heavy tank. One hundred units of the vehicle were assembled in 1919 at Rock Island Arsenal. The vehicle remained in U.S. Army service until 1932 when they were placed in storage. In 1936, the requirement for heavy tanks was canceled and the vehicles were eventually scrapped. (Patton Museum)

PILOT MODEL APPROVED

It was originally intended that the pilot version of the T1 would be fitted with a Hydramatic transmission; however, at the last minute it was decided to try installing a General Electric (GE)-developed electric drive and steering mechanism in the pilot vehicle. Reflecting the component change, the vehicle was now designated as the T1E1.

As the GE electric drive and steering mechanism was not ready in time, the pilot model of the T1 was delivered in August 1941 with a Twin Disc torque convertor and designated the T1E2. Without crew or ammunition onboard, the pilot vehicle weighed in at 55 tons. It was 20 feet 1 inch long, 10 feet 3 inches wide, and 10 feet 3 inches high. The maximum armor thickness was 5 inches (127mm) on the front of the turret.

Plagued by numerous unresolved design issues, the pilot T1E2 tank was debuted by the Ordnance Department at the Baldwin Locomotive Works the day after the Japanese attack on Pearl Harbor in an obvious morale-boosting event for the country. The vehicle, which featured both a cast-armor turret and hull, was then sent off for testing to APG. A number of modifications to the T1E2 reflected changes suggested by both the testing conducted at APG and British Army combat experience.

External changes to the T1E2 included the deletion of the rotor-mounted .50 caliber machine gun at the rear of the turret and doing away with one of the two fixed, forward-firing .30 caliber machine guns in the vehicle's front hull. The vehicle commander's cupola borrowed from the medium tank M3 was also removed. It was replaced with a rotating two-piece circular split hatch as installed on the early M4 series tanks.

With the United States now at war, a decision was made to release the T1E2 for large-scale series production despite its many design shortcomings. To meet the large production numbers of heavy tanks anticipated, it was proposed that some of them feature a welded-armor hull and be powered by four General Motors 6-71 diesel engines connected to two Hydramatic transmissions. In this configuration, the vehicle was designated the T1E4. Another version of the heavy tank with a welded-armor hull, but retaining the original Wright G-200 engine, was designated the T1E3.

Ordnance Committee action recommended in April 1942 the standardization of the T1E2 and T1E3. The T1E2 soon became the heavy tank M6, and the T1E3 the heavy tank M6A1. This standardization was approved the following month along with funding for 1,084 units of the M6 and M6A1. However, the senior leadership of the Armored Force saw no requirement for that many heavy tanks and cut the number authorized for procurement down to 115 units, although this number was later bumped up to 230 units by the U.S. Army Services of Supply.

The heavy tank M6 shown here weighed 63 tons and was 27 feet 8 inches long with its main gun forward. The vehicle had a width of 10 feet 3 inches, and a height of 9 feet 10 inches. Plagued by design shortcomings that were never resolved and with no interest from the user community, it never saw combat. (TACOM Historical Office)

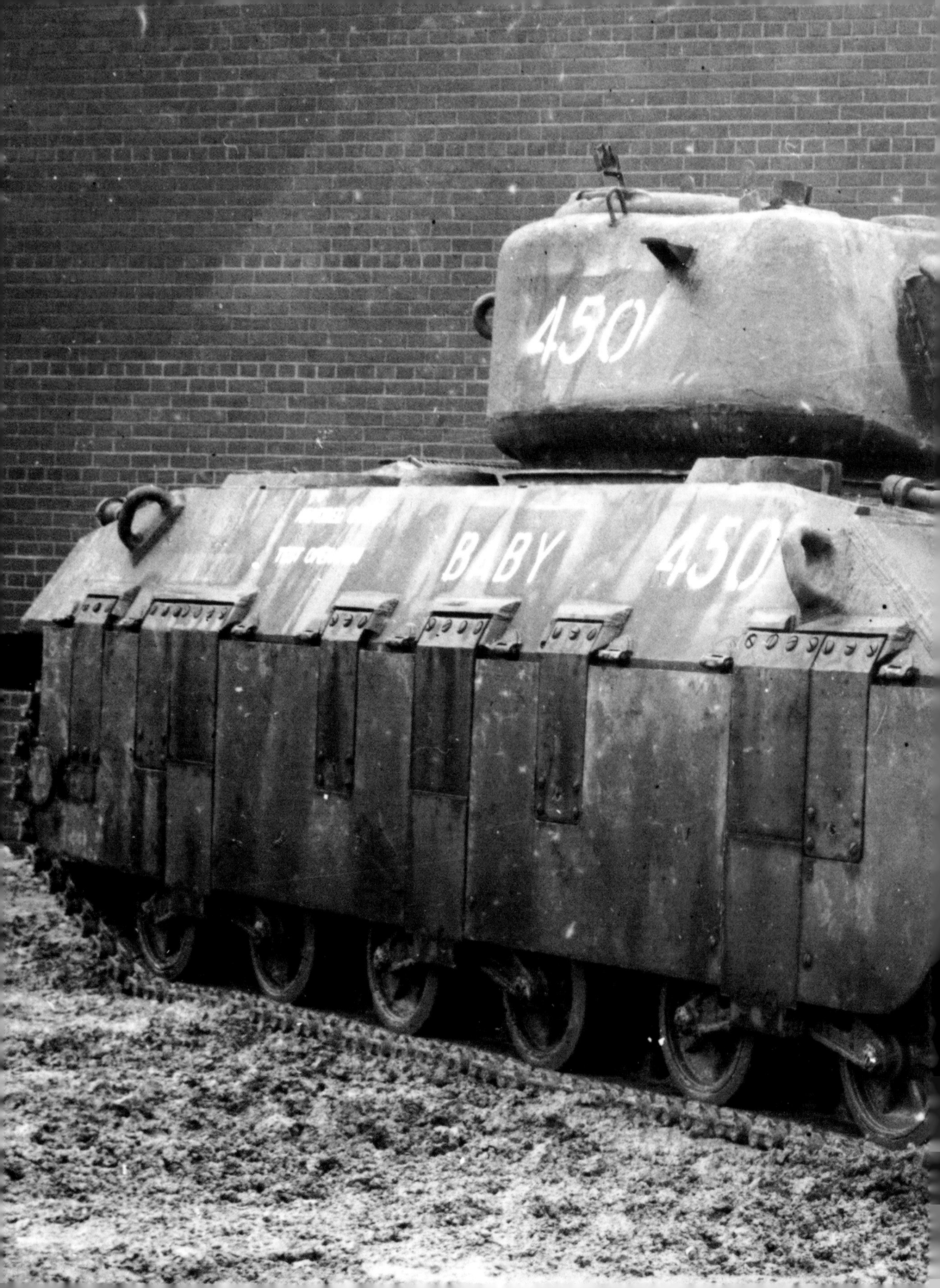
4501
BABY
450

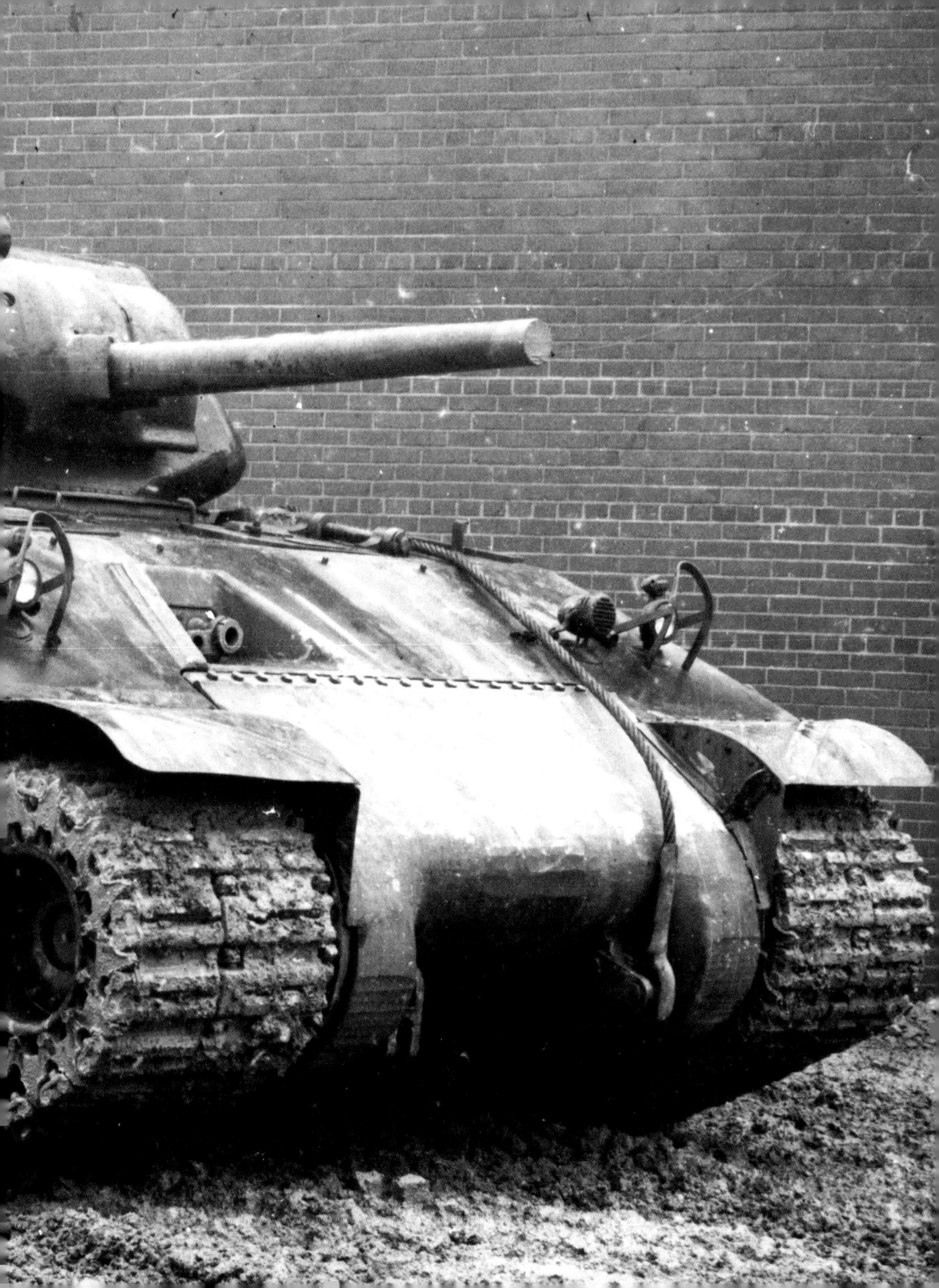

By this time, the Ordnance Department probably realized that the user community was not that interested in the M6 or M6A1. What the Ordnance Department really wanted at this stage was the approval to have 115 units of the T1E1 built solely for an extended test program. There was a measure put forth to have the T1E1 standardized as the M6A2 but this was not approved. With series production of the M6 and M6A1 slated to begin in late 1942, the Ordnance Department decided that these vehicles could be offered to the British Army under Lend-Lease.

It was in December 1942 that the first series production M6 came off the assembly line. That same month the commanding general of the Armored Force, Lieutenant-General Jacob Devers, wrote a letter to Lieutenant-General Lesley J. McNair, the Commanding General of the AGF, in which he stated: "Due to its tremendous weight and limited tactical use, there is no requirement in the Armored Force for the heavy tank. The increase in the power of the armament of the heavy tank does not compensate for the heavier armor." Devers would go on to recommend in the same letter: "The heavy tank program [should] be discontinued and the material already used in the program be diverted to essential production or scrapped."

Devers reflected the popular opinion of the Armored Force at the time that it made more sense to ship overseas two medium tanks in the space that one heavy tank would take up. The U.S. Army Services of Supply concurred with Devers' thoughts on heavy tanks and the M6 series was eventually terminated. In the end, only 40 series production units of the heavy tank were built, this being divided among eight M6s, 12 M6A1s, and 20 T1E1s. The Ordnance Department, with some foresight, believed that there would be occasions when the greater firepower and thicker armor of a heavy tank would offset any other considerations, something the Armored Force would later realize, at the cost of the lives of many U.S. Army tankers.

Despite the cancellation of the M6 series program, the Ordnance Department continued testing the vehicles, hoping the Armored Force would have a change of heart. Test results were generally dismal, as in an Armored Force report dated July 1943, which offered the following conclusions:

1. The firepower was not commensurate with weight and size of the vehicle in view of the designs now being developed.
2. The fire control equipment was obsolete.
3. The crew was unable to man the armament because of the location of controls, crew, and seats.
4. The fighting compartment had insufficient ventilation.
5. The transmission was unsatisfactory and would require a complete redesign to obtain satisfactory shifting.
6. The accessories required daily or more frequent maintenance and were inaccessible.

Because the Armored Board decided that the 3-inch gun M7 was not suitability potent for the weight and size of the M6 series tanks, the Ordnance Department decided to test mount the 90mm gun T7 in the first production T1E1 in early 1943. Firing tests completed in April 1943 showed the gun could be successfully installed in the M6 series tanks. However, with only 40 units authorized, it made no sense to go forward with the project.

BACK FROM THE GRAVE

Some folks at the Ordnance Department did not give up completely on the M6 series. In the summer of 1944 there was a proposal made by the Ordnance Department that 15 of the 20 T1E1s be up-armored and up-gunned with a new, high-velocity 105mm main gun, designated the T5E1. The five remaining unmodified vehicles would be used for spare parts. Tentative approval was given for the project and the vehicle so modified was designated the M6A2E1. The intention was that these vehicles assist in overcoming German defensive systems.

The AGF leadership was not excited about the M6A2E1. They pushed the matter up the chain of command until August 2, 1944, when the U.S. Army Service Forces sent a cablegram to General Dwight D. Eisenhower, the Supreme Allied Commander of the Allied Expeditionary Force. Eisenhower responded on August 18, 1944, in which he stated: "We do not want at this time the fifteen M6 that were offered, as they are not considered practicable for use in this theater now." This response was enough to have the project quickly canceled.

Strangely enough, after the proposed M6A2E1 project was rejected, the Ordnance Department received permission to actually assemble two of the vehicles. They were not interested in the actual up-armoring of the M6A2E1. Their only interest was in the up-gunning aspect of the project, which involved grafting the massive 105mm main-gun-armed turret from the heavy tank T29 onto the existing chassis of an M6 series tank. This line of development came to an end when the Ordnance Committee classified the M6, M6A1, and T1E1 obsolete on December 14, 1944.

ASSAULT TANK

In March 1942, a requirement for a special assault tank (also known as a heavy tank) was agreed upon by representatives of the Chief of Ordnance, APG, and the British Tank Mission in the United States. It was the latter who had expressed interest in acquiring a large number of such vehicles when and if built. In contrast, the Armored Force had no interest in taking into service a large number of heavy tanks at that time. However, it was agreed upon by both countries that they would each build two pilot tanks, with the British version based on the Mark VIII Cruiser "Cromwell" tank and the American version based on the M4 series tank. The Ordnance Committee issued military characteristics for the proposed heavy tank in May 1942, now designated the assault tank T14.

The first T14 pilot showed up at APG in July 1943, with the second pilot showing up the following month. Crew layout followed that of the standard M4 series tank with a five-man crew. The armament of the T14 was the same as the first-generation M4 series tank with a turret-mounted 75mm gun M3 and a coaxial .30 caliber machine gun. There was a bow-mounted .30 caliber machine gun in the front hull and provisions made to mount another machine gun on the turret roof. The staff at APG were told to prepare layout drawings for the possible mounting of other weapons in the turret, including everything from a 90mm gun to a 105mm howitzer.

Maximum armor thickness on the cast-armor front glacis of the T14 was 4 inches (101mm). The sloping side-armor hull plates were 2 inches (50mm) thick. The vehicle's HVSS suspension system, which rode on 25.75-inch-wide tracks coming

from the M6 series tank, was protected by 0.5-inch (13mm) armor skirts that folded out for maintenance. The front of the cast-armor turret on the T14 was 3 inches (76mm) thick and sloped at 30 degrees, with the vertical sides and rear wall of the turret being 4 inches (101mm) thick.

The T14 weighed 42 tons, with power provided by a gasoline-powered, liquid-cooled engine referred to as the Ford GAZ V-8. The top speed was 24 miles per hour on level roads. Testing of the two pilot tanks at APG revealed a host of design problems, including a tendency to throw tracks. Standard vehicle maintenance proved difficult due to the inaccessibility of many components. Overall mobility of the vehicle was considered unsatisfactory, and the staff at APG recommended that no further consideration be given to procuring the T14 for the U.S. Army. Testing of the pilot vehicles would continue until December 1944 when the project was officially canceled.

A STOPGAP

Even before the Normandy landings in June 1944, the U.S. Army anticipated that as Allied forces pushed out of France and into the German homeland, they would encounter increasingly organized and fortified German defensive positions. Therefore, there would be a need for a tank with both a heavier gun and thicker armor to attack these prepared positions.

The U.S. Army hoped that the heavy tank M26 would be available by late 1944. But as this would not be the case, a decision materialized to use a drastically up-armored second-generation M4A3(75)W tank in the assault role. An agreement to manufacture 254 such tanks quickly appeared, and the Ordnance Committee designated the tank as the M4A3E2 in March 1944.

Recent research conducted by noted Sherman tank expert Joe DeMarco has uncovered that in at least one wartime US Army report, located in the American National Archives, the M4A3E2 is referred to as the "Jumbo." However, most wartime reports do not mention that name.

Construction of the M4A3E2 proceeded rapidly, and the first units entered into combat in Northwest Europe in late 1944. American tankers greeted its arrival with great enthusiasm because they appreciated the extra protection they received from the tank's very thick armor protection levels and quickly asked for more of them. The normal role for the M4A3E2 was that of point tank (leading a column) as it could better absorb the projectiles fired from German tank and antitank guns. However, the crews of the assault tanks were under no illusions about their vehicles being proof against the enemy weapons they encountered – merely that they had a bit more protection than the standard M4 series tanks.

The M4A3E2 featured 4 inches (101mm) of armor on its welded armor glacis and 3 inches (76mm) on the sides of its upper welded armor hull. There was no additional armor on the side plates of the lower hull or on the rear of the tank. The vehicle's final-drive armor casting was 5.5 inches (140mm) at its thickest horizontal point, with 4.5 inches (114mm) on its sloping surfaces. The cast-armor gun shield on the front of the M4A3E2 turret was 7 inches (178mm) thick with the rounded turret sides being 6 inches (152mm) thick.

The M4A3E2 rode on the VVSS. Due to the vehicle's increased weight, now approximately 42 tons, extended end connectors were fitted on the outside of the tracks to maintain a ground pressure of about 14 pounds per square inch. There was a price to pay for all that extra armor on the M4A3E2, as seen in this U.S Army report dated October 24, 1944: "One thing that users must realize is that in rough cross-country operation the front volute springs will fail if permitted to 'bottom' violently."

Power for the M4A3E2 came from the standard Ford V-8 GAA tank engine, but the final-drive gear ratio was changed to compensate for the increased weight to provide for a maximum speed of 22 miles per hour.

Though the original intent was to mount a 76mm main gun, the M4A3E2 received the standard 75mm main gun since it fired a more potent HE round than the 76mm gun. There was enough internal room in the tank's lower hull to stow 106 of

The original plans called for the M4A3E2 to be fitted with a 75mm main gun. As can be seen from this photograph, some in the ETO were later re-gunned with a 76mm main gun, which is evident by the longer gun tube. (Patton Museum)

the 75mm main gun rounds in wet storage containers. The M4A3A2 mounted the same secondary armament found on other M4 series tanks, which included two .30 caliber machine guns and one .50 caliber machine gun on the turret roof. During their deployment in Northwest Europe, some of the M4A3E2s had their 75mm main guns replaced with the 76mm gun.

IMPROVISED M4A3E2s

With only 254 M4A3E2 tanks built, the demand quickly outstripped the available supply then in Northwest Europe. To satisfy the need for additional units Lieutenant-General George S. Patton, commander of Third Army, ordered in early 1945 that as many M4A3(76)W tanks (with or without HVSS) not already up-armored with add-on armor plates be quickly modified to the M4A3E2 standard with additional armor.

As a result of Patton's order, many disabled welded-armor hull M4 series tanks had their glacis and upper-hull armor cut off with acetylene torches and welded onto operational M4A3(76)W tanks. The armor from captured German tanks was also used in this up-armoring effort.

Third Army's up-armoring program employed local civilian firms to complete the work. In less than a month's time, it is estimated that about 100 M4A3(76)W tanks went through the up-armoring process ordered by Patton. Third Army also received a number of disabled M4 series tanks from Seventh Army for use in the up-armoring project.

ANOTHER APPROACH

Based on development work done on the medium tank T23, two additional versions were proposed mounting the 90mm gun T7 (later re-designated as the 90mm gun M3) originally tested on the M6 series. These were the medium tank T25 and the medium tank T26. Developmental work on the T25 came to a halt in September 1944 in favor of the T26. The War Department approved production of ten T26s on May 24, 1943 as part of a larger production order for the T20 series medium tank series.

Following on the heels of the Allied invasion of France, someone saw a need for a heavily armored and up-gunned heavy tank M6. That resulted in the M6A2E1 heavy tank, seen here with a new turret and a 105mm main gun. It was rejected by General Dwight D. Eisenhower as being impractical for use in the ETO. (TACOM Historical Office)

The T26 boasted thicker armor than that found on either the T23 or T25 and could be the fighting equal of the German Tiger E heavy tank armed with an 88mm main gun, but weighing less. On September 13, 1943, the Ordnance Department requested that 500 additional T26s be built. This proposal was vetoed by Lieutenant-General Lesley McNair who continued to be opposed to all heavy tanks.

On November 13, 1943, Lieutenant-General Jacob Devers, former head of the Armored Force, and now commanding general of U.S. Army forces in the ETO, requested production of 250 units of the T26 based on a British Army-inspired request. McNair's biased objections to the T26 can be seen in a November 30, 1943 memo to the Chief of Staff of the U.S. Army, General George C. Marshall. In that memo, McNair states:

> There can be no basis for the T26 tank other than the conception of a tank versus tank duel – which is believed to be unsound. Both British and American battle experience has demonstrated that the antitank gun ... is the master of the tank... Tank destroyers can support an armored division or other unit in whatever degree is necessary to protect them against hostile tanks, leaving friendly tanks themselves free for their proper mission [exploitation].

Trying to stay ahead in the gun/armor race with the German tank designers, a U.S. Army development program began in spring 1942 to field a better armored and armed replacement for the M4 series of medium tanks. That new tank started off life as the wooden mockup referred to as the medium tank T20, seen here. (Patton Museum)

On December 7, 1943, Major-General Joseph McNarney of the War Department queried Devers about whether his request was based on operational requirements or was just something he felt that was needed, as McNair continued to have strong objections to the production of the T26. Devers confirmed on December 10, 1943 that indeed his request for the T26 was based on operational requirements.

On December 16, 1943, Major-General Russell Maxwell, the Assistant Chief of Staff, G4 (logistics) in the War Department General Staff, ordered that McNair arrange for production of 250 units of the T26 as requested by Devers. Five days later, General Marshall became personally involved with the T26 project when he cabled Devers about the decision to build the vehicle and suggested a nine-month delay before production would commence. On January 15, 1944, Marshall inquired of Eisenhower if he believed there was a need for the T26, which he quickly confirmed.

As developmental work on the T26 went forward there was the inevitable weight gain. To bring down the weight of the T26, its heavy 1.9-ton electric transmission and steering mechanism was replaced with the much lighter and more conventional Torqmatic transmission and controlled differential on some pilot units. This component swap resulted in those pilot T26s being affected becoming the T26E1. The Ordnance Department stated in February 1944 that they believed the first production units of the T26 could roll off the assembly line by October 1944.

Testing of the T26 by the Armored Board, however, would not begin until fall 1944, alongside the T26E1 and an improved version designated the T26E3. Test results showed that the electric transmission and steering mechanism in the T26 was too complicated for the average U.S. Army mechanic to work on, so the T26 was dropped, and developmental work continued on the T26E1 and T26E3. All versions of the medium tank T26 were re-designated as the heavy tank T26 on June 29, 1944.

By the time series production began in November 1944 at the Fisher Tank Arsenal, it was the T26E3 version that was produced and not the T26E1. Beginning in March 1945, production of the T26E3 also took place at the Detroit Tank Arsenal. It was that same month that the T26E3 was standardized as the heavy tank M26.

DESCRIPTION

The M26 weighed 46 tons combat loaded and had a maximum armor thickness of 4.5 inches (114mm) on its vertical cast-armor gun shield. The near-vertical sides of the tank's power-operated cast-armor turret were 3 inches (76mm) thick. The cast-armor glacis on the M26 was 4 inches (101mm) thick and sloped at 46 degrees. The lower portion of the cast-armor front hull on the M26 was 3 inches (76mm)

thick and sloped at 53 degrees. Along the vertical sides, armor thickness was 3 inches (76mm) and consisted of welded- armor plates.

As with the M4 series tanks, the M26 had a three-man turret crew, with the vehicle commander and gunner on the right side of the main gun and the loader on the left. Unlike the M4 series tanks, there were dual driver controls in the M26 with the driver in the left front hull and the assistant driver/bow gunner in the right side of the front hull. This feature was made possible by the Torqmatic transmission, which did not have the manual clutch and gear shift lever of the M4 series tanks' manual transmission. The crew was provided with an interphone system to talk to each other. Like the M4 series tanks, the driver and assistant driver/bow gunner on the M26 could raise their seats to look out over the top of the front hull when not in combat. The vehicle had storage space for 70 main gun rounds.

The most heavily armored offshoot of the T20 series evolved into the medium tank T26 and was armed with a 90mm main gun. The electric transmission originally intended for the T26 proved too heavy and was replaced with a Torqmatic transmission, resulting in a new designation for the vehicle as the medium tank T26E1, seen here. (TACOM Historical Office)

OFFICE CHIEF OF ORDNANCE-DETRO
NEG. No. 6388   DATE 5-23-45   DEVELOPMENT
Tank, Heavy, M26. 3/4 right front view.

The M26 rode on an Ordnance Department-developed torsion bar suspension system with 12 individually sprung dual road wheels, six on either side of the hull. The vehicle used an all steel track designated the T81, which had a width of 24 inches, or an improved version designated the T81E1. The eight-cylinder, Ford GAF liquid-cooled, gasoline-powered engine produced 500hp at 2,600rpm and provided the vehicle a maximum sustained speed on level roads of 25 miles per hour and a dash speed on level roads of 30 miles per hour.

The cruising range of the M26 on level roads was approximately 100 miles, similar to that of the M4 series tanks. Unlike the M4 series tanks, the transmission and the controlled differential on the M26 were located in the rear hull of the vehicle with the engine, and could be removed as a single unit. As the GAF engine in the M26 was virtually the same as the GAA mounted in the much lighter M4A3 first- and second-generation tanks, the M26 was underpowered for its weight. However, this was seen as a necessary tradeoff at the time to put the vehicle into service as quickly as possible.

FIREPOWER

The 90mm gun M3 mounted in the M26 was 15.5 feet long. Like its German late-war opponents, the American tank gun was fitted with a muzzle brake to reduce both recoil and the effects of obscuration when firing from a stationary position.

The 90mm gun M3 firing an early version of the standard M82 APC-T round had a muzzle velocity of 2,650ft/sec and at 500 yards could penetrate 4.7 inches (119mm) of armor. At a range of 2,000 yards it could still penetrate 3.8 inches (97mm) of armor. An improved version of the M82 APC-T round had a muzzle velocity of 2,800ft/sec and could penetrate 5.1 inches (129mm) of armor at 500 yards and 4.2 inches (107mm) of armor at 2,000 yards, but few improved M82 rounds reached the field before the fighting in Europe concluded.

In addition to the two versions of the M82 APC-T round, there was another tank-killing round, designated the M304 hypervelocity armor-piercing tracer (HVAP-T) solid shot. It was nicknamed "Hyper-Shot" by the tank crews. While the complete

Testing of the medium tank T26E1 pilots went well, and with a number of modifications the vehicle was designated as the medium tank T26E3, and 250 were initially ordered. By March 1945 the vehicle was standardized as the heavy tank M26. It had authorized storage space for 70 main gun rounds. (TACOM Historical Office)

OVERLEAF
The heavy tank M26 had a maximum armor thickness of 4.5 inches (114mm) on its vertical cast-armor gun shield. The cast-armor glacis on the M26 was 4 inches (102mm) thick and sloped at 46 degrees. Pictured is an unrestored M26. (Michael Green)

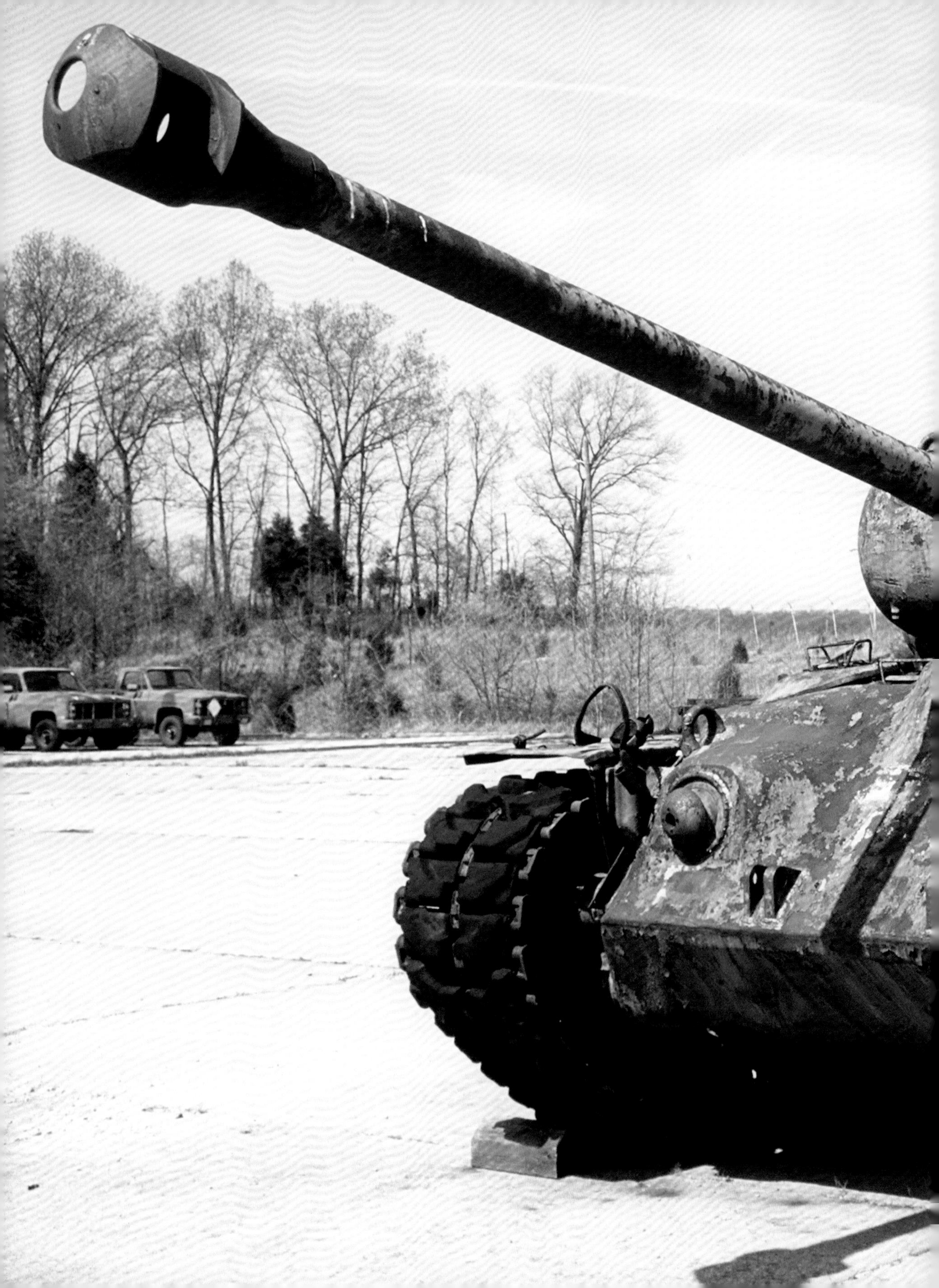

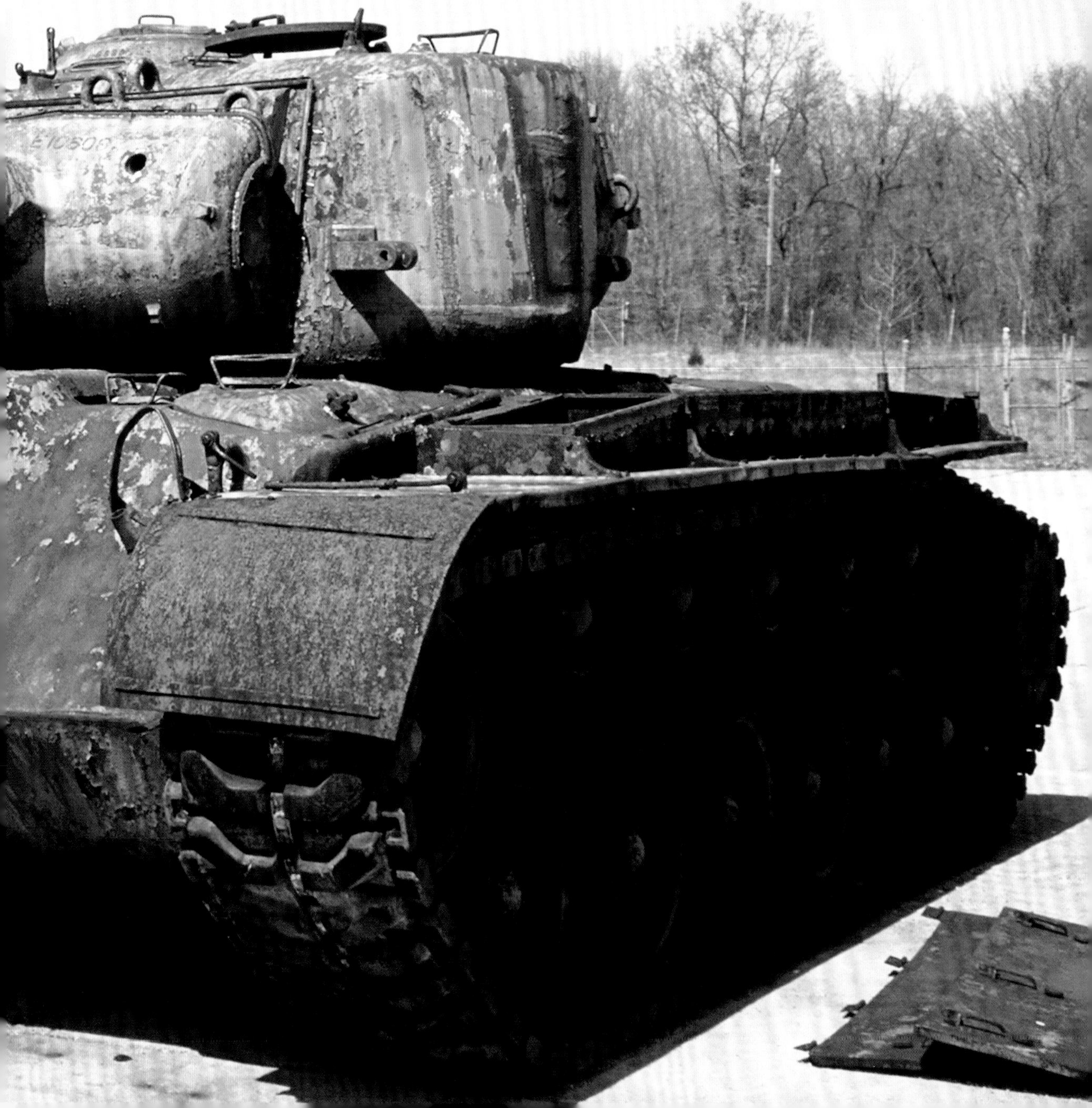

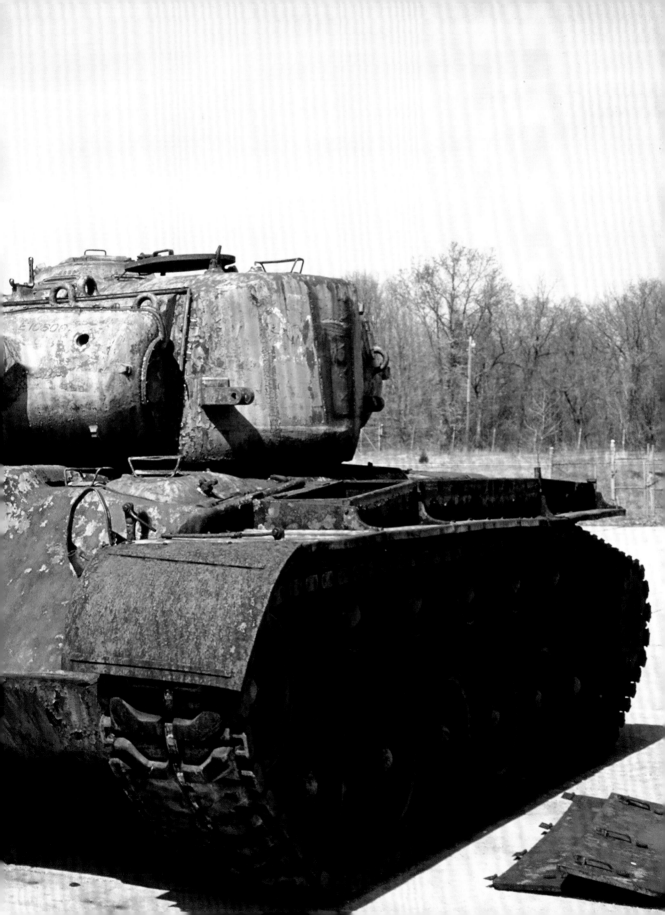

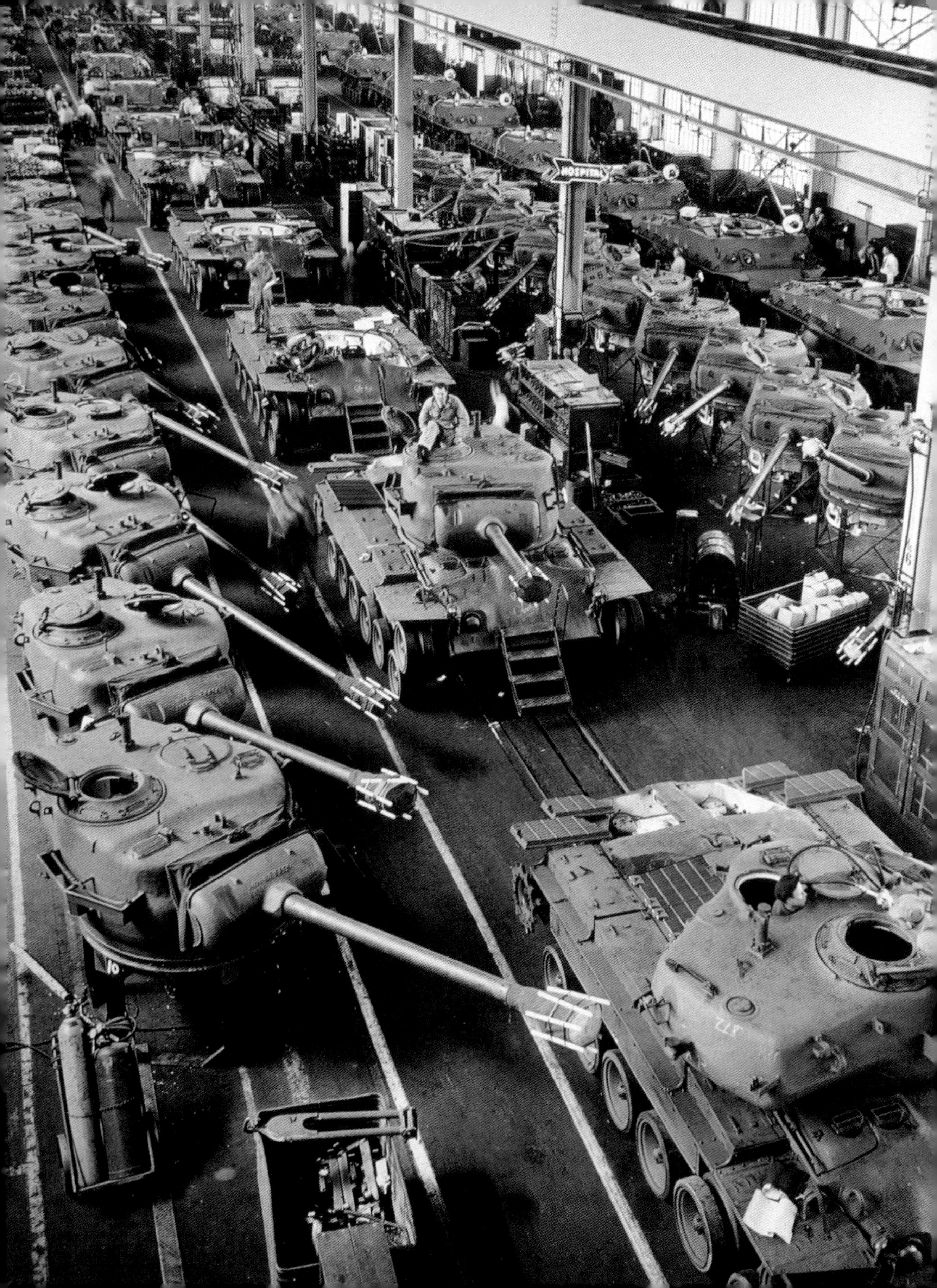
HOSPITAL

HVAP round weighed 37.13 pounds, the lightweight 16.80-pound projectile consisted of a dense penetrator core of tungsten carbide with an aluminum outer body, nose, and windshield.

With a muzzle velocity of 3,350ft/sec, the projectile portion of the M304 HVAP-T could penetrate 8.7 inches (221mm) of armor at 500 yards (almost double that of the original M82 APC-T main gun round). It penetrated up to 7.9 inches (201mm) of armor at 1,000 yards, and at 2,000 yards, 6.1 inches (155mm).

As the M304 HVAP-T round had difficulty dealing with the highly sloped glacis of the Panther tank at ranges of less than 450 yards, another round was developed to address the issue. It was referred to as the T33 AP-T shot and was based on the substitute standard mono-bloc M77 projectile that went through an additional reheating treatment to improve its hardness and was fitted with a thin metal windshield to improve its drag characteristics. The 2,800ft/sec muzzle velocity of the T33 shot projectile was far less than that of the M304 HVAP-T, but it could penetrate the Panther glacis at 1,100 yards. Like the M304 HVAP-T round, the T33 AP-T rounds were available only in small numbers.

As with the M4 series of tanks, the M26 had a coaxial .30 caliber machine gun mounted in its turret and another one in the front hull operated by the assistant driver/bow gunner. There were also provisions for fitting a .50 caliber machine gun on the turret roof of the M26.

By the time the war ended in the ETO, there were 310 heavy tank M26s in place, with 200 of them issued to field units. Of those issued to field units, it was only the first 40 tanks, which arrived in February 1945 that saw much combat action. The M26 pictured was heavily damaged by two enemy HE rounds. (Patton Museum)

OPPOSITE

Heavy tank M26s are seen on the Fisher Tank Arsenal production line during winter 1944–45. The vehicle commander's cupola on the tank was developed for the second generation of M4 series medium tanks and later retrofitted to many rebuilt first-generation M4 series medium tanks. (Patton Museum)

COMBAT USE

The desire by U.S. Army tankers fighting in Northwest Europe to have a vehicle in service that was at least an equal in fighting effectiveness to the late-war German tanks they were encountering in battle appears in a quote from this March 1945 U.S. Army report titled *United States vs. German Equipment*:

> No comment can be made on the M26 tank itself, as none have been seen by any of this crew as yet, however, a more heavily armored vehicle would be very desirable, and if the M26 tank lives up to its publicity releases, it should be the answer to the German Mark V [Panther] and Mark VI [Tiger] problem. The 90mm gun, as used by the T.D. units [M36 Tank Destroyers] has proved to be very satisfactory as far as we have seen.

In response to this type of demand, 20 T26E3 tanks out of the initial production run of 40 vehicles were rushed overseas, arriving at the Belgium port of Antwerp in January 1945. These tanks were part of a technical mission, codenamed "Zebra," intended to assist the rapid introduction of the new tank as well as several other items of new equipment to the rigors of combat. To evaluate how the 20 T26E3s would fare in combat against German late-war tanks, the U.S. Army 3rd and 9th Armored Divisions each received ten vehicles.

A relatively late arrival to the Western European Theater of Operations was the U.S. Army T26E3 tank. The vehicle pictured served with Company B, 19th Tank Battalion, 9th Armored Division, at Remagen, Germany, in March 1945. Unlike the more gaudy camouflage paint schemes seen on many late-war German tanks and self-propelled guns, the T26E3 is in its nondescript base olive drab. The vehicle's registration number was done in yellow on a ventilation hump located on the front hull. The national insignia is seen on the front upper hull and rear engine deck. Also seen on the rear engine deck is an air identification panel. (© Osprey Publishing Ltd.)

The first T26E3 tank-on-tank combat action took place on the evening of February 26, 1945, when a T26E3 tank from the 3rd Armored Division, nicknamed "Fireball," was knocked out. The American tank had been trying to break through a roadblock and its position was given away by a fire among the rubble that silhouetted the vehicle's turret to a nearby, but unseen, German Tiger Ausf. E heavy tank that put three rounds into the M26 at a range of 100 yards.

The first German 88mm projectile penetrated the Browning .30 caliber coaxial machine gun port on the M26, killing the vehicle's gunner and loader, while the second projectile destroyed the tank's muzzle brake. The strike on the muzzle brake caused the propellant of a chambered round to detonate, with the projectile portion exiting the gun barrel and the explosion causing the barrel of the 90mm main gun to swell. The third projectile strike gouged a chunk of steel out of the right side front of the vehicle's turret and ripped off the open vehicle commander's hatch. Upon conclusion of firing on the American tank, the German tank backed up in the darkness and managed to immobilize itself on a pile of rubble and was abandoned by its crew.

The following day, another T26E3 tank from the same division avenged "Fireball" by destroying a German Tiger Ausf. E heavy tank with four shots at a range of about 900 yards. The initial round was an M304 HVAP-T that destroyed one of the German tank's front hull-mounted final drives, immobilizing the vehicle. The second round was a T33 AP-T that struck and penetrated the bottom of the tank's thick gun shield and caused an internal explosion that rendered the vehicle inoperable. Two follow-on HE rounds did not cause any additional damage to the German tank.

A restored heavy tank M26 owned by a private collector in the United States. Reflecting the fact that the vehicle was rushed into service, a number of minor design issues were uncovered in the field as was common to all tanks. Many of these were addressed during the vehicle's production run that lasted until October 1945. (Bob Fleming)

A single pilot example of the heavy tank T26E4 was sent to Germany just before the war in the ETO ended. It mounted a new, more powerful T15E1 90mm main gun. To increase its armor protection levels, appliqué armor was mounted on the front hull and turret of the vehicle in the ETO. (Patton Museum)

In addition to destroying the German Tiger tank, The other T26E2 tank engaged and knocked out two German Pz.Kpfw. IV medium tanks that same day at a range of 1,200 yards with one round of T33 AP-T each. These were followed by two rounds of HE that killed the crews as they attempted to evacuate their respective vehicles.

During the fighting for the German city of Cologne on March 6, 1945, a T26E3 now designated the M26, of the 3rd Armored Division, destroyed a German Panther tank with three shots. This "tank-versus-tank" incident was captured on film by a U.S. Army Signal Corps cameraman and is often seen on television shows about the war in Europe. The best known combat action in which the M26 tanks took part – but did not include any tank-versus-tank action – occurred on March 7, 1945, when four M26s of the 9th Armored Division aided in the capture of the Ludendorff railroad bridge over the Rhine River at the German town of Remagen.

By the end of March 1945, 40 additional M26 tanks arrived in Western Europe. These were assigned to the American Ninth Army, with 22 going to the 2nd Armored Division and 18 to the 5th Armored Division. In early April 1945, the 11th Armored Division of Lieutenant-General George S. Patton's Third Army received 30 M26s. However, with the war in Europe winding down, there were no additional tank-versus-tank combat actions between the M26 and German tanks.

With the war in Europe ending, greater attention was paid to the fighting in the Pacific. A Marine Corps document dated April 1945 titled *Iwo Jima, 4th Tank Battalion Report* made this recommendation based on combat experience gained during the fighting on Iwo Jima:

> Tank, Army, medium, M4A3 should be replaced by Tank, Army, heavy, M26 (also known as Tank, Army, medium T26, T26E1 and General Pershing). M4 series tanks are extremely vulnerable to 47mm AT [antitank] fire, magnetic mines, shaped charges and field artillery. This is especially true in operations against a well manned, heavily fortified position or in a slow moving situation over difficult terrain where the M4 loses its maneuverability. The 75mm M3, Tank, the primary armament of the M4 series tank is not effective against well-constructed reinforced concrete positions. The M4 series tank, with its increased weight from many modifications and its narrow track and bogie-volute suspension system has too much ground pressure to successfully negotiate loose sand or heavy going.
>
> The M26 presents the following advantages over the M4. It is shorter, wider and lower, presenting a lower silhouette; it weighs forty-four tons, the additional weight being caused by increased armor. Since it is now evident that M4 series tanks cannot safely be loaded in LCMs [landing craft, mechanized], this increased weight would not affect the use of the M26 in amphibious operations.

On the island of Okinawa, Japanese 47mm antitank guns were taking a heavy toll of thinly armored M4 series tanks. As a result, it was decided to send 12 M26s to that theater of operations. The M26 tanks arrived on Okinawa after fighting concluded in July 1945. The M26s were then considered as playing an important role in the planned follow-up invasion of Japan; however, the Japanese surrender in August 1945 ended that mission.

The M26 was officially designated the "General Pershing" after World War II in honor of General of the Armies John J. "Black Jack" Pershing, who was commander of the American Expeditionary Forces (AEF) in France during World War I. In May 1946, the M26 was reclassified as a medium tank, as the U.S. Army was envisioning developing much larger and heavier tanks. Wartime and postwar tankers simply referred to the vehicle as the "M26" or the "26."

Following World War II, the U.S. Army revised its weight classifications for tanks. Light tanks could weigh up to 25 tons. Anything between 26 and 55 tons was classified as a medium tank. Heavy tanks ranged in weight from 56 to 85 tons. Tanks weighing over 86 tons would be classified as super heavy tanks.

M26 VARIANTS

The Ordnance Department decided they needed a version of the M26 that mounted a main gun equal in performance to that of the 8.8cm Kw.K. 43 fitted in the German Tiger Ausf. B heavy tank (often referred to as the Tiger II) first encountered by American tankers during the Battle of the Bulge in December 1944. The gun they installed in a modified T26E1 was designated the 90mm gun T15E1 and fired one-piece, fixed ammunition. Another version of the same gun firing two-piece, separated ammunition was designated the T15E2, and was fitted to a second pilot tank.

The modified T26E1 with the T15E1 gun was designated the T26E4 in March 1945 with authorization for 1,000 units to be built. In the end, only the original pilot of the T26E4 made it to the ETO before the end of the war in Europe. Upon arrival in Europe it was re-designated as the T26E4, temporary pilot number 1. Among modelers, the vehicle is referred to incorrectly as the "Super Pershing." It was rushed into service to see if it could be placed into action against the German Army Tiger Ausf. B in order to compare the respective vehicles' combat capabilities.

When firing an HVAP-T round (designated the T44 shot) from the T15E1 gun at a muzzle velocity of 3,750ft/sec the projectile could penetrate 9.6 inches (244mm) of armor at 500 yards, and at 1,000 yards 8.7 inches (221mm) of armor. At 1,500 yards, it could penetrate 7.7 inches (196mm) of armor, and at 2,000 yards it could still penetrate 6.8 inches (173mm) of armor. The barrel of the T15E1 was 21.5 feet long and needed a double external equilibrator mounted on the top of the vehicle's turret to balance the gun in its mount, and a heavy counterweight welded onto the rear of the turret bustle to balance the turret in traverse.

Once the T26E4, temporary pilot number 1 arrived in Western Europe in March 1945, the front of the vehicle's gun shield was up-armored by the 3rd Armored Division with an approximately 80mm chunk of armor flame-cut from the glacis of a captured German Panther tank. The glacis of the T26E4 and the bottom front hull plate on the vehicle were provided more protection by welding on two large steel armor plates 1.6 inches (41mm) thick.

The T26E4's first combat action occurred on April 4, 1945, when it engaged and destroyed what was perceived as a German tank or self-propelled gun at a range of 1,500 yards. On April 21, 1945, it supposedly engaged in a short-range duel in the German town of Dessau with an enemy tank identified as a Tiger Ausf. B. However, current research into the location of all known Tiger Ausf. B tanks indicates that none were near Dessau on this date, and the location of the ammunition more closely corresponds to that of a Pz.Kpfw. IV.

Formerly on display for many years at the now closed Patton Museum of Armor and Cavalry was this impressive-looking vehicle, initially classified as the heavy tank T28. It was later designated as the 105mm gun motor carriage T95 in February 1945. In June 1946 the vehicle was renamed the super heavy tank T28. (Michael Green)

The end of the war in Europe resulted in all but 25 units of the T26E4 being canceled. As these vehicles eventually entered the U.S. Army inventory, a decision was made that the preferred postwar tanks would fire one-piece, fixed ammunition, and not the two-piece, separated ammunition fired from the 90mm main gun on the T26E4. As a result, those T26E4s not used as range targets were scrapped.

M26 ASSAULT TANK

Based on the successful employment of the M4A3E2 assault tank in Northwest Europe, it was decided to develop an assault tank version of the M26. It was proposed that such a vehicle have almost 5 inches (125mm) of armor on its cast-armor sloped front glacis and a cast-armor turret that was 8 inches thick (203mm) all the way around. To compensate for the increased weight of the proposed vehicle, it was decided to attach 5-inch end connectors to the existing 23-inch-wide T80E1 track. In February 1945, the Ordnance Committee designated the proposed M26 assault tank as the T26E5.

In March 1945, the Ordnance Department decided that the T26E5 needed even thicker armor than originally anticipated. The sloping glacis was now 6 inches (152mm) thick, and the vertical gun shield on the front of the tank's turret was 11 inches (279mm) thick. The front of the tank's turret on either side of the gun shield was 7.5 inches (191mm) thick, with 3.5 inches (89mm) on the vertical sides of the turret. The rear of the turret was to be 5 inches (127mm) thick to help balance the turret in traverse. The additional armor pushed up the weight of the T26E5 to 51 tons.

Testing showed that the mobility of the T26E5 was almost equal to that of the standard M26 on level roads. However, like the M4A3E2, the suspension system failed when driven over rough ground at higher speeds. Production of the T26E5 began in June 1945, with 27 units built before the program was canceled with the official end of World War II.

Power for the super heavy tank T28 seen here came from a single gasoline-powered, liquid-cooled Ford GAA engine that produced 500 gross horsepower at 2,600rpm, providing the vehicle with a maximum speed on a level road of 8mph. Cruising range of the vehicle was about 100 miles. (Patton Museum)

GETTING BIGGER

Despite the Armored Force's lack of interest in heavy tanks in the early part of World War II, the Ordnance Department went ahead and initiated another heavy tank project in September 1943. Their studies indicated that after the planned invasion of Western Europe, the U.S. Army would eventually come up against the German western border defensive system, known as the West Wall and also referred to as the "Siegfried Line." With this in mind, the Ordnance Department envisioned the need for a new heavy tank that had a main gun capable of penetrating thick concrete. That weapon had just been developed, and was designated the 105mm gun T5E1.

The Ordnance Department envisioned mounting the 105mm gun T5E1 in a tank protected by 8 inches (203mm) of cast frontal armor. Power for this proposed heavy

tank would come from the GE electric drive system as mounted in the heavy tank T1E1. The Ordnance Department anticipated a need for 25 of these new heavy tanks and believed they could be made ready in as little as eight to 12 months for the invasion of Western Europe.

The AGF were not enthralled with the proposed building of a small fleet of new heavy tanks and suggested that only three examples be built, and instead of an electric drive system, they have a standard mechanical transmission. After a bit more discussion, the number of new heavy tanks armed with the T5E1 105mm main gun was bumped up to five units. They were to be designated the heavy tank T28. As the latest plans called for the vehicle to have at least 12 inches (305mm) of cast frontal armor, it was estimated the completed weight of the vehicle would be 95 tons.

Unlike the M6 series tank, the proposed T28 would not have a rotating turret. The main gun was mounted in the front hull with a limited traverse of 10 degrees either way. There was no coaxial machine gun for this vehicle. It was intended that the T28 have a crew of four men. The driver was located on the left of the main gun in the front hull, with his vision provided by a cupola equipped with vision blocks and an overhead periscope. The gunner was located in the right front hull and provided with both a telescope and periscope sight.

The five-man heavy tank T30 seen here on display at the former U.S. Army Armored School then located at Fort Knox, Kentucky, was basically a heavy tank T29 armed with a 155mm main gun. Reflecting the larger and heavier main gun, it weighed in at 70 tons. (Dean and Nancy Kleffman)

The vehicle commander for the proposed T28 was located behind the gunner and had an overhead cupola, while the loader was located behind the driver's position. The only secondary armament on the vehicle was to be a .50 caliber machine gun fitted to a ring mount above the vehicle commander's cupola.

Reflecting the fact that the T28 was turret-less, it was decided in March 1945 to no longer refer to it as a heavy tank. It was therefore re-designated as the 105mm GMC T95. There seemed to be no rush placed on putting the T95 into service for the

invasion of Western Europe, or for that matter, the U.S. Army's eventual encounter with the Siegfried Line, which first occurred in August 1944. It can only be surmised that the lack of interest by the AGF might have something to do with this.

Another problem with getting the T95 constructed was finding a contractor who had the spare capacity to build something as massive as the completed vehicle was to be. Finally, the Pacific Car and Foundry Company stepped up to the plate and agreed to take on the project in May 1945. With the end of the war in the Pacific in August 1945, the number of pilot T95 vehicles was reduced from five to two units.

The first completed pilot arrived at APG in December 1945. The second pilot left the factory floor in January 1946 and was sent first to Fort Knox, Kentucky, and then on to the Engineer Board located at Yuma, Arizona, where it was used for testing floating bridges. The vehicles were 36.5 feet long, approximately 15 feet wide, and 9 feet 4 inches tall.

To reduce the T95's ground pressure to a reasonable figure, it came with four sets of tracks, two on either side of the vehicle, with the outer set being removable when shipment by rail was required. With the outer set of tracks removed, the width of the vehicle dropped to 10 feet 4 inches. On paved roads, the second set of removable tracks could be joined together as a trailer and towed behind the vehicle. In June 1946, the T95 was re-designated as the super heavy tank T28. Testing of the vehicle continued until 1947, and shortly thereafter the project was terminated.

TOO LATE FOR THE PARTY

Beginning in summer 1944, reports of even more heavily armored German tanks and self-propelled guns generated interest within the Ordnance Department in the development of a heavy tank that would be the equal of whatever the German Army fielded. By September 1944, the Ordnance Committee recommended the development and building of four pilots for a new heavy tank. Of the four new heavy tanks, two would be armed with a turret-mounted 105mm gun T5E1 and be designated the T29, while the other two would be armed with a turret-mounted 155mm gun T7 and be designated as the T30. The turrets for these heavy tanks were all made of cast armor.

Of the two new proposed heavy tanks, the Ordnance Committee recommended in March 1945 that 1,200 units of the six-man T29 be acquired. The following month, that number was reduced to 1,152 units. Also in April 1945, the building of four additional pilots of the T29 was approved. At the same time, it was also decided that one of the four pilots of the T29 be armed with the 120mm gun T53 and be designated the heavy tank T34. As events transpired, the Japanese surrender in August 1945 brought the T29 program to a halt with only a single example of the 70-ton vehicle completed and a second example partially completed.

DETROIT ARSENAL
NEG. NO. 14316   DATE 17 March 1947   DEVELOPMENT & ENGINEERING
Tank, Heavy T29.

CHAPTER FIVE

TANK DESTROYERS

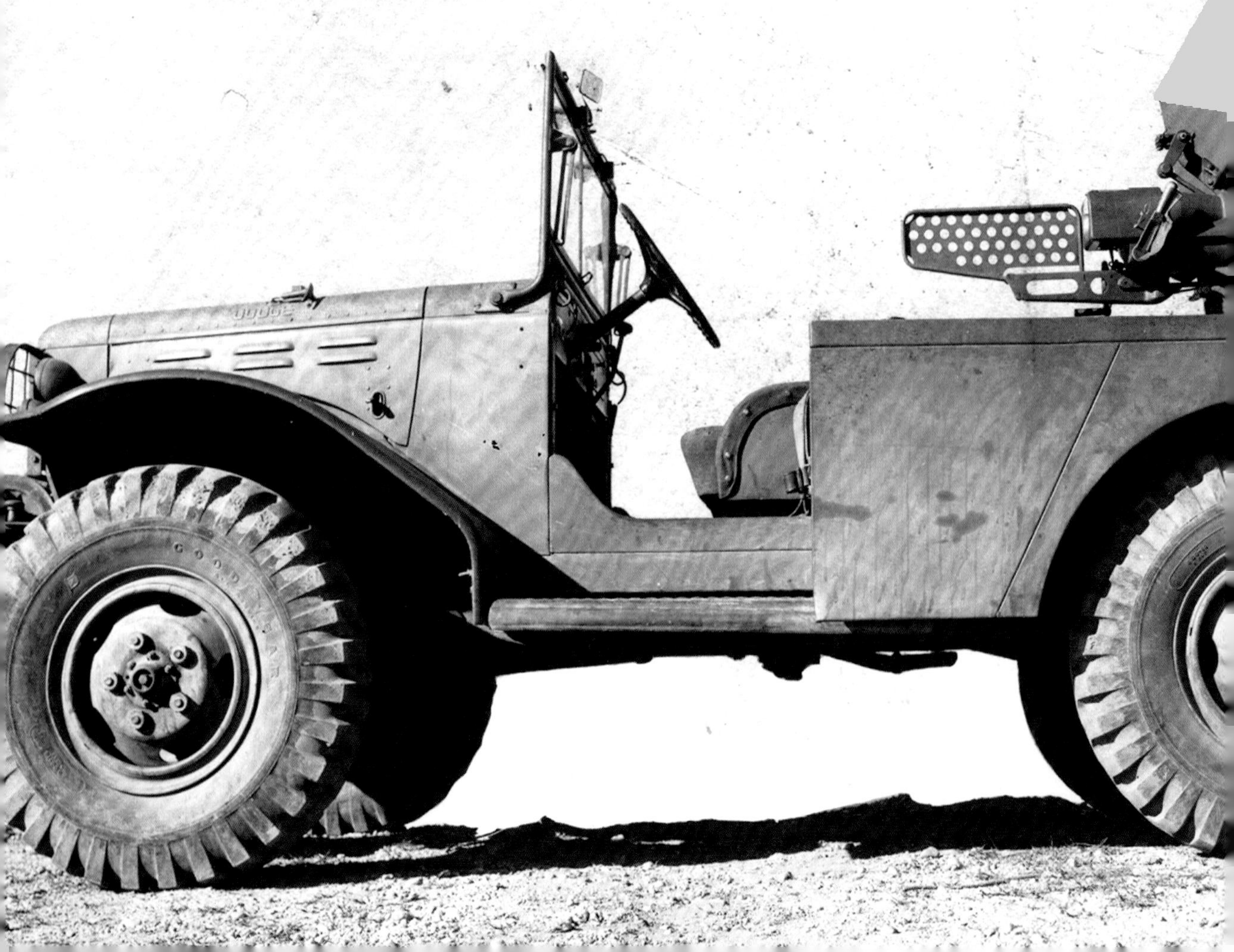

The German Army's stunning defeat of France in summer 1940 within the span of six weeks was generally (and incorrectly) attributed by most to the Germans fielding thousands of tanks that overwhelmed the French Army's defenses. In reality, the French Army and the British Expeditionary Force (BEF) in France fielded more tanks, and in some cases, superior tanks to those of the Germans.

The surprisingly quick collapse of the French Army had more to do with a litany of endemic political and military reasons that included incompetent leadership at its most senior levels. However, the image of massive fleets of German tanks rolling over all they met seemed to have infected some within the U.S. Army senior leadership with a deep fear in those dark, early days of World War II.

In response to the perceived German armored juggernaut that would no doubt be encountered by the U.S. Army ground forces on some future battlefield, the Army's senior leadership began deliberations that extended into early summer 1941 on how best to deal with this future threat.

Some believed that the best defense against enemy tanks was another tank. Others firmly believed that American tanks should be reserved for exploitation duties only as was the accepted U.S. Army doctrine at the time. In a doctrinal miscalculation, it was eventually concluded that highly mobile antitank guns, held in pooled reserves behind the front lines, would use offensive maneuvers to deal with any German armored thrust. U.S. Army tanks of the armored divisions would only deal with those enemy tanks encountered during the exploitation phase of an offensive operation. Sadly, what many in the senior ranks of the U.S. Army failed to grasp in 1940 and 1941 was that German Army tactics with its armored divisions employed in France were not based on all-tank formations. Rather, their armored divisions practiced combined-arms warfare, which included tanks working in conjunction with

The 37mm gun motor carriage M6 was the U.S. Army's first stopgap tank destroyer. The gun-shield protected weapon had 360 degrees of traverse but in reality could only be fired over the rear of the vehicle cargo body. The 0.75-ton Dodge weapon carrier chassis upon which the weapon was mounted could reach speeds over 50mph. (Patton Museum)

aviation elements, self-propelled artillery, antitank guns, engineers, and mechanized infantry. The U.S. Army did not fully understand this reality until they first met the German Army in North Africa in early 1943.

THE BEGINNING

U.S. Army thoughts on stopping mass German tank attacks with mobile antitank guns first appeared in a War Department Training Circular dated September 23, 1940. In that document it was urged that a minimum number of antitank guns be positioned in fixed forward defensive positions, with the majority held back in a mobile, offensive-oriented reserve. Prior to that time, the U.S. Army antitank doctrine was strictly defensive in nature, relying on antitank guns in the front lines to deal with any enemy tank-led offensive action.

In May 1941, the Chief of Staff of the U.S. Army, General George C. Marshall, stated that he was convinced that mobile antitank units should be organized, and he directed that work on the forming of such an institution begin. In early 1941, Marshall stated: "It occurs to me that possibly the best way to combat a mechanized force would be to create anti-mechanized units on self-propelled mounts, with emphasis on visibility (for the gunner), mobility, heavy armament, and very little armor."

On November 27, 1941, the War Department set up the Tank Destroyer Command at Fort Meade, Maryland. Its job was to serve as the developmental clearing house for all tank destroyer-oriented doctrine and equipment, and to provide a central training site for all tank destroyer personnel and their units. Eventually, the Tank Destroyer

One of the many paths of development in the search for a suitable armored full-tracked tank destroyer for the U.S. Army resulted in the gun motor carriage M5 armed with a 3-inch gun. Designed and built by the Cleveland Tractor Company (Cletrac) in 1941, it proved a dismal failure during testing in 1942, and only one example, seen here, was ever assembled. (Patton Museum)

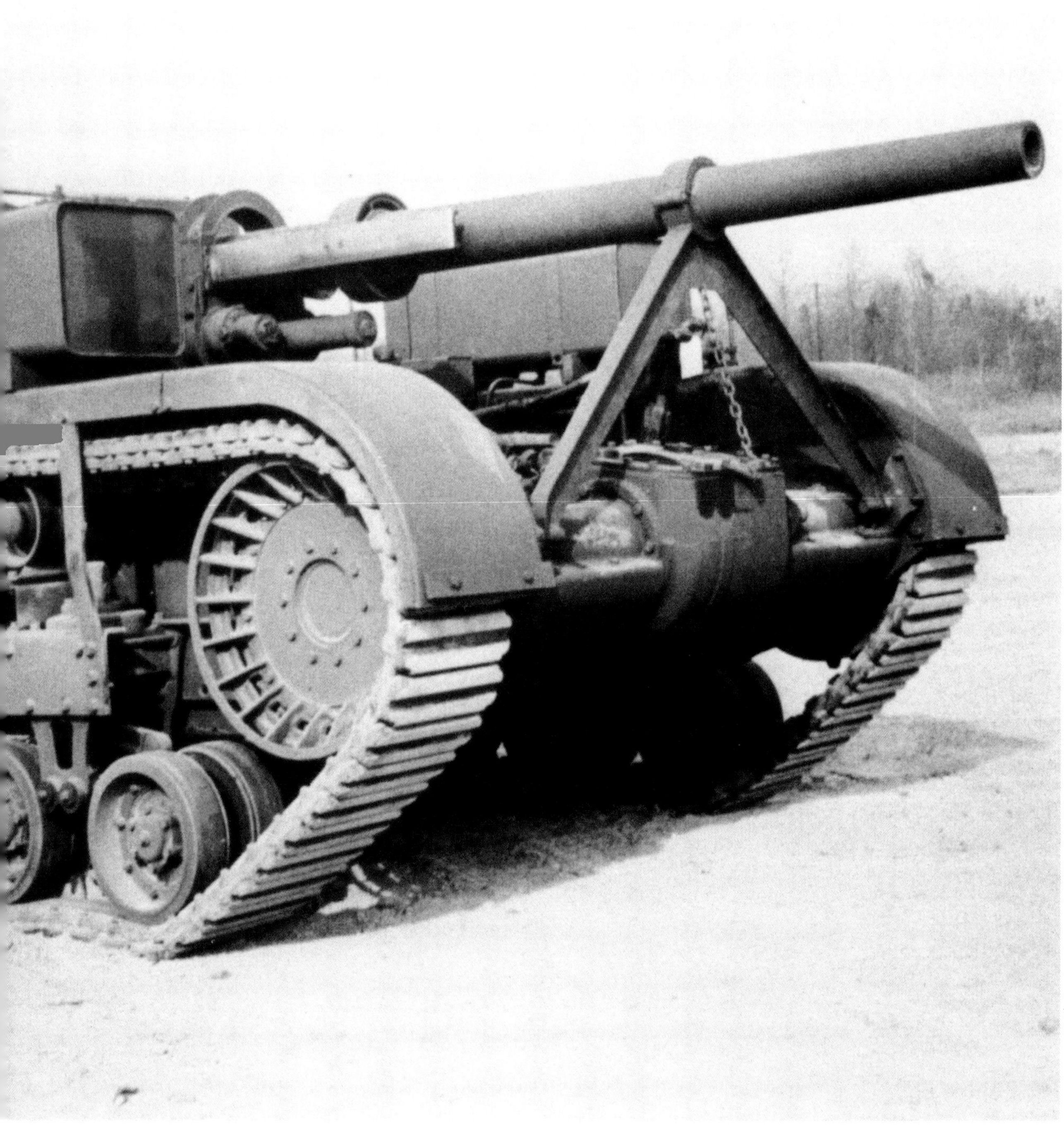

Command found a larger and much more spacious home at Camp Hood, Texas, now known as Fort Hood, for training tank destroyer units.

The two men who were crucial to the forming of the Tank Destroyer Command were Lieutenant-General Lesley J. McNair, commander of the Army Ground Forces (AGF), and Lieutenant-Colonel Andrew D. Bruce (later promoted to lieutenant-general). They strongly differed in opinion as to what types of tank destroyers were required by the growing U.S. Army. McNair, a former artillery officer, preferred towed antitank guns, while Bruce, a former infantry officer, wanted to see antitank guns mounted on self-propelled, armored tracked vehicles. Marshall's interest in self-propelled guns helped Bruce overcome McNair's objections.

The Tank Destroyer Command at Camp Hood, Texas was downgraded to the Tank Destroyer Center in August 1942, reflecting a sharp decline in its power and prestige within the U.S. Army. Some of the center's training activities were curtailed as early as October 1943 because the need for tank destroyers in the field was not nearly as high as originally anticipated; in the crucible of combat, the concept was proving to be unsound.

THE SEARCH BEGINS

Bruce had no illusions about quickly fielding the optimum self-propelled tank destroyer that met all his expectations. Instead, he knew that an expedient self-propelled tank destroyer was needed in the short term until American industry could provide him with a more suitable vehicle.

The Ordnance Department approved a vehicle concept in December 1940 proposed by the Cleveland Tractor Company. The firm had suggested mounting a suitable antitank gun in a fixed forward-firing position (with limited traverse) on a modified chassis of a small, 7-ton, unarmored, full-tracked, aircraft-towing tractor the company built, which they called the "Cletrac." The Ordnance Department approved the concept and mandated that the weapon mounted on the Cletrac be a modified version of the 3-inch antiaircraft gun M3.

It was Brigadier-General Gladeon M. Barnes of the Ordnance Department (later promoted to major-general) who had pushed for employing the 3-inch gun as an antitank weapon in December 1940. Not everybody in the U.S. Army believed that a 3-inch gun was needed, as is evident from an Ordnance Committee meeting, dated December 1940, in which the Infantry Board at Fort

The inadequacies of the 75mm gun motor carriage M3 became very clear to the U.S. Army during the fighting in North Africa and Sicily in 1943 and they were withdrawn from service. However, the vehicle remained in Marine Corps service as seen here until the end of World War II in the special weapon battalions of Marine Corps divisions. (Patton Museum)

Benning, Georgia, stated: "In view ... of the lack of information as to the need for a weapon with the great penetrating ability of the subject gun, the Chief of Infantry cannot agree that there is a need for antitank material of such great weight and consequential poor mobility."

The initial prototype/pilot vehicle that mated the Cletrac chassis with the 3-inch gun M3 was designated the 3-inch Gun Motor Carriage (GMC) T1 and featured a welded armored gun shield. There was some limited armor protection provided for the driver's and assistant driver's positions. Despite numerous teething problems with the prototype, the Ordnance Department recommended that the vehicle be standardized as the 3-inch GMC M5. With the War Department's approval, the production of 1,580 vehicles was authorized in early January 1942. The vehicle's weight had now risen to 12 tons.

Bruce was against the 3-inch GMC M5 from the very beginning and called it the "cleak track" because he thought it was so poorly designed. Bruce's opinion of the vehicle was confirmed after a series of tests conducted at APG in July 1942. The tests proved conclusively that the 3-inch GMC M5 was clearly unsuitable as a tank destroyer and dashed the hopes of the Ordnance Department, which had expressed a great deal of faith in the vehicle's potential prior to the APG tests.

An Ordnance Department historian, after testing at APG by a crew from the Tank Destroyer Board, described the condition of the 3-inch GMC M5: "The sides were dished in, the gun supports buckled, the suspensions were out of line, the travel lock folded, and the gun mount loosened." After reviewing the test results, McNair would write a letter on July 10, 1942 in which he had to admit to Bruce that the 3-inch GMC M5 looked "pretty hopeless." The Army Ground Forces recommended the cancellation of the vehicle in August 1942.

Despite the end of the line for the 3-inch GMC M5, the Ordnance Department had other irons in the fire. In September 1941, they proposed mounting the 3-inch gun M7, developed for the heavy tank M6 series, on the open-topped chassis of the light tank M3, designating it the 3-inch gun motor carriage (GMC) T20. However, it quickly became clear that the chassis could not support a weapon that powerful and it was canceled in October 1941. Other follow-on projects, including the T50, T56, and T57 GMCs, involved mounting the 3-inch gun M7 on a light tank chassis, but all were canceled for the same reason.

HALF-TRACK-BASED TANK DESTROYER

The first vehicle to enter into field service of the new Tank Destroyer Force was the 10-ton pilot 75mm GMC T12, cobbled together by a team of Ordnance Department personnel at APG in summer 1941. The pilot T12 consisted of a standard, open-topped half-track personnel carrier M3 modified to mount a forward-firing, World War I vintage 75mm light field piece designated the 75mm gun M1897A4. The half-track personnel carrier M3 had just entered into service and there were a great many surplus M1897A4 guns in storage.

The T12 pilot was used by the Autocar Company as a template to fulfill a contract from the Ordnance Department for 36 additional units of the vehicle. That order was subsequently boosted to 86 units, all of which were delivered to the U.S. Army between August and September 1941.

Of the 86 T12s provided to the U.S. Army, 50 were hastily shipped off to the Philippines to bolster America's defenses in that country against future Japanese aggression. Some of the remaining T12s went to APG for additional testing by the Ordnance Department, with the others provided to the newly formed 93rd Antitank

Battalion for field testing. The testing process resulted in a number of modifications to the T12, including a new gun shield. The vehicle was standardized as the 75mm GMC M3 in October 1941.

VEHICLE DETAILS

It was originally intended that the 75mm GMC M3 be staffed by a crew of four men: driver, assistant driver (radio operator), gunner, and loader. After field testing, it was decided that the vehicle was better served by a crew of five, with the addition of a gun commander. Field testing also showed that a .50 caliber machine gun, fitted onto a pedestal at the rear of the vehicle, was not needed, and it and its onboard ammunition storage containers were eliminated and replaced with two sheet metal storage boxes. The maximum armor thickness on the GMC M3 was 0.625 inches (15.9mm) on the welded armor gun shield.

The gunner aimed the 75mm main gun on the GMC M3 with the telescopic sight M33, based on the telescopic sight from the 37mm gun M3. The weapon itself was fitted to the upper parts of the M2A3 gun carriage, which sat on a steel base mounted on the rear of the half-track frame. However, as the number of M2A3 gun carriages dwindled, an earlier version still in the inventory, designated the M2A2 field carriage, was used. The Ordnance Committee re-designated the M2A2 field carriage as the 75mm gun mount M5, and vehicles fitted with this gun mount were referred to as the GMC M3A1.

There was storage space for 59 rounds of 75mm main gun ammunition in the GMC M3/M3A1. The main gun could be traversed 19 degrees left, and 21 degrees right on the GMC M3, but on the GMC M3A1, the main gun could be traversed 21 degrees left or right. Maximum elevation for the main gun on the GMC M3/M3A1 was 29 degrees, with depression being limited to 7 degrees.

Series production of the 75mm GMC M3/M3A1 took place between February 1942 and February 1943 with a total run of 2,116 vehicles including the 86 T12s. The M3/M3A1 were reclassified as limited standard in March 1944 and declared obsolete in September 1944.

PROPOSED CHANGES

By early 1943 the inventory of 75mm M1897A4 cannons for fitting to the GMC M3/M3A1 was running low. As the vehicles were expected to remain in production (at that time) for the foreseeable future, it was decided to see if a 75mm gun M3 as fitted to the first generation of M4 series tanks could be modified to replace the M1897A4 on the GMC M3/M3A1. Testing quickly affirmed that the concept would work, and the modified 75mm gun M3 was designated as the 75mm gun T15 in gun mount T17.

When the 75mm gun T15 and gun mount T17 were fitted to a modified half-track personnel carrier M3, it became the 75mm GMC T73. However, by the time the T73 was ready for series production, a new generation of fully tracked tank destroyers appeared and the T73 project was canceled. At the same time, it was decided to convert 1,360 units of the already existing GMC M3/M3A1s back into the half-track personnel carrier M3A1.

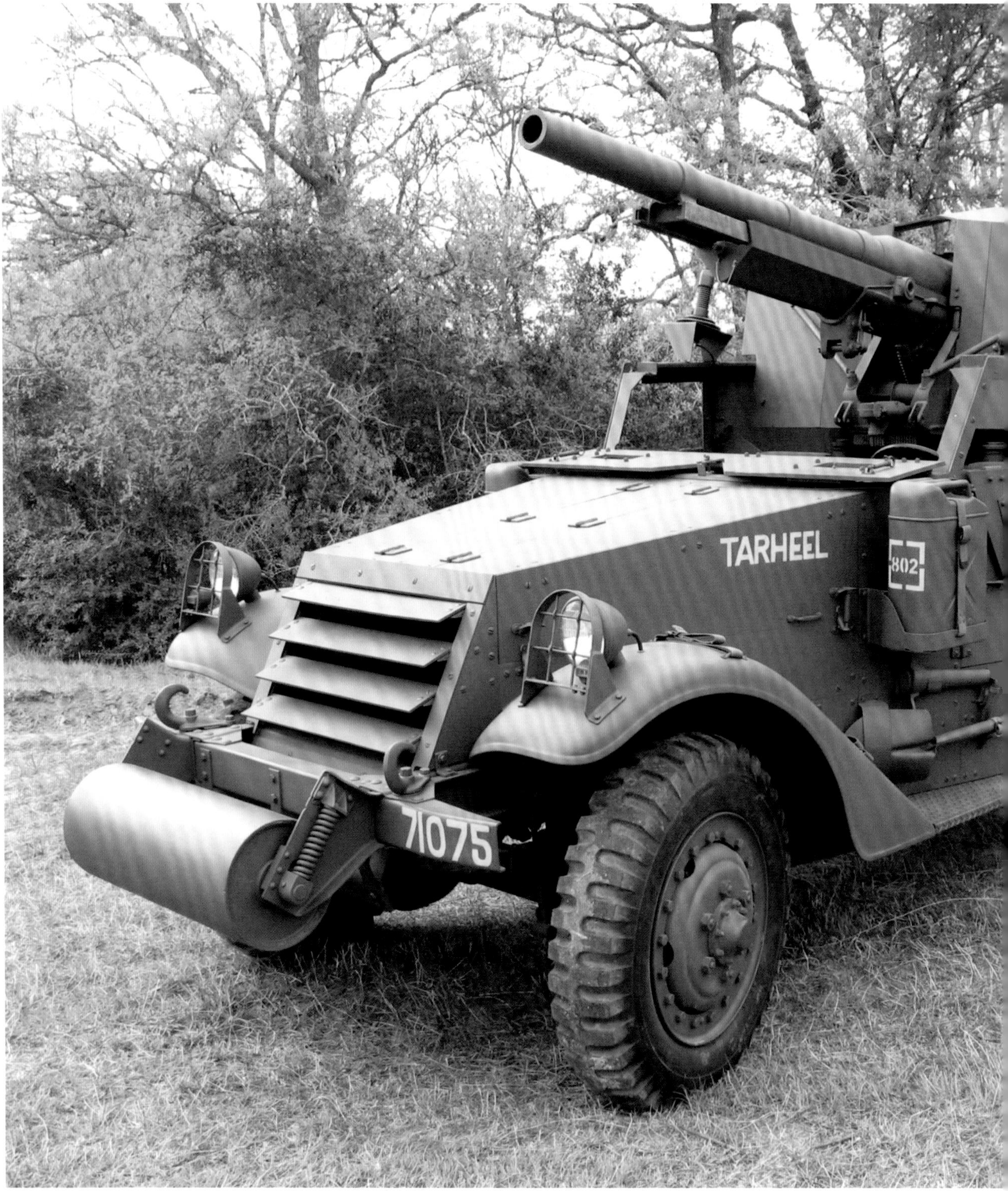
TARHEEL
802
71075

FOR LEND-LEASE

Another half-track-based tank destroyer was the GMC T48 armed with an American-built but British-designed 6-pounder (57mm) gun. In U.S. Army service the weapon was referred to as the 57mm gun M1 and differed from the British-built version as it featured a longer and thinner barrel. It was mounted in the rear of a modified half-track personnel carrier M3 on a conical structure, designated the 57mm gun mount T5.

Like the 75mm gun M1897A4 in the GMC M3/M3A1, the 57mm gun in the T48 was a forward-firing weapon with limited traverse left or right. The towed version of the weapon was offered to the Tank Destroyer Force, but was rejected because there were reports from the fighting in North Africa that the weapon's AP-T projectile shattered on the welded armor of German tanks. Also disliked was the fact that the British had not designed an HE round for the weapon.

From the beginning, the T48 was intended not for the U.S. Army Tank Destroyer Force, but for the British Army under Lend-Lease only. However, of the 962 built by the Diamond T Motor Car Company between December 1942 and May 1943, only 30 went to the British Army, with 650 going to the Red Army. Of the remaining vehicles, one went overseas with the U.S. Army, and the remaining 281 T48s were eventually converted back to the half-track personnel carrier M3A1 configuration. As the T48 was not intended for American military service, the Ordnance Department listed it as a limited procurement item.

Pictured is a newly made 75mm gun motor carriage M3 at a World War II reenactment. The vehicle was recreated by a Texas-based military vehicle dealer. The vehicle and weapon are authentic, with the gun mount and armored gun shield being accurate reconstructions of the original. (Michael Green)

The 57mm gun motor carriage T48 was American-built, but primarily intended for use by the British Army, which received them under Lend-Lease. As events unfolded, the Red Army took the bulk of those built, assigned them the designation SU-57, and employed them until the end of the war in Europe. (Patton Museum)

INTO COMBAT WITHOUT A PRAYER

Another expedient tank destroyer was the 37mm GMC M6, which was nothing more than a standard 37mm antitank gun M3 mounted in the rear cargo bay of a three-quarter-ton 4x4 Dodge WC-55 truck. The only protection afforded the four-man crew was a large 0.25-inch-thick (6mm) welded armor gun shield. The 37mm gun could only be fired over the rear of the vehicle, as the muzzle blast would have harmed the driver and assistant driver sitting in the front of the vehicle and no doubt destroyed the vehicle's windshield.

The GMC M6 was standardized in February 1942, with 5,380 units built between April and October 1942. It was originally intended solely as a training vehicle, but some were sent to fight in North Africa, with a few actually seeing

combat. An observer of the M6 in North Africa, in a U.S. Army report dated March 1943, stated: "The sending of such a patently inadequate destroyer into combat can be best termed a tragic mistake."

In the 1943 AAR *Operations of the 1st Armored in Tunisia*, Major-General E. N. Harmon stated: "The 37mm self-propelled gun, mounted on a ¾-ton truck, is positively worthless and has never been used in this division." Most of the M6s sent to North Africa had their weapons dismounted, and in some cases, the 37mm antitank guns were mounted on half-tracks.

The M6 was reclassified as limited standard in September 1943 and declared obsolete in January 1945.

MEDIUM-TANK-BASED TANK DESTROYERS

To provide the new Tank Destroyer Force with a vehicle mounting a weapon powerful enough to deal with any German tanks it might encounter in battle, the Ordnance Department proposed an interim full-tracked, self-propelled vehicle in September 1941. The concept consisted of a 3-inch gun M3 in a forward-firing position with limited traverse on the open-topped chassis of a medium tank M3. The Ordnance Committee approved the project in October 1941 and designated it the 3-inch GMC T24.

The pilot model of the T24 showed up at APG in November 1941 for testing. The Japanese attack on Pearl Harbor and America's entry into World War II the following month prompted the U.S. Army to authorize the Ordnance Department to mount 50 of the 3-inch M1918 antiaircraft guns (the predecessor to the 3-inch antiaircraft gun M3) on the open-topped chassis of the medium tank M3 as a stopgap measure. As testing progressed in early 1942, the Ordnance Committee had the vehicle designated the 3-inch GMC M9. The Tank Destroyer Force, however, had no interest in the vehicle.

The parameters of what constituted a viable self-propelled tank destroyer were determined in a Special Armored Vehicle Board meeting held from October 13 to 15,

OVERLEAF

The 3-inch gun motor carriage T24 seen here was created in late 1941 and based on the open-topped superstructure of an M3 medium tank. The gun had very limited traverse and elevation. Only the single example pictured was ever built and it was rejected as being too tall at 8 feet 7 inches. (Patton Museum)

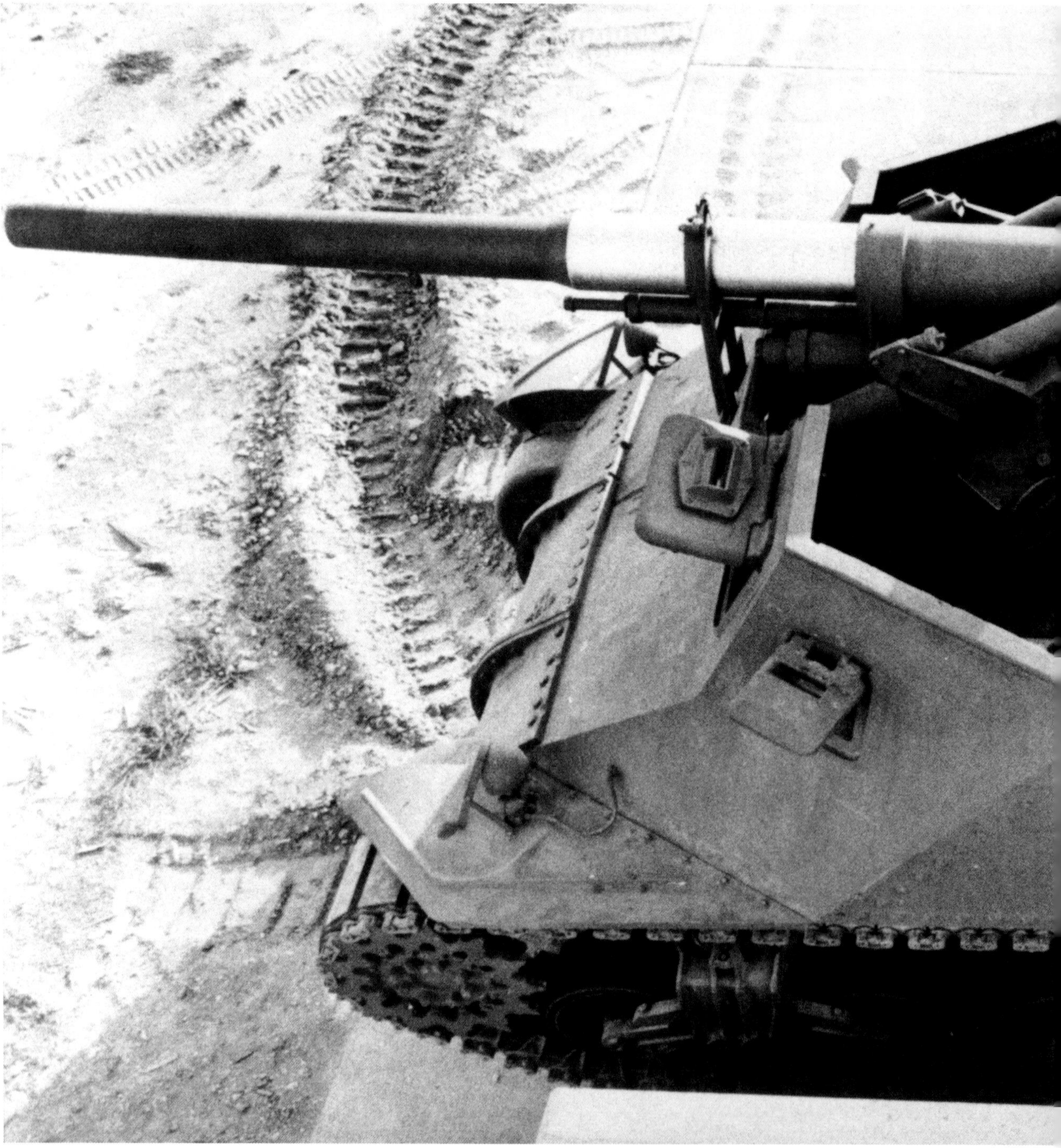

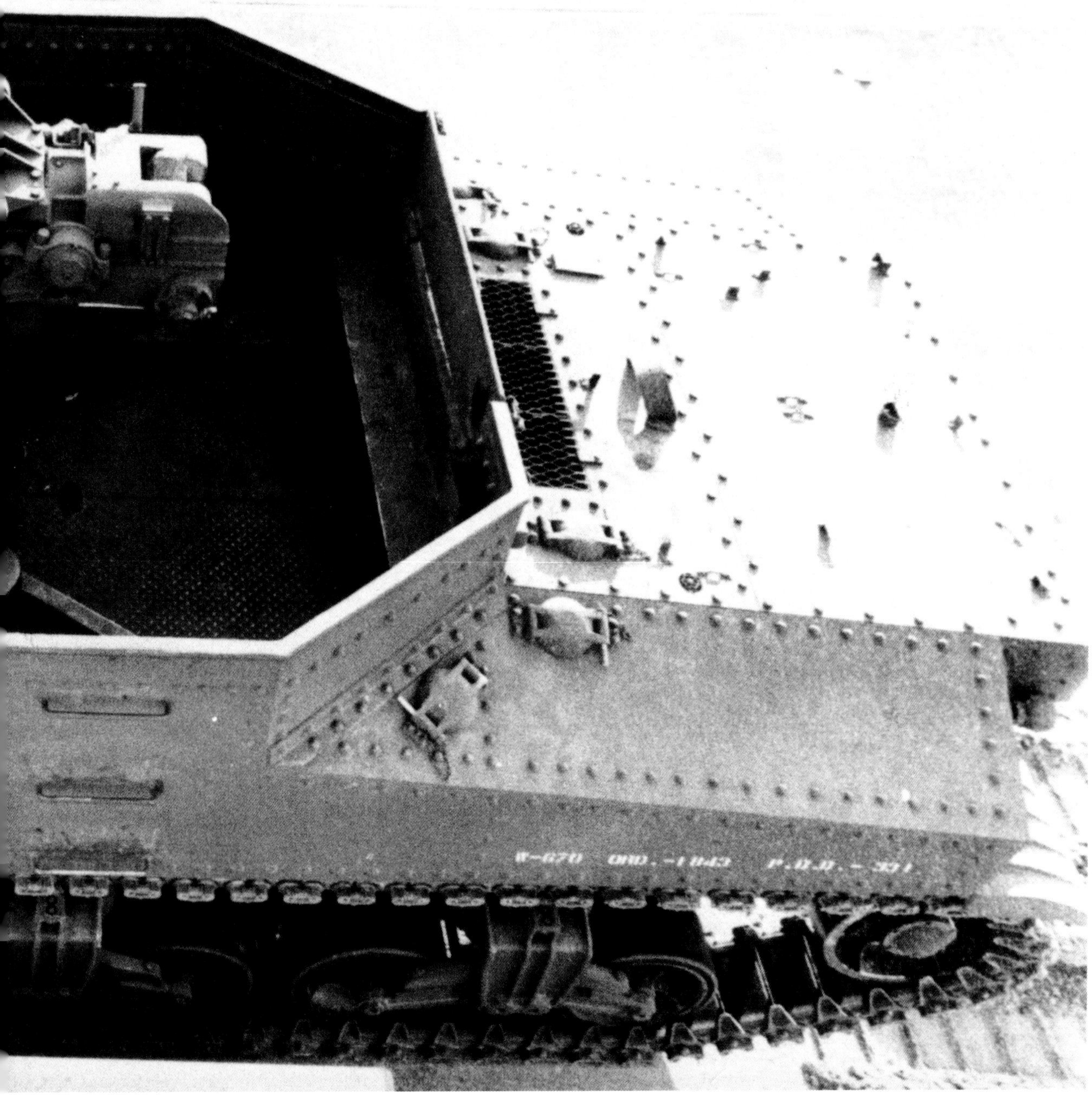

1942. Its conclusions were published on December 5, 1942 and appeared in a U.S. Army research report dated 1941–42, titled *Anti-Tank Defense – Weapons and Doctrine*:

> A suitable tank destroyer gun motor carriage is critically needed. The tank destroyer is essentially a gun carriage, not a tank. The first requirement is a gun adequate to knock out the stoutest enemy tank. The gun carriage must have; cross-country mobility fully equal to that of a tank (which makes a tracked vehicle essential); top speed on roads of 45–50mph; top weight of 20 tons; traverse for the gun of at least thirty-five degrees total; space for the crew, equipment, and forty rounds of ammunition; armor protection against small arms and fragments.

GETTING CLOSER TO A SOLUTION

What had convinced the Tank Destroyer Force not to accept the 3-inch GMC M9 was the development of another full-tracked tank destroyer, supposedly more capable, based on the new M4 series tank. It featured a turret-mounted 3-inch gun M7, an adaptation of a prewar antiaircraft gun. It was the weapon of choice for Lieutenant-Colonel Bruce because its muzzle velocity of 2,600ft/sec with the standard APC M62 round suggested it had the penetrative power needed to destroy current German tanks. Sadly, the penetrative powers of the 3-inch gun M7 proved inadequate against late-war German tanks and self-propelled guns.

The first drawings for this proposed tank destroyer designated it as the 3-inch GMC T35 in November 1941. A wooden mockup of the proposed T35 appeared in January 1942. By this time it had already been decided to base the new vehicle on the welded armor chassis of the medium tank M4A2 with thinner armor on the sides and rear of the vehicle's upper hull, and an open-topped cast-armor turret.

Unlike the M4A2, the T35 lacked either a coaxial or bow machine guns, reflecting its main purpose of killing enemy tanks and not enemy infantry. There were provisions made to mount a .50 caliber machine gun on the top of the vehicle's open-topped turret.

User feedback from the fighting in the Philippines in December 1941 indicated angled welded armor plates had an advantage in deflecting projectiles, so the Tank Destroyer Board requested that the new tank destroyer feature sloped armor on its upper hull sides and rear. The vehicle with this feature was designated the 3-inch GMC T35E1. It was this version of the vehicle that was standardized in May 1942 with a number of modifications ordered.

The most noticeable external change on the T35E1 was the replacement of the original open-topped cast-armor turret with an open-topped welded-armor turret,

which offered superior ballistic protection. In another weight-saving measure, the welded armor glacis on the T35E1 was only 1.5 inches (38mm) thick compared with the 2-inch (51mm) welded armor glacis on the M4A2 and the T35. The thickest armor on the T35E1 was the cast-armor gun shield that was 2.25 inches (57mm) thick.

Concerns about the reduction of armor thickness on the T35E1 led to the adoption of bosses on the turret and upper hull of the vehicle for the mounting of auxiliary welded armor plates if the need arose. There was never an official, factory-designed and built, auxiliary armor kit designed or made for the T35E1. On later series production examples of the vehicle, the bosses were retained only on the glacis. Some units added their own improvised armor arrangements to their vehicle in a belated effort to improve their protection from enemy antitank weapons.

Following on the heels of the unsuccessful 3-inch gun motor carriage T24, the 3-inch gun motor carriage T35 and T35E1, based on the chassis of the M4A2 medium tank, appeared in April 1942. The big advantage with the T35 seen here, and the T35E1, was the mounting of the 3-inch gun in an open-topped turret arrangement with 360 degrees of traverse. (Ordnance Museum)

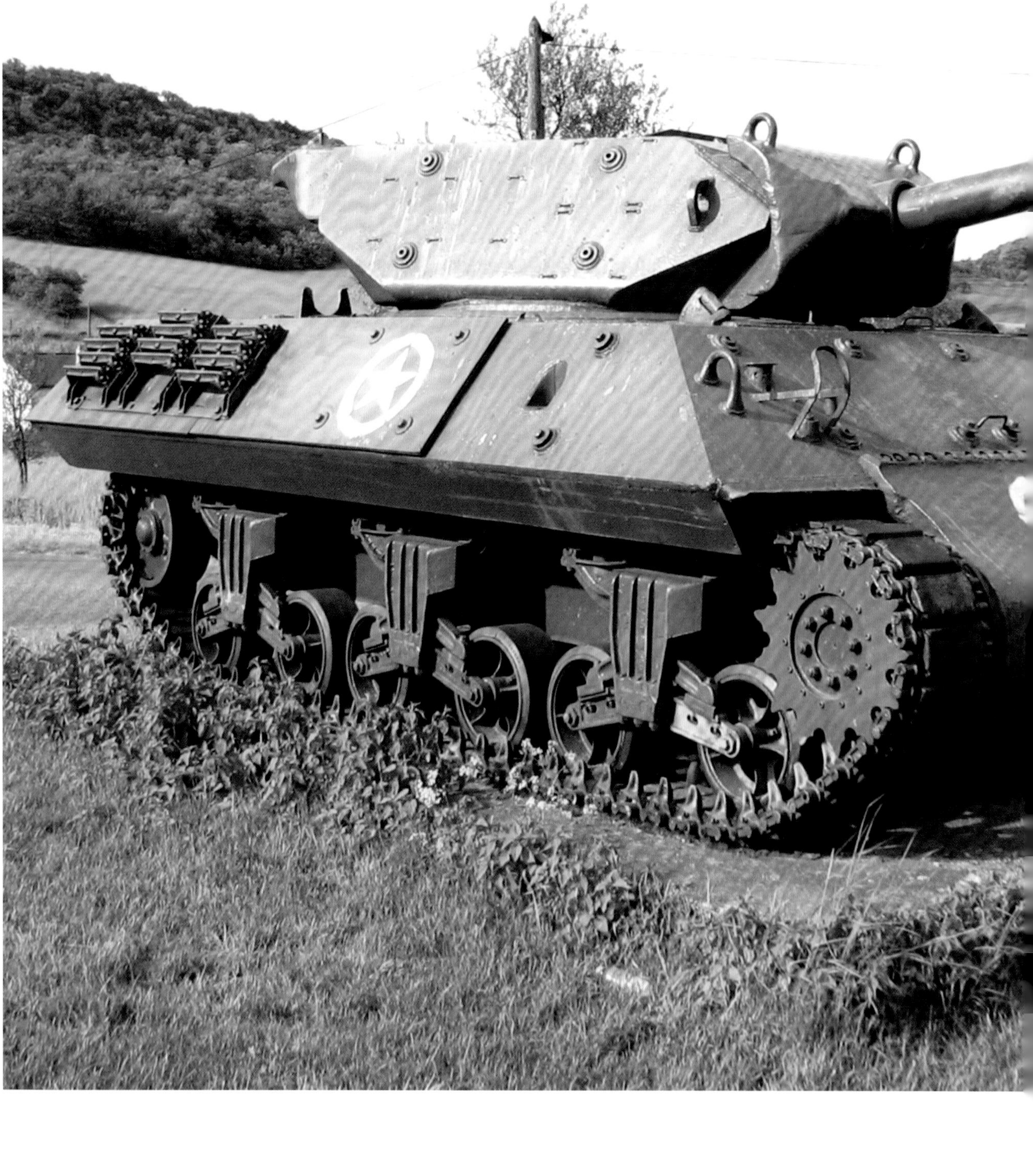

A NEW TANK DESTROYER APPEARS

The T35E1 was designated the 3-inch GMC M10 in June 1942. At this time, the T35 project was canceled. Series production of the 32.6-ton M10 began at the Fisher Tank Arsenal in September 1942 and lasted until December 1943, with a total of 4,993 units assembled. So pressing was the perceived need for tank destroyers at this time it was assigned a higher priority in production than the M4 series tanks.

Bruce was in favor of the 3-inch gun M7, but not the vehicle in which the Ordnance Department wanted to mount it. This can be seen in a letter he wrote to Brigadier-General W. B. Plamer of the Armored Force on January 26, 1942, in which he stated that the vehicle "weighs too much and is too slow." Bruce felt that the M10 was another expedient vehicle forced upon him by the Ordnance Department and McNair, and would delay the fielding of the ideal tank destroyer. The Armored Force agreed with him and felt the M10 should not be built as it offered little in capabilities over the M4A2 besides being just a little bit lighter and faster.

With the Ordnance Department worried about having a sufficient number of the M4A2 chassis to modify into the M10, it was decided that another version of the vehicle be built based on the M4A3 tank chassis. That version of the M10 was designated the 3-inch GMC M10A1, and the Ford Motor Company and the Fisher Tank Arsenal constructed 1,713 units between October 1942 and December 1943, with the last 300 completed without turrets. It turned out that there were more than enough M10s built to satisfy the overseas needs of the Tank Destroyer Force and the M10A1s were retained in the United States for training duties.

In June 1942 it was decided that the 3-inch gun motor carriage T35E1 best met the U.S. Army's requirements for a new tank destroyer, and it was standardized as the 3-inch gun motor carriage M10. Those tank destroyers based on the modified chassis of the M4A3 medium tank were designated the M10A1. Pictured is a slightly weatherworn M10 on display in France as a monument vehicle. (Pierre-Olivier Buan)

Belonging to the French Tank Museum at Saumur is this restored and running 3-inch gun motor carriage M10 in the markings of the Free French Army during World War II. The rounded bosses seen on the sides of the vehicle's turret and hull were intended for add-on armor plates that were never manufactured. (Christophe Vallier)

M10 FIREPOWER

The M10 saw service not just in North Africa, but also in the ETO and the PTO. Comments on the vehicle's armor and armament arrangement appeared in an article titled "Brassing off Kraut" by Major Edward A. Raymond, in the October 1944 issue of *The Field Artillery Journal*:

> Use of TDs to take advantage of their armor alone is unconventional, since the superstructure is open and made of armor in places only 1½ inches thick. M10s have an equally obvious weakness in armament, when it comes to fighting against infantry infiltration. The M10 turret must be rotated to permit forward fire of the .50 cal. machine gun and even then it cannot be depressed enough to fire at infantry close-in. There is no flexible machine gun over the assistant driver's seat as in a tank, nor is there one coaxially mounted with the [gun] tube. There are not even fighting slits in the turret.

Some M10 units added additional machine guns to their vehicles to make up for the lack of a coaxially mounted machine gun. The AAR of the 640th Tank Destroyer Battalion that fought in the Philippines from January through March 1945, describes this arrangement:

> In addition to the .50 caliber machine gun normally mounted on the turrets of the M10s a provisional .30 caliber machine gun was placed on the turret immediately in front of the gun commander where it could be employed by him against foot troops or to mark targets. By using tracer ammunition the target can be marked for 3-inch fire thus conserving ammunition. With the .50 caliber mounted in the rear and the .30 caliber in front it gave the crew all around protection from enemy ground troops. In the Luzon campaign this arrangement proved highly successful.

The same AAR also described the effectiveness of the 3-inch gun on the M10 in combat:

> As no threat of enemy armored attack developed the firing companies remained attached each to an infantry regiment for direct fire support to the ground troops. The principal targets attacked were caves, pillboxes, and machine gun nests. The 3-inch gun proved to be very effective against these targets and the companies destroyed many. Often the M10s were used for direct fire against Jap artillery of various calibers, destroying many. However, often this heavy armor was employed for missions normal to infantry weapons. At times the M10 tank destroyers were used continuously while the M7s [105mm self-propelled howitzers] and 57mm antitank guns of the infantry were idle.

The open-topped turret of the M10 made it extremely vulnerable to Japanese close-in antitank tactics, and infantry had to be assigned to the vehicles for added protection when engaged in combat operations.

As seen in this quote from the March 1945 U.S. Army report *United States vs. German Equipment*, by Technical 5 Ernest B. Forster, the 3-inch gun M7 on the M10 was not up to tackling late-war German tanks: "While at Amperveiler I saw three dug in tank destroyers with three inch naval guns open fire on two Mark Vs [Panthers] at a range of 800 yards, resulting in two ricochets on the German tanks and two tank destroyers knocked out, the third one withdrawing. The tank destroyer men held the element of surprise but naval guns are not capable of knocking out the Mark V."

DESIGN ISSUE

Due to the haste in which the M10/M10A1 was rushed into service, a number of design issues remained unresolved. These had to be corrected on the production lines or in the field.

One of the more important issues that had to be tackled was the heavy weight of the 3-inch gun M7 that seriously unbalanced the turret of the vehicle, especially on slopes. To address this problem a number of solutions were tried. The first attempt involved moving the vehicle's track grouser stowage to the upper rear turret plate, which did not work. Next was something that was referred to as the "quick-fix" counterweight that could be either fitted to the upper rear turret plate at the Fisher Tank Arsenal, or improvised in the field based on factory drawings. Some M10 units made counterweights to their own designs.

The quick-fix involved two identical counterweights, weighing 2,400 pounds together, and made using lead, cast iron, or steel. The quick-fix counterweights were all externally identical except for the lead version that was not as thick as the others, and also had a pair of steel bands attached to each of the two counterweights for added strength. They were designed to be attached to the existing track grouser racks and then bolted to the upper rear turret plate.

The M10 3-inch GMC pictured belonged to the 776th Tank Destroyer Battalion, in North Africa in March 1943. As the vehicle's base olive drab proved to be too dark for the light-colored terrain in which it was serving, it tended to stand out. This led to the vehicle crews applying a wash of the local mud to their tank destroyers to lighten the shade of the olive drab. The application of the local mud has partially obscured the national insignia done in yellow of the vehicle shown. Also partly obscured is the vehicle's nickname done in white chalk, and a tactical marking, also done in white. There is no registration number visible on the vehicle. As an early-production vehicle the M10 pictured has the quick-fix counterweights located on the rear of the turret. (© Osprey Publishing Ltd.)

Because the Fisher Tank Arsenal quick-fix counterweights didn't solve the problem of the M10's unbalanced turret, their engineers went back to the drawing board and came up with another solution that involved two larger, cast-iron, wedge-shaped counterweights with a combined weight of 3,700 pounds that attached to the upper rear turret plate. Unfortunately, this still did not solve the problem, and the Fisher Tank Arsenal folks came up with a final counterweight design that weighed 2,500 pounds and was attached to a redesigned upper rear turret plate. This final counterweight design was called "duck-bill" due to its shape.

Despite the relatively poor showing of the U.S. Army's tank destroyers during the fighting in North Africa, there had been a huge commitment made in men and equipment to the concept, which resulted in the continued employment of them for the remainder of World War II. Pictured is an M10 3-inch GMC belonging to the 636th Tank Destroyer Battalion, at Salerno, Italy, in September 1943. This particular vehicle was commanded by Sergeant Edwin Yosts, which succeeded in knocking out five German medium tanks, a German half-track, and a pillbox in the span of 25 minutes during an enemy counterattack on September 14, 1943. The base olive drab has been over-painted with irregular patches of black. The national symbol in white is surrounded by a segmented white circle. (© Osprey Publishing Ltd.)

The British Army was not inspired by the penetrative abilities of the main gun on the 3-inch gun motor carriage M10 supplied to them under Lend-Lease. They had the original American-supplied main gun on many of their late-model M10s replaced with the more powerful 17-pounder Mk II gun as seen here, and designated either the M10 17-pounder or M10C. (Pierre-Olivier Buan)

FOREIGN SERVICE

Of the 4,993 units of the M10 built, a total of 1,648 went to the British Army under Lend-Lease. Those received with the Fisher Tank Arsenal-designed and built wedge-shaped counterweights on the rear upper plate of the turret were referred to as 3-inch self-propelled mount (SPM) M10 Mk I by the British Army. The M10s shipped to England that had the larger duck-bill counterweights were designated the 3-inch SPM M10 Mk II.

The British Army was not impressed with the penetrative powers of the 3-inch gun M7 mounted in the M10. They decided to re-arm later production units of the M10, mostly the 3-inch SPM Mk II, with the more powerful, British-designed and built 17-pounder (76.2mm) Mk V gun, a modified version of the 17-pounder Mk II gun already in service.

The up-arming process of the M10s began in May 1944 and continued until April 1945, with 1,017 units completed. Some additional units may have been constructed in field workshops. These re-armed M10s would be referred to by the British Army as the "M10C" or the "M10 17-pounder." Some of the M10s and up-armed M10Cs would go on to serve with various British Commonwealth forces, such as the Canadian, New Zealand, and South African armies. The British Army would also supply the M10 and M10C to Polish Forces fighting for the Western Allies.

Despite the widespread belief that the M10Cs were officially nicknamed "Achilles," this was not the case and does not show up in British Army documents from World War II.

As the gun mount of the M10 was designed from the beginning to accept other weapons, it required only two additional lugs to mount the 17-pounder Mk V gun in the vehicle's turret. Because the barrel of the 17-pounder Mk V was thinner than the barrel of the 3-inch gun M7, it was necessary to come up with a special casting that was welded over the M10's original gun shield to reduce the opening to the correct size to fit the smaller diameter 17-pounder Mk V gun. To help balance the long gun barrel, a counterweight was added just behind the muzzle brake.

Besides the British Army, the Red Army received 52 M10s under Lend-Lease, with another 155 units of the M10 going to the Free French Army under Lend-Lease. The Free French Army received another 100 or so M10s from U.S. Army stockpiles when attached to the Sixth U.S. Army Group in 1944/1945.

THE IDEAL TANK DESTROYER

Bruce finally got his ideal tank destroyer in July 1943 when the first 76mm GMC T70 rolled off the production line of the Buick Motor Car Division of the General Motors Corporation. The vehicle was standardized in March 1944 as the 76mm GMC M18. It had a cast-armor turret and a welded-armor hull. The thickest armor on the M18 was 1 inch (25mm) on the cast-armor gun shield. This thin armor remained a major concern to the crews of the vehicle throughout World War II. Additionally, in order

The planned replacement for the M10 3-inch GMC was the T70 76mm GMC standardized in March 1944 as the M18 76mm GMC. The vehicle shown belongs to the 894th Tank Destroyer Battalion at the Anzio, Italy, beachhead in May 1944. Its base olive drab has been over-painted with black splotches. National symbols in white, surrounded by a segmented white circle, have been applied to both the hull and turret. This was done to aid in long-range identification. Combat experience had shown that at a distance, the national symbol minus the segmented circle could be mistaken for the white Balkan crosses sometimes seen on German tanks. (© Osprey Publishing Ltd.)

to keep the height of the vehicle to a minimum, yet still allow depression of the gun, the turret was completely open-topped, except for a canvas "convertible top," which would be fitted in inclement weather.

The 19.5-ton M18 was the first vehicle in the U.S. Army inventory to ride on the Ordnance Department-developed torsion bar suspension system. The combination of light weight and the torsion bar suspension, plus a gasoline-powered, air-cooled radial engine that produced 400hp at 2,400rpm, meant the M18 had a maximum speed on level roads of 60mph, making it the fastest tracked vehicle of World War II.

The name "Hellcat," often used to describe the M18, was never an official military designation and does not show up in U.S. Army documents from the period. It was merely a public relations nickname thought up for the vehicle by the builder.

The main armament of the five-man M18 was one of three different versions of the 76mm gun, designated the M1A1, M1A1C, and M1A2. These were the same series of main guns mounted in the second generation of M4 series tanks. There was no secondary armament on the vehicle except a .50 caliber machine gun fitted to a ring mount on the top of the open-topped turret.

Pictured is an M18 76mm GMC, of the 805th Tank Destroyer Battalion, in Northern Italy in April 1945. The vehicle is in the base olive drab, with the national symbol in white on the turret, surrounded by a segmented white circle. The explanation for the segmented circle, rather than a continuous circle, can be attributed to the stencil cutouts. Obviously, only those portions of the stencils that were cut out were painted in. Also visible on the vehicle shown are the tactical numbers painted onto a white background. In addition, the Tank Destroyer Center's emblem, a black panther crushing a tank in its jaws on a circular orange background, appears on the vehicle's hull.
(© Osprey Publishing Ltd.)

This picture is looking into the open-topped turret of an M10 17-pounder belonging to the Military Vehicle Technology Foundation. Missing of course are the main gun rounds stored around the inside turret walls as is indicated by the various brackets assemblies. Visible are the vehicle commander's and gunner's seats. (Chris Hughes)

Lieutenant-General McNair, ever the diehard tank destroyer enthusiast, expressed some gratification with the fielding of the M18 in a letter from him to Bruce dated October 25, 1943, which is paraphrased in this passage from the 1946 U.S. Army report *The Tank Destroyer History Study No. 29*: "He [McNair] further expressed confidence in the T70 tank destroyer as bidding fair to becoming an outstanding weapon of the self-propelled type and that for the first time we had weapons which were suited for tank destroyer purposes, and that they had inflicted serious damage to German armor."

Because of the rush to put the M18 into production, a host of design faults quickly became evident. Most of these initial defects were addressed on the production line; however, additional problems surfaced during testing. It became such a serious issue that the Ordnance Committee recommended that the first 684 units of the M18 built be returned to the factory for upgrading. In the end, 640 of these early-production M18s returned to Buick for upgrading were converted into the open-topped armored utility vehicle M39.

The M39 was originally envisioned as having many different roles: a reconnaissance vehicle to replace the armored utility car M20, prime mover for towed antitank guns, or as a replacement for the armored personnel carrier M3/M3A1. It was armed with a single .50 caliber machine gun in a ring mount located at the front of the vehicle's open-topped crew compartment. As a personnel carrier, it could carry eight passengers along with its three-man crew. The M39 might have seen some service in the last couple of months of the war in Europe; however, there is no pictorial evidence of this.

Like the other tank destroyers before it, the open-topped design of the M18 and lack of machine guns meant that it was very vulnerable to aggressive Japanese antitank tactics, which often involved close-in attacks by infantrymen equipped with explosives.

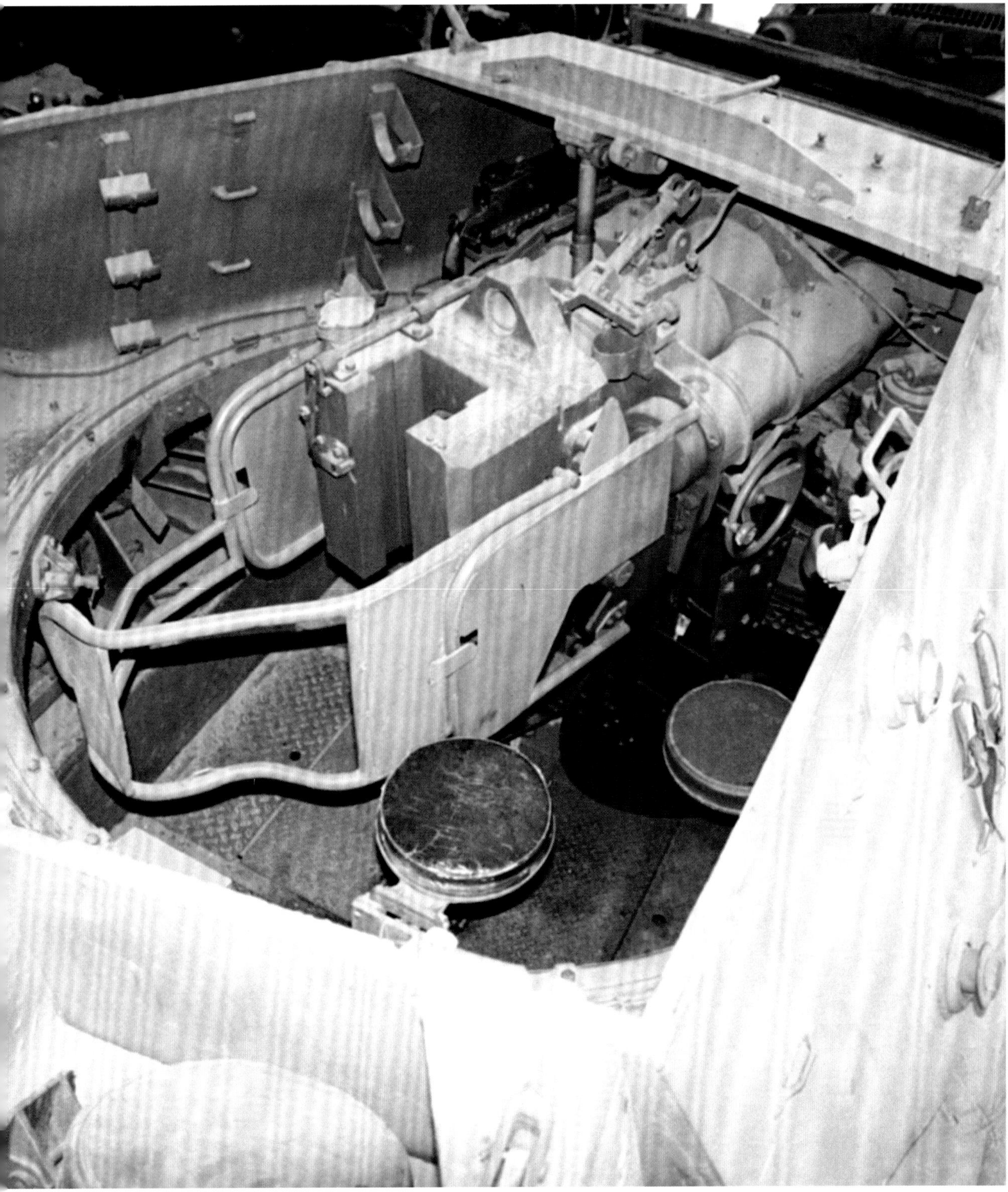

Among the many armored fighting vehicles restored by the talented crew of vehicle restoration specialists (both paid and volunteer) at the Virginia Museum of Military Vehicles is this 76mm gun motor carriage M18. Before being standardized in March 1944 it was designated the 76mm gun motor carriage T70. (Michael Green)

OPPOSITE

A bit dirty and wet, this 76mm gun motor carriage M18 is taking part in a World War II reenactment event. The vehicle was 21 feet 10 inches in length with its gun pointed forward. It had a width of 11 feet 1 inch and a height over its turret roof-mounted antiaircraft machine gun of 10 feet 5 inches. (Bob Fleming)

INTO SERVICE

Originally, it was intended that almost 9,000 M18s would be built, with 1,600 meant for Lend-Lease. That number was never met, as the U.S. Army's senior leadership favored towed antitank guns as "tank destroyers" in the summer of 1943, and potential Lend-Lease customers had expressed no interest in acquiring the vehicle.

It had also become painfully clear to all by summer 1944 that the 76mm main gun on the M18 was not up to dealing with late-war German tanks. As a result, only 2,507 units of the vehicle were ever completed. The War Department document *Report of the New Weapon Board*, dated April 27, 1944, describes the reaction of tank destroyer personnel in Italy to the shipping of the M18 to their theater of operations in the spring of 1944:

> Reviewing officers felt that the introduction of these vehicles into the theater should be with troops trained on them in the zone of the interior [the United States] and in no case should the M18s be issued as a replacement for the M10s… It was noted by several officers that the silhouette and suspension of this tank destroyer vehicle are very similar to a number of the German self-propelled mounts, and a complete recognition course on the M18 would be necessary for units already in the field [to prevent friendly fire incidents].

Not everybody disliked the M18. According to a Third Army AAR dated May 9, 1945, which addressed the vehicle's mobility, the M18 was "the finest piece of tracked equipment in the U.S. Army." First Army was more concerned with protection than Third Army and hence rated the M18 much lower in preference.

M18s in action

Despite their firepower disadvantage against late-war German tanks, the M18 occasionally proved its usefulness as is recounted in a U.S. Army research report dated 1949–50 and titled *Employment of Four Tank Destroyer Battalions in the ETO*. The report described a brief encounter around Arracourt, France in mid-September 1944, when the U.S. Army's 4th Armored Division, supported by the 704th Tank Destroyer Battalion, armed with the M18, encountered some tanks of the German Fifth Panzer Army.

As the hill was approached, Lt. Leiper, who was still in front with his jeep, was startled to see the muzzle of a German tank gun sticking out through the trees at what seemed to be less than 30 feet away! He immediately gave the dispersal signal and the many months of continuous practice proved its worthiness as the platoon promptly deployed with perfect accord.

The lead tank destroyer, commanded by Sgt. Stacey, had evidently seen the German tank at the same time as Lt. Leiper, and opened fire immediately. Its first round scored a direct hit, exploding the German tank. The flames of the burning tank revealed others behind it in a V-formation and Sgt. Stacey's next round hit a second German tank, but immediately afterwards he had his own tank destroyer knocked out by fire from a third German tank. This enemy Mark IV was taken under fire by the No. 2 tank destroyer, and was destroyed. The maneuver and fire of the 3rd tank destroyer got another German tank as it tried to back out of the unhealthy situation and a fifth enemy tank was destroyed almost immediately thereafter.

The entire affair was over in a matter of minutes, and as soon as the shooting had stopped, Lt. Leiper ordered the platoon to the area to make sure the enemy tanks were all out of action and to be certain that there were no more there.

The box score for the short action stood at five German tanks destroyed, and one tank destroyer knocked out of action. The tank destroyer had been hit on an angle along the base of its gun barrel and through the gun shield. The ricocheting round had bounced around the interior of the tank.

It was clear to some by 1942 that the introduction of more heavily armored tanks by the German military rendered the main gun in the 3-inch gun motor carriage M10 obsolete. In response, work began on the development of a more potent tank destroyer, seen here, that was designated the 90mm gun motor carriage M36. It was based on the chassis of the M10A1. (Pierre-Olivier Buan)

THE M10 REPLACEMENT

The anticipation that the German Army would field new, better armored tanks that the 3-inch gun M7 mounted on the M10 would be unable to deal with pushed the Ordnance Department to develop and field another tank destroyer. The original designation for the pilot vehicle was the 90mm GMC T71. It was standardized in July 1944 as the 90mm GMC M36.

The heart of the new 31.5-ton M36 was its powerful 90mm gun M3 fitted in the combination gun mount M4. It was this gun that proved to be the only tank destroyer weapon reliably able to penetrate the thick armor of late-war German tanks at normal combat ranges.

The 90mm gun M3 mounted in the M36 was the same weapon that went into the heavy tank M26. The effectiveness of the 90mm gun can be seen in a First Army report dated September 1944, which stated that an M36 destroyed a Panther tank with a lucky first shot at 3,200 yards.

Lucky shots could not be counted on in battle, as is recounted by Lieutenant Colonel John A. Beal, commander of the 702nd Tank Destroyer Battalion, in *United States vs. German Equipment*: "the M36 90mm gun with present ammunition will not penetrate the front slope [glacis] of a Mark V [Panther] at greater than 800 yards range as shown by the repeated ricochets in the Puffendorf-Ederen battle." In the same report, he was asked if the M36 could meet the Panther tank on equal terms. His answer was no, due to the German tank's advantages in armor, firepower, and sighting equipment. He went on to say that the M36 could only take on a Panther tank by striking through stealth and prepared ambushes and never in a head-on duel.

Lieutenant-Colonel Beal goes on in the same report to highlight the armor protection disadvantage suffered by the crews of his M36s: "Upon interviewing all of my gun commanders and from my own personal experiences, we have all agreed that our frontal armor plate on the M36 is far too light. The Mark V [Panther] and Mark VI [Tiger] tanks can safely stand off from 1,500 to 1,600 yards and knock out our M36s at will, achieving penetrations in any part of the hull or turret."

Major Philip C. Calhoun of the 3rd Battalion, 66th Armored Regiment, commented in the same report: "The 88 and 75 on the Mark V [sic] are superior to the 90mm, partly because of higher velocity and flatter trajectory, this making it more possible for them to hit what they point at; and partly because of the muzzle brake which we have seen on both the Mark V and Mark VI, thus allowing them to observe their fire better than our 90mm gunners or destroyer commanders can."

The Armored Board had recommended that a muzzle brake and long primer ammunition be provided for the 90mm main gun on the M36 to reduce the effects of target obscuration due to the flash and muzzle blast when fired. However, few M36s ever had the muzzle brake installed until very late in the production run.

During the development cycle of what became the 90mm GMC M36, it was decided to do away with the troublesome turret design of the M10 and come up with a new, open-topped, rounded cast-armor turret for the vehicle that looked very different from the angled construction of the welded-armor turret of the M10/M10A1. Like the turret on the M10, the turret of the M36 came with a rear turret counterweight. However, the counterweight on the M36 was an incorporated part of the vehicle's cast-armor turret design, and not an add-on feature as the counterweights developed for the M10/M10A1.

Rather than build brand new M36s, the Fisher Tank Arsenal was requested in November 1943 to convert 500 of the M10A1s it was then building into M36s. Because the process of building the M10A1 was so far along on many of the vehicles, only 300 units of the M10A1 chassis were available for conversion into the M36. These units were converted to the M36 configuration between April and July 1944.

Due to demand for additional M36s, it was decided to return 500 M10A1s already in the field or in ordnance depots to the factory for conversion into M36s. As the Fisher Tank Arsenal was working at maximum capacity at the time, conversion work went to the Massey Harris Company, which was supplied the turrets for the M36s from the Fisher Tank Arsenal. These units were completed between June and December 1944.

After the invasion of France in June 1944, the painful realization hit the Army Ground Forces that none of their existing tanks or tank destroyers could be counted on to destroy late-war German tanks and a loud cry arose for more M36s. In July 1944, the request went out that all tank destroyer battalions equipped with the M10 be converted to the M36. This was a complete turn-around by almost all concerned, as the Ordnance Department had developed the M36 on its own initiative. The Tank Destroyer Center felt that the M36 was just another expedient vehicle being forced on them like the M10, diverting resources away from their beloved M18.

The first M36s reached Western Europe in August 1944. To meet the demand for even more M36s, the American Locomotive Company was contracted to convert another 413 M10A1s into M36s, which they accomplished between October and December 1944. To build up the M36 numbers, the Fisher Tank Arsenal installed the turret of the M36 on the modified hulls of 187 second-generation M4A3s. The vehicle was then designated as the 90mm GMC M36B1 and classified as substitute standard in October 1944.

Even as the war in Europe was entering its closing stages, there remained a call for more M36s. To meet this demand, in May 1945 the Montreal Locomotive Works in Canada began to convert 200 M10A1s into M36s. As the supply of M10A1s was exhausted, it was decided to convert M10s into M36s. Vehicles so converted were designated the M36B2. The American Locomotive Company undertook this project beginning in May 1945 and produced 672 of the M36B2 by the end of 1945.

Another 52 M10s were also converted into M36B2s by the Montreal Locomotive Works by the end of 1945, bringing the total number of M36s built to 2,324 units. The vehicles were classified as substitute standard in March 1945.

The main gun on the 90mm gun motor carriage M36 was the only U.S. Army vehicle-mounted weapon that had the penetrative power to deal with late-war German medium and heavy tanks with some degree of success. The M36 had authorized stowage for 47 main gun rounds. (Christophe Vallier)

INTO COMBAT FOR THE TANK DESTROYER FORCES

The first combat test of the Tank Destroyer Force took place in the Far East in December 1941, through January 1942 when the Japanese military invaded the Philippines and encountered GMC M3s. Of the 50 T12s sent to defend the island chain before America's official entrance into World War II, 48 of them were divided into three provisional self-propelled artillery battalions. They took part in the fighting for the main island of Luzon, and eventually withdrew to the Bataan Peninsula.

In August 1942, 12 Marine Corps GMC M3s took part in the fighting for the island of Guadalcanal. A description of one of the combat actions they engaged in comes from an article prepared by Brigadier-General H. T. Mayberry, Commandant of the Tank Destroyer School, for the December 1943 issue of the *Military Review* magazine:

> Despite the poor terrain conditions for armored warfare on Guadalcanal, the United States Marines used the M3 tank destroyer half-tracks on various missions in that theater. Under orders to support defense of the Matanikau River against a Japanese counterattack, the Marines moved their M3s into firing positions. Twelve Japanese tanks led the assault across a sand-spit at the mouth of the river just at dusk. At a range of less than a hundred yards, the self-propelled 75s destroyed ten of the tanks. The leading tank tried to maneuver and escape the devastating fire. It ran off the sand spit and into the river, disappearing beneath the surface of the water. The last tank succeeded in crossing the spit but was destroyed while trying to make its escape down the beach.

The true test of the validity of the GMC M3, the M10, and the entire tank destroyer doctrine took place in North Africa in early 1943 and all were found wanting. This was mostly because the tank destroyer units in that theater were poorly used. They were widely spread out and making it nearly impossible to mass them together according to their training and doctrine.

Neither did the German Army mass their tanks together in North Africa for offensive operations except on a couple of occasions. On one of those occasions, in March 1943, the intact 601st Tank Destroyer Battalion equipped with the GMC M3 supported by a company of M10s from the 899th Tank Destroyer Battalion managed to blunt an attack by the 10th Panzer Division and were reported to have destroyed 30 German tanks for a loss to themselves of 35 tank destroyers.

Part of the problem with self-propelled tank destroyers employed in North Africa taking such high losses was summed up by the Historical Section of the Army Ground Forces in *Antitank Defense Weapons and Doctrine*, a U.S. Army research report dated 1951–52: "The trouble in North Africa was that the tank destroyers, instead of firing from concealed positions, maneuvered too freely during combat. Instead of being aggressive in their reconnaissance and preparatory dispositions, they were aggressive in the face of tanks themselves, and suffered severe casualties because of their virtual lack of armor."

Adding to the problem with the self-propelled tank destroyers was the fact that senior commanders were unfamiliar with the approved doctrine for employing these vehicles and were loath to leave these vehicles massed behind the rear lines for a German tank offensive that may or may not occur. As a result, the self-propelled tank destroyers were dispersed among divisions for added firepower throughout the front lines, similar to the "penny packet" concept that proved so disastrous for British and French armored units fighting the Germans in France in 1940.

For this and other reasons, the self-propelled tank destroyers in North Africa did little to impress the senior commanders in theater. Generals George S. Patton, Omar N. Bradley, and Lloyd R. Fredendall all expressed their dislike of the offensively oriented concept of self-propelled tank destroyers. According to *Notes on Combat Experience during the Tunisian and African Campaigns*, a U.S. Army report issued in late 1944 by Major-General E. N. Harmon:

> There is no need for tank destroyers. I believe the whole organization and development of the tank destroyer will be considered a great mistake of the war. In the first place, the doctrine originally promulgated, of the tank destroyer seeking out and pursing the tank was a fallacy which caused the destruction of many lives and much equipment before it was corrected. The tank destroyer M10, now in use, has proven of great benefit simply because it contained a 3-inch gun that was the best gun for coping with the Tiger and Panther tanks. Had this gun or a more powerful gun been installed in a tank, there would have been no need for the tank destroyer.

Major-General John P. Lucas stated in a report dated September 8, 1943, after witnessing the fighting in Sicily (July–August 1943): "the tank destroyer has, in my opinion failed to prove its usefulness… I believe that the doctrine of an offensive weapon to 'slug it out' with the tank is unsound." Lucas went on to suggest that purely defensive, towed antitank guns made more sense. All of these negative responses to the concept of offensively oriented tank destroyers in North Africa and later in Sicily caused a number of tank destroyer battalions originally issued with the M10 to be replaced with the 2.5 ton, towed 3-inch antitank gun M6, which was standardized in November 1943. Over half the tank destroyer battalions slated for use in the June 1944 invasion of France were towed.

COMBAT IN WESTERN EUROPE

Soon after the invasion of France, the conversion of self-propelled tank destroyers to towed antitank guns was seen as a big mistake, because the towed tank destroyer battalions were almost completely useless in the very restricted terrain of Western France. This turned the U.S. Army's interest back to 100 percent self-propelled tank destroyer battalions by December 1944.

However, fighting in Western France clearly demonstrated that the 3-inch gun M7 in the M10 tank destroyer – like the 75mm main gun in the first generation of the M4 series or the 76mm main gun in the second generation of M4 series tanks – lacked the penetrative powers to destroy the Panther and Tiger tanks they were encountering on the fields of battle. Fortunately for the M4 and M10 crews, the German Army never fielded that many Panther and Tiger tanks in Western Europe, except on a handful of occasions.

U.S. Army tank destroyer units fighting in Italy never saw the massed, tank heavy, offensive operations for which they were intended. What the typical American tanker or tank destroyer crewman faced was just the opposite extreme, as is attested to in this passage by Lieutenant-Colonel P. H. Perkins, tank battalion commander in

As the supply of M10A1 hulls was exhausted for conversion into the 90mm gun motor carriage M36, it was decided to use the unmodified hulls of M4A3(75)W medium tanks as a platform upon which to mount the 90mm main armed turret of the M36, as seen in this wartime photograph. In this configuration the vehicle was designated the 90mm gun motor carriage M36B1. (Patton Museum)

Postwar, some units equipped with the 90mm gun motor carriage M36 featured the new M31 90mm main gun that had a muzzle brake and bore evacuator as seen on the vehicle pictured. When fitted with a muzzle brake, and with the main gun pointed forward, the vehicle was 24 feet 5 inches in length. Vehicle width was 10 feet. (Bob Fleming)

Italy, which appeared in an undated wartime U.S. Army publication referred to as *Combat Lessons Number 1*: "The standard German attack here consists of three or four tanks in line in the lead. They are followed by infantry in trucks at four to five hundred yards. The rest of the tanks follow the infantry. When fire is drawn the infantry dismounts. The leading tanks mill about, fire and withdraw. We have never seen the reserve tanks committed."

As in North Africa and Italy, senior U.S. Army commanders in Western France refused to keep the self-propelled tank destroyer battalions behind the lines in pooled reserves to react to any German tank offensive. Like in North Africa, they often pushed the self-propelled tank destroyers into the frontlines where their added firepower was often welcomed by the infantry units they assisted, as is seen in this quote from a tank destroyer officer from a document dated April 26, 1946, titled *Report on Study of Organization, Equipment, and Tactical Employment of Tank Destroyer*

A picture of the only restored armored utility vehicle M39 in the United States. It was originally intended that the vehicle be used to tow antitank guns. Following World War II, the towed antitank guns disappeared from U.S. military service, and the M39 saw use as an armored personnel carrier among other roles. (Patton Museum)

Units: "The appearance and knowledge that self-propelled tank destroyers were at hand was a major reason that the infantry attained success and victory… The towed guns can be just as brave and trained but they never give much 'oomph' to the fighting doughboys when the 'chips are really down.'"

However, this infantry support role came at a price for the crews of the self-propelled tank destroyers. Pushed into the frontlines in roles they were never designed for, they quickly attracted heavy enemy fire with which their thin armor and open-topped design couldn't cope. According to a U.S. Army Ground Forces report dated October 30, 1944, some infantry commanders preferred to employ tanks rather than the open-topped tank destroyers, because they were better protected from sniper fire and hand grenades. Enemy artillery and mortar fire were also serious dangers to the turret crewmen of the self-propelled tank destroyers.

The AAR of the 808th Tank Destroyer Battalion, dated May 1–4, 1945, recounted the trials and tribulations that their M36s were subjected to when supporting infantry in an urban environment:

The encircling [tank] destroyer proceeded about halfway to his new position under heavy sniper fire and machine gun fire, than it was charged by six enemy rocket grenadiers. In attempting to back into position from which to ward off the attack, the destroyer driver dropped the vehicle into a large shell crater, temporarily immobilizing it. The assistant driver and loader left the destroyer and made their way back to the tank destroyer platoon

Throughout World War II, the Ordnance Department conducted numerous experiments with its vehicles by mixing and matching equipment from different vehicles to see what might show some promise. Here we see the turret from a 90mm gun motor carriage M36 mounted on a 76mm gun motor carriage M18 chassis. (Ordnance Museum)

leader, to whom they reported the incident. The tank destroyer platoon leader ran under heavy enemy fire to the shell crater to determine the extent of the damage. Before he arrived, however, the gunner had fought off the grenadier attack, killing six of the enemy; firing both caliber .50 machine gun and 90mm HE at point blank ranges; and the gunner, destroyer commander, and the driver had managed to extricate the destroyer from the shell crater and proceed to their destination to wait in readiness to attack the enemy tank.

When not supporting the infantry or armored units, the self-propelled tank destroyer often saw service in the indirect-fire role as backup artillery. This first took place during the fighting in North Africa and continued during the fighting in Sicily and Italy. An example of their use in the indirect-fire role comes from the records of the U.S. Army's VIII Corps fighting in Normandy, France in the summer of 1944, which state that 87 percent of the main gun rounds fired by their self-propelled tank destroyers were expended in the indirect-fire support role.

The last big test of the tank destroyer concept took place in December 1944 and ended in January 1945 with the powerful German offensive operation known as the Battle of the Bulge. Despite the large number of German tanks taking part in the operation, estimated to have numbered 1,800 vehicles, the U.S. Army tank destroyer battalions that took part in helping to stop the German offensive were, as in almost all other previous encounters, spread out over a large area and were never massed in sufficient numbers to be employed in the doctrine envisioned for them.

This is not to say that U.S. Army self-propelled tank destroyers did not play an important part in that two-month-long struggle to push the Germans back to their original starting positions. The official U.S. Army history of the operation published by the U.S. Army Center of Military History, titled *The Ardennes: Battle of the Bulge*, by Hugh M. Cole states: "The mobile, tactically agile, self-propelled, armored field artillery and tank destroyers are clearly traceable in the Ardennes fighting as over and over again influencing the course of battle."

Despite the contributions made by the U.S. Army tank destroyer battalions during the Battle of the Bulge, they were quickly disbanded after World War II. The reason why can be summed up in various conclusions reached by the Stillwell Board on May 29, 1945, one of which is quoted here:

In its original concept, the purpose of the tank destroyer was to place upon the battlefield a highly mobile and powerful antitank gun not then available in a tank. Whereas the typically thin-skinned, highly gunned vehicle known as the tank destroyer will always be able to carry more powerful armament for the same overall weight than the corresponding tank, this inherent advantage does not justify the continuation of the development of this class of fighting vehicles in view of the present and future potentialities of tank armament, mobility, and maneuverability. Therefore development of the tank destroyer should be terminated.

CHAPTER SIX

ARMORED CA

ARMORED CAR · T4
ORDNANCE DEPARTMENT
BUILT BY
ROCK ISLAND ARSENAL

RS

K. 818a. 32

In 1928 the U.S. Army began looking for an armored car. The first candidate vehicle was referred to as the T1, and was based on a General Motors Pontiac car chassis, equipped with four wheels, but only two of which were driven – abbreviated as (4x2). Armed with a single, pedestal-mounted .30 caliber machine gun, the only armor on the vehicle was a removable 0.25 inch (6mm) armored plate that replaced the front windshield.

There was also another armored car based on a Pontiac car chassis, also (4x2), known as the T3. Like the T1, it had a removable 0.25-inch (6mm) armored plate that replaced the front windshield. In contrast to the T1, it also had 0.25-inch-thick (6mm) steel armored louvers in front of its radiator to protect the vulnerable cooling system. Earlier iterations of these types of vehicles lacked any armored protection whatsoever.

The light armored car T1 was followed by the medium armored car T2, based on the General Motors (4x2) La Salle car chassis. In the tricky semantics of the time, these vehicles were considered as "cross-country cars" or "scout cars" by the U.S. Army. Unlike the T1 and T3, the T2 had all-around armor protection with a maximum thickness of 0.125 inches (3mm).

The original version of the 2.4-ton T2 featured gun ports on either side of the vehicle with an armored roof that could be raised for better visibility. Subsequent versions of the three-man vehicle appeared with an open-topped, manually operated turret armed with a .30 caliber machine gun. None of the experimental T1, T2, and T3 armored cars/scout cars were standardized by the U.S. Army.

With the poor economic conditions at the time, the U.S. Army's search for a suitable armored car during the interwar period inspired a large number of organizations, and even individuals, both inside and outside the military, to develop a wide range of vehicles that might be suitable for series production. However, in the end, the only

The first standardized armored car for the U.S. Army in the period between World War I and World War II was the T4 pictured here, which was standardized as the M1. Turret armament consisted of a .50 caliber machine gun and a coaxial .30 caliber machine gun. A second .30 caliber machine gun could be mounted on the roof of the turret for antiaircraft purposes. (Patton Museum)

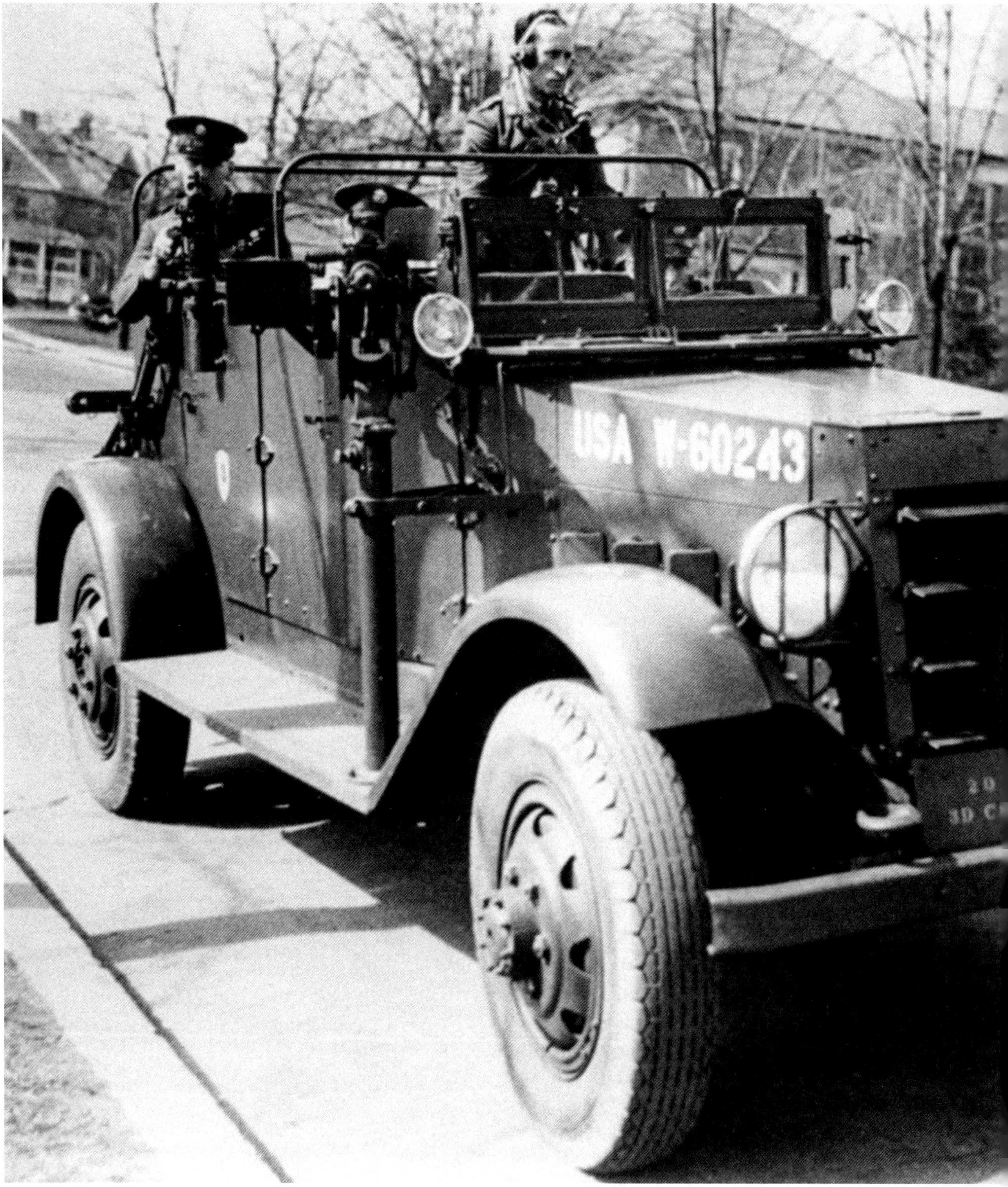
USA W-60243

armored car standardized by the U.S. Army occurred in 1931. It was an Ordnance Department-designed (6x4) vehicle originally referred to as the T4, and later as the armored car M1.

Twenty units of the 5-ton M1 were built by the James Cunningham, Son and Company by 1938. Like the later models of the T2, it had all-around armor protection with a maximum armor thickness of 0.5 inches (13mm). The three-man M1 also featured an enclosed turret with overhead protection armed with a .50 caliber machine gun as its main armament and a coaxial .30 caliber machine gun. There were also provisions for mounting another .30 caliber machine gun on the top of the vehicle's turret for antiaircraft use.

SCOUT CARS

In another confusing change in semantics, the U.S. Army decided in 1937 that it would no longer use the term "armored car," and that all vehicles intended for the reconnaissance role would be labeled as "scout cars." The first of the new scout cars was standardized as the scout car M1, a different vehicle than the armored car M1 described previously. The open-topped M1 was a four-wheel-drive vehicle protected by a maximum armor thickness of 0.5 inches (13mm). Armament on the vehicle consisted of two .50 caliber, and two .30 caliber pedestal-mounted machine guns. The Indiana Motor Truck Company, a subsidiary of the White Motor Company, built 76 units of the 3.8-ton M1 for the U.S. Army between 1934 and 1937.

The scout car M1 was quickly followed in production by another very similar looking vehicle that was standardized as the scout car M2 in 1938. A total of 22 units of the (4x4) M2 were built between 1935 and 1938. Improvements to the vehicle's design resulted in a new designation, the M2A1, which was later changed to M3. One hundred units of the M3 were built by the White Motor Company between 1936 and 1939. The machine-gun armament of the scout car M3 was no longer pedestal mounted, but fitted onto a skate rail that ran around the inside of the vehicle, allowing greater flexibility in delivering fire onto the enemy.

After experimenting in the early 1930s with a number of modified civilian cars for possible use as reconnaissance vehicles, the U.S. Army standardized the scout car M1 seen here in 1934. It was designed from the ground up by the White Motor Company as a military vehicle using as many commercial truck components as possible. (Patton Museum)

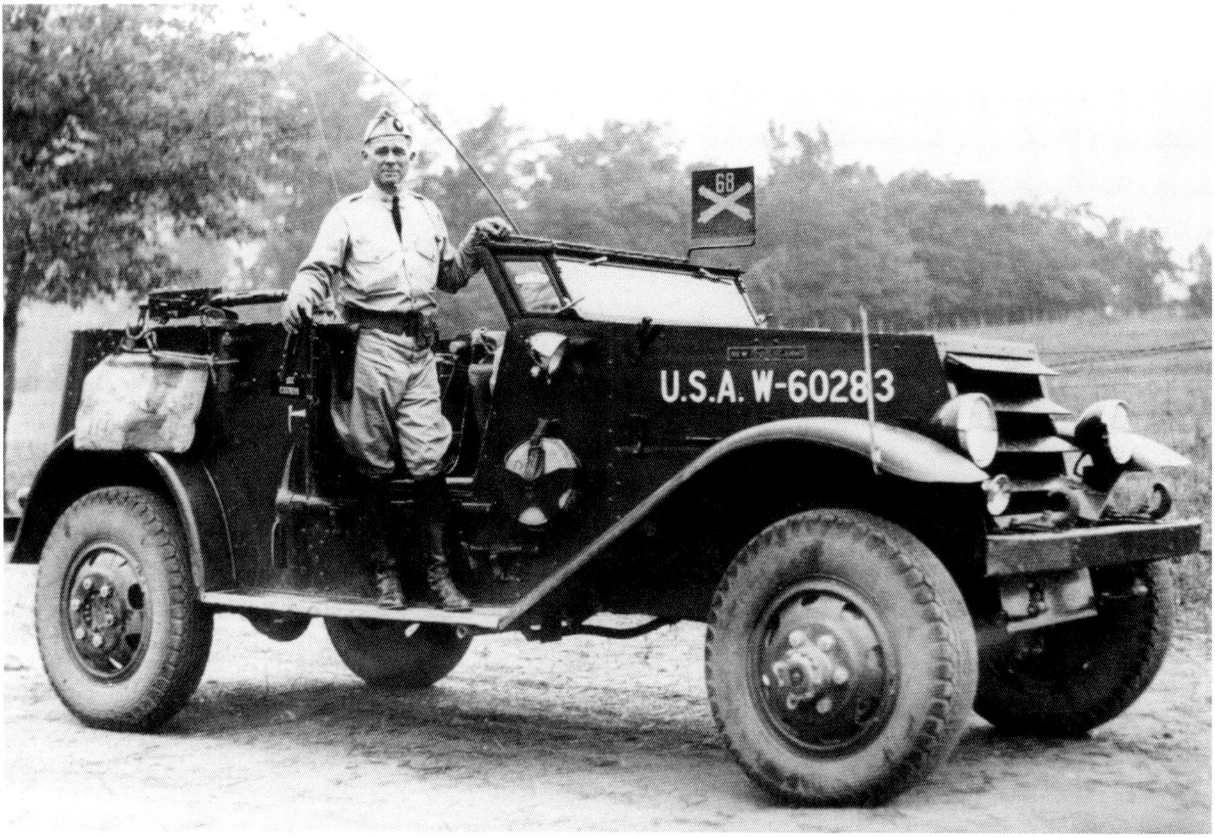
68
U.S.A. W-60283

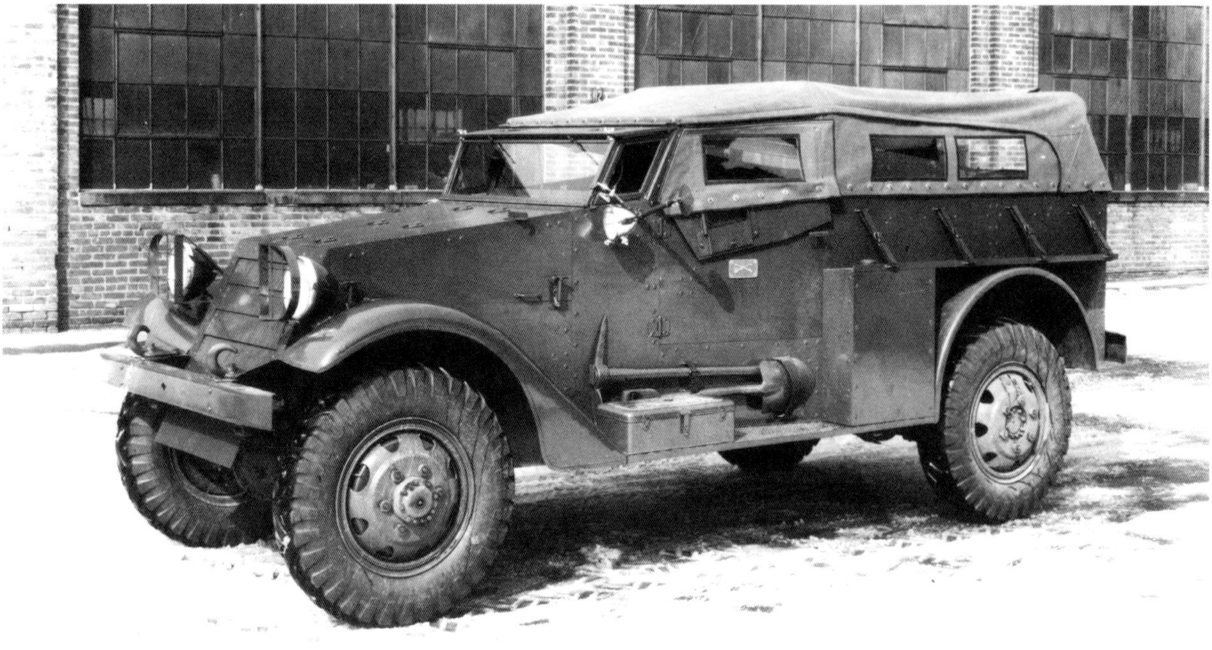

By 1939, an improved version of the scout car M3 appeared that was referred to as the M3A1. The vehicle weighed 4.3 tons and featured a wider body. Unlike its predecessor, there was no rear door and the front bumper was replaced by a spring-loaded un-ditching roller to ease the vehicle in and out of shallow ditches. The armor plate on the M2A1 and the M3/M3A1 was 0.25 inch (6mm) thick FHA and fabricated and mounted on the chassis of the vehicle by the Diebold Safe and Lock Company.

From 1939 to March 1944, 20,894 units of the M3A1 rolled off the factory floor. During the fighting in North Africa, the M3A1 served in U.S. Army armored reconnaissance battalions where their lackluster off-road performance made them extremely unpopular as they could not keep up with the unit's machine-gun-armed quarter-ton jeeps. Their lack of overhead armor protection made them vulnerable to everything from enemy hand grenades to artillery.

Besides service with the U.S. Army in World War II, the Marine Corps ordered hundreds of the scout car M3A1 in early 1942. However, once the Marines' attention turned to the invasion of enemy-occupied islands in the Pacific, it was decided that the quarter-ton jeep was a much more suitable reconnaissance vehicle, and the Marine Corps inventory of M3A1s was discarded.

A total of 11,440 units of the M3A1 were allocated for Lend-Lease. Of that number, 6,987 went to the British Army, 3,310 to the Red Army, and 104 to the Nationalist Chinese Army. In British Army service, the M3A1 was commonly referred to as the "White" after the builder of the vehicle.

British Army testing of the M3A1 found its off-road ability dismal, and the armor protection more for morale purposes than for true physical protection. Despite these

In 1938 the U.S. Army standardized the scout car M2, seen here acting as a command vehicle for an artillery unit as indicated by the pennant with the crossing cannons. As with the very similar-looking scout car M1, it was based on commercial truck components. Maximum speed of the M1 was 50mph. (Patton Museum)

Following on the heels of the scout car M2 appeared the improved scout car M2A1 (later changed to scout car M3), seen here with a canvas top and windshield installed. Both the scout car M2 and M3 were built by the White Motor Company. The M3 had a maximum speed on level paved roads of up to 60mph. (Patton Museum)

Not content with the scout car M3, the Ordnance Department continued to work on improving the design of the vehicle. This resulted in the building of the scout car M3A1 in 1939. To provide more room for the crew and weapons, the rear body of the M3 was widened. Pictured is an M3A1 belonging to the former Military Vehicle Technology Foundation. (Michael Green)

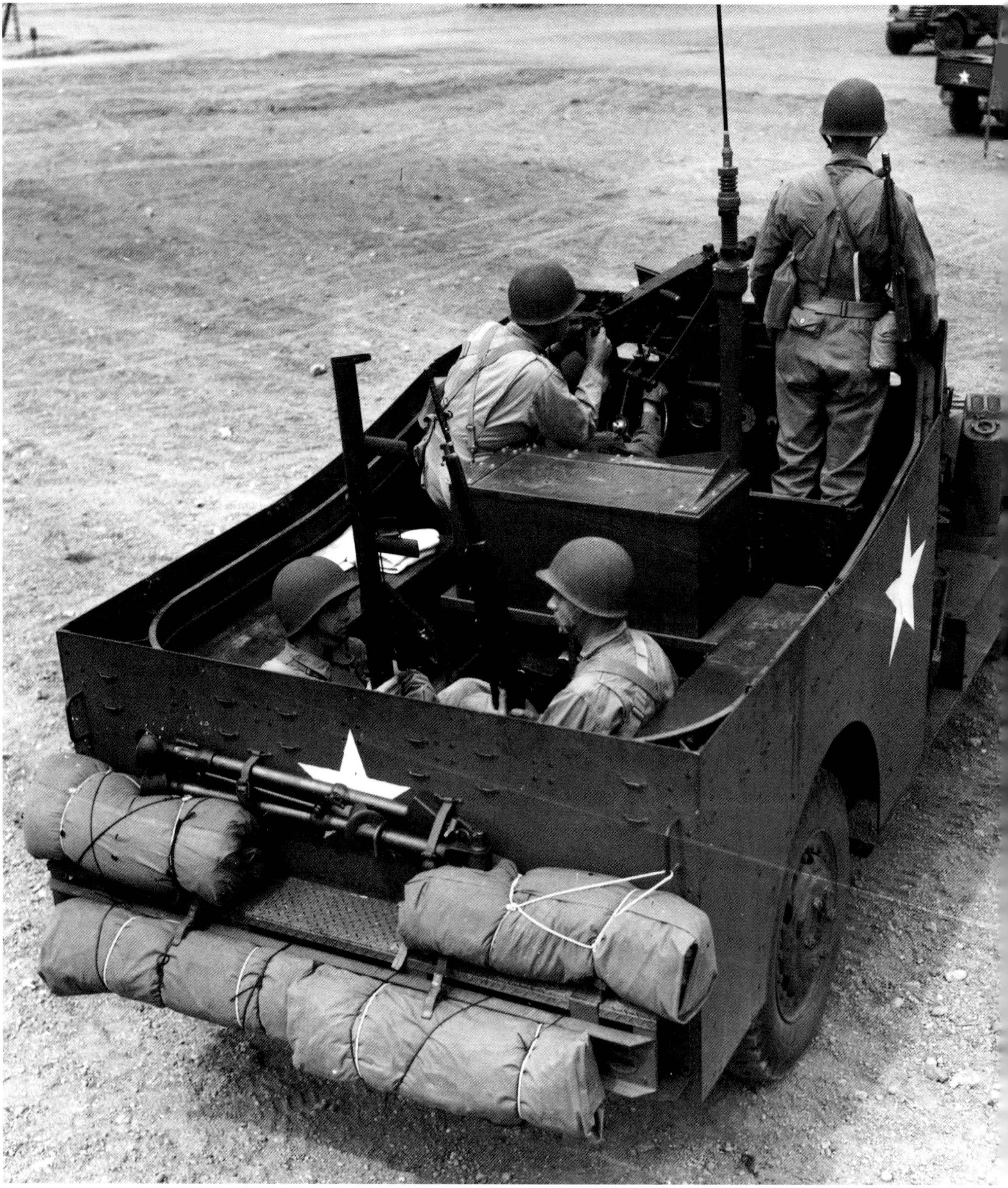

flaws, it proved a popular vehicle in the field. It served in a wide variety of roles with the British Army, ranging from reconnaissance to armored ambulances.

Considered for the scout car role but eventually rejected in late 1941 was a (4x4) vehicle by the Ford Motor Company designated the observation post tender T2. Another vehicle that did not pass muster was a six-wheel-drive vehicle based on a large number of jeep components that came from Willys-Overland Motors, Inc. It was named the scout car T28 but was canceled in late 1942.

Other vehicles that did not make it into series production were two private venture scout cars constructed by the Chrysler Corporation; one was a (4x4) and the other a (6x6). There were also a number of (4x4) armored jeep designs proposed for the scout-car role that were referred to as the T25 series. As the vehicles proved unable to carry the additional weight of armor without comprising their handling characteristics, none ever made it out of the testing stage.

MEDIUM ARMORED CARS

Besides their interest in acquiring American-made tanks in early 1941, the British Purchasing Commission also made it known that they were looking for both medium and heavy armored cars and issued specifications for such vehicles. The successful employment of large armored cars in the reconnaissance role by the British Army during the fighting in North Africa also generated interest among some in the U.S. Army for a large armored car in the same role. The U.S. Army semantic distinction between a scout car and an armored car seemed to have been forgotten by this time.

In this overhead photograph of a U.S. Army scout car M3A1, the skate rail that circled the interior of the vehicle's open-topped hull can be seen. The skate rail first appeared on the scout car M3 for mounting machine guns. Unlike the scout car M3, the scout car M3A1 did not have a rear hull door. (Patton Museum)

The Trackless Tank Corporation developed a large, eight-wheel armored vehicle with independently sprung wheels. It was intended that it could serve as a platform for a variety of roles, ranging from a self-propelled antitank gun to a self-propelled howitzer. As seen here during U.S. Army testing, it was designated the armored car T13. (TACOM Historical Office)

ARMORED CAR T13   11-14-42

The British Purchasing Committee (based in the United States) submitted to American industry in early 1941 the specifications for a desired new medium armored car. Both Ford and General Motors responded to the British request. Ford came up with a six-wheel armored car seen here, designated the T17. (TACOM Historical Office)

As a result of their combined interest in large armored cars, it was decided that the British and the American armies would merge their requirements into vehicles that met both their needs. A proposed medium armored car was labeled the T17, and a proposed heavy armored car as the T18. In response, Ford Motor Company offered a (6x6) medium armored car designated the T17. The Chevrolet Division of General Motors came up with a (4x4) medium armored car referred to as the T17E1.

The larger Ford T17 was tested, and it was decided that both its size and 16-ton weight made it impractical as a reconnaissance vehicle for the U.S. Army. Despite this finding, the Ordnance Department authorized 250 to be built for the British Army, who had already named it the "Deerhound." However, none of the five-man T17s would ever be shipped overseas to see service with the British Army. Instead, the gasoline-engine-powered vehicles were retained in the United States and issued to military police units without their 37mm M6 main guns.

Testing of the 15.3-ton Chevrolet T17E1 went reasonably well after some minor design issues were addressed and the vehicle was authorized for series production for the British Army. The British Army name for the five-man vehicle was the "Staghound." It was not the most popular armored car in British Army service as its size and weight often became an issue when going off-road. Maximum armor thickness on the vehicle was 1.25 inches (31.7mm) with the turret made of cast armor and the hull of both cast and welded armor. Power for the vehicle was provided by two gasoline engines.

Chevrolet constructed a total of 2,844 units of the T17E1 between October 1942 and December 1943. Eight T17E1s remained in the United States for testing purposes. At one point, the U.S. Army considered fielding the T17E1 as the medium armored car M6, but the vehicle was never standardized for U.S. service.

An important variant of the T17E1 was an antiaircraft version designated the T17E2. It mounted two .50 caliber air-cooled machine guns in a power-operated, open-topped one-man welded armor turret. Built at the request of the British Army, in whose service it was referred to as the "Staghound AA," Chevrolet assembled 1,000 units of the vehicle between October 1943 and April 1944. As the T17, T17E1, and T17E2 were all originally intended to serve only in the British Army, the Ordnance Department classified them as limited procurement items.

At one point, the British Army requested a version of the Staghound armed with the turret from the 75mm howitzer motor carriage M8. It would have been designated the T17E3 but series production was never begun and the project was later canceled.

The truck-type suspension components in the armored cars T17 and T17E1 limited their off-road mobility. In pursuit of something better, the Ordnance Department began looking at independently sprung wheels for its armored cars.

The General Motors candidate vehicle to satisfy the British Purchasing Committee interest in a new medium armored car rode on four wheels. Originally designated the T17E1 by the U.S. Army for test purposes, it was never standardized for American military use. In British Army service the vehicle was named the "Staghound." An example is seen here in Belgian Army markings. (Michel Krauss)

As the U.S. Army was interested in comparing the mobility effectiveness of medium armored cars riding on conventional truck-type suspension systems as found on the T17 and T17E1, they authorized developmental work on similar armored cars riding on independently sprung wheels. This resulted in the medium armored car T19 shown here. (TACOM Historical Office)

The Ordnance Committee authorized development of another (6x6) vehicle by Chevrolet, referred to as the medium armored car T19, to test and prove this concept.

Very similar in appearance to the T17 and T17E1, the T19 was fitted with a T17 turret and sent off for testing. A modified version was soon developed and designated the T19E1. Despite improvements in off-road mobility, it was clear to all that the vehicle was too large and heavy for the reconnaissance role and would be too costly to build. As a result, the project was canceled. There was also a pilot vehicle built using the chassis of the T19E1 and mounting an open-topped turret fitted with a 75mm M3 main gun. The vehicle was referred to as the 75mm gun motor carriage T66 but it never went beyond the pilot stage.

At the same time the U.S. Army authorized the development of the Ford medium armored car T17 and the General Motors medium armored car T17E1, permission was given to begin work on a heavy, eight-wheel armored car, seen here, designated the T18 for testing purposes. The vehicle was never standardized for American military service. (TACOM Historical Office)

HEAVY ARMORED CAR

It fell to the General Motors Truck and Coach Division to design and build an (8x4) heavy armored car, designated the T18. Like the T17E1, it was originally armed with a turret-mounted 37mm M6 main gun and a .30 caliber coaxial machine gun. However, by the time the first pilot vehicle was ready for testing in July 1942, it was clear that the T18 was under-armed. It was therefore decided to up-gun a second pilot with the American license-built version of the British Army 6-pounder (57mm) antitank gun, designated the 57mm gun M1. The up-gunned T18 was referred to as the T18E2.

It was originally envisioned that the T18, with a maximum armor thickness of 2 inches (51mm), would weigh no more than 18 tons. However, this early estimate was totally unrealistic, and by the time the T18E2 pilot vehicle was constructed, the vehicle's weight had risen to 26.5 tons. The turret was cast armor and the hull constructed of cast and welded armor. Power for the vehicle came from two gasoline-powered truck engines.

Production of the T18E2 was authorized for the British Army, who attached the name "Boarhound" to the vehicle. Despite the initial British Army request for 2,500 units of the vehicle, by the time series production began in December 1942, the need for such a vehicle had disappeared in North Africa, and in the end, only 30 units were completed by May 1943. None saw field service with the British Army. The U.S. Army never expressed interest in the vehicle, and the Ordnance Department classified it as a limited procurement item.

ALMOST TAKEN INTO SERVICE

The medium armored car T19E1 pictured here was a modified version of the T19. An external difference between the two vehicles was the smaller and lighter turret on the T19E1. Neither medium armored car was standardized for service with the American military as they were considered too large and heavy as reconnaissance vehicles. (TACOM Historical Office)

The T18E2 was not the only large American (8x8) armored car designed and built during World War II. In January 1941, the Trackless Tank Corporation submitted a prototype of an (8x6) armored vehicle, for review by the Ordnance Department and the Armored Force Board. It was powered by a diesel engine and so impressed those who oversaw its testing that a contract was awarded for 17 more examples of the vehicle. Of those 17 vehicles, 13 were fitted with a light-tank-type turret armed with a 37mm M6 main gun, and a coaxial .30 caliber machine gun. These vehicles were designated the armored car T13.

The remaining four vehicles were configured as tank destroyers and referred to as the 3-inch gun motor carriage T7, a project that never went past the prototype stage. By early December 1941, the U.S. Army was considering mounting a 105mm howitzer on the Trackless Tank Corporation vehicle, an idea that did not go past the concept stage.

With the Japanese attack on Pearl Harbor and the rush to re-arm, the Ordnance Committee suggested that 1,000 units of the T13 be contracted out even before it was standardized. This number was later dropped to 500 vehicles. Two T13 pilot vehicles, built by REO Motor Car Company under a subcontract with the Trackless Tank Corporation, arrived at Fort Knox, Kentucky in summer 1942 for testing. This time around, the vehicles were looked at in a much harsher light, and their weight of roughly 16 tons soon became an issue, as well as numerous design problems, resulting in the cancellation of the T13 project.

Another eight-wheeled armored car that was being developed at the same time as the T13 was referred to as the T20. It looked like the T13, but was designed by Reo Motors and featured a gasoline engine that drove a generator that in turn supplied power to electric motors in each of the vehicle's eight large wheels. The Ordnance Committee wanted to order 500 T20s, but it was soon obvious, even in its mockup stage, that it would be both overweight and not able to meet the performance standards set by the U.S. Army. It was canceled in August 1942 before a single prototype was built.

LIGHT ARMORED CARS

The Ordnance Department began work in July 1941 on a (6x6) vehicle referred to as a light tank destroyer. It was to be armed with a turret-mounted 37mm M6 main gun and a coaxial .30 caliber machine gun. The concept was well received by the Tank Destroyer Command, and development of this wheeled vehicle was approved in October 1941. Three firms vied for the contract to build the vehicle, based on an Ordnance Department layout configuration, including the Ford Motor Corporation, the Fargo Division of the Chrysler Corporation, and the Studebaker Corporation.

Even as development of the light tank destroyer moved forward, it became clear to all concerned that the main armament on the vehicle would no longer be a viable antitank weapon by the time it would be fielded. However, there was a need in the new tank destroyer battalions for a reconnaissance vehicle, and the Cavalry Branch was desperate to replace their less-than-satisfactory machine-gun-armed scout car M3A1s. So it was decided to progress with the development of what was no longer referred to as a light tank destroyer, but as a light armored car, beginning in March 1942.

Of the three companies interested in building the new light armored car, it was the Ford Motor Corporation that managed to supply the Ordnance Department with the first pilot vehicle in March 1942, designated the T22. Testing of the four-man vehicle went well enough that the Armored Force Board felt it did not need to test the other contenders' products, and decided in April 1942 that the T22 with some modifications would be suitable for everybody's purposes as a reconnaissance vehicle. The vehicle's turret was made of cast armor and the chassis of welded armor, with a maximum armor thickness of 1 inch (25.4mm) on the front of the vehicle.

A production contract for 5,000 units of an improved version of the T22, designated the T22E2, went to Ford in May 1942; however, unresolved contract issues between the U.S. Army and Ford caused series production of the vehicle to be delayed until March 1943. The vehicle was standardized as the light armored car M8 in June 1942.

Series production of the M8 continued until May 1945 with a total of 8,523 units built. Originally intended as a vehicle for the Tank Destroyer Command, most of the M8s constructed saw service with U.S. Army mechanized cavalry units during World

=DETROIT ARSENAL=
NEG. NO. 8404   DATE 16 Nov 1942   DEVELOPMENT & ENGINEERING
Armored Car, T21.

The U.S. Army's interest in a wheeled tank destroyer armed with a 37mm main gun resulted in three firms submitting vehicles for consideration. These included the Ford Motor Company, the Chrysler Corporation, and the Studebaker Corporation, who presented the vehicle pictured here, originally designated as the 37mm gun motor carriage T21. (TACOM Historical Office)

War II. From a February 24, 1944 War Department manual titled *Cavalry Reconnaissance Troop (Mechanized)* comes this description of the M8 and its role:

Armored cars are the basic command and communication vehicles. The light armored car, M8, is a 6x6 vehicle, weighs 16,400 pounds with equipment and crew, and is capable of cruising from 100 to 250 miles cross country or 200 to 400 miles on highways without refueling. On a level, improved road, it can sustain a speed of 55 miles per hour. Each armored car is equipped with a long-range radio set to assist in the exercise of command or for the purpose of relaying information received from subordinate elements to higher headquarters, and a short-range radio set for communications within a platoon, reconnaissance team, or headquarters.

After World War II, the mechanized cavalry was simply referred to as the "cavalry." Under Lend-Lease, the British Army received 496 units of the M8, which they referred to as the "Greyhound." The Free French Army received a total of 689 M8s under Lend-Lease, and the Brazilian Army received 20.

Due to interest by the Tank Destroyer Command, a turret-less version of the M8 was built and referred to as the armored utility car M20. It served in a number of roles, including command car, personnel carrier, and cargo carrier. It could carry from two to six men. Production of the M20 began at Ford in July 1943, and continued until June 1945 with a total of 3,791 units built. Some World War II-era M20s were provided with a ring mount over the open-topped crew compartment for mounting a .50 caliber machine gun.

VEHICLE OPINIONS AND VARIANTS

As the M8 employed standard truck-type suspension components, its off-road capabilities were limited and resulted in some unhappiness among those using the vehicle. Major-General Robert Grow, commander of the 6th Armored Division, so disliked the M8 that he replaced them within his division and even as his own command vehicle with the full-tracked 75mm howitzer motor carriage (HMC) M8.

An interesting comment about this shortcoming of the light armored car M8 and the Armored Utility Car M20 appeared in this AAR of the 640th Tank Destroyer Battalion that fought in the Philippines between January and March 1945: "The M8s and M20s in reconnaissance companies proved to be very good reconnaissance vehicles over terrain where a road net existed, however over difficult terrain their ability to reconnoiter for track vehicles is very limited."

In the end, the U.S. Army selected the Ford Motor Company six-wheel contender, originally designated the 37mm gun motor carriage T22, as its newest tank destroyer. However, by the time the vehicle was standardized by the U.S. Army it was classified as the light armored car M8. An example of an M8 is seen here. (Christophe Vallier)

Another impression comes from a March 1945 U.S. Army report titled *United States vs. German Equipment*. In that report, Lieutenant Colonel John A. Beall, Commander 702nd Tank Destroyer Battalion, comments on the M8 and M20:

1. The armor on the M20 and the M8 is unsatisfactory and gives the crew little or no protection against any caliber AT [antitank] weapon.
2. The crew is exposed to artillery and mortar fire since no overhead protection is provided.
3. Grenades can be easily thrown into the vehicle by enemy infantry.
4. The .50 caliber MG [machine gun] on the M20 can be manned only by exposing the entire body and cannot be depressed for ground firing without extreme danger for the gunner.
5. The 37mm gun on the M8 is wholly inadequate as an anti-vehicle weapon.

By contrast, in a U.S. Army report dated October 7, 1944, Lieutenant -Colonel Michael Popowski, who commanded M8s in the Italian Theater of Operations, shares these positive comments: "During my combat experience I saw only one instance

Originally envisioned as a tank destroyer, the M8 light armored car found a new role as a reconnaissance vehicle with the U.S. Army and others under Lend-Lease. The M8 pictured belonged to Company C, 82nd Armored Reconnaissance Battalion of the 2nd Armored Division, during Operation *Cobra*, which took place in France between July and August 1944. Appearing in base olive drab the vehicle has been over-painted with black paint in an irregular pattern. Tactical markings appear in both yellow and white. On the rear engine deck is an air identification panel. (© Osprey Publishing Ltd.)

where the armored car was not able to go where tanks went… Some of its capabilities over a tracked vehicle are: quietness, range, maintenance, and weight, which are all important in reconnaissance."

Some comments on needed modifications to the M8 in the Italian Theater of Operations appeared in this extract from a late 1944 report written by Major-General E. N. Harmon titled *Notes on Combat Experience during the Tunisian and African Campaigns*:

1. Added protection on the floor to protect the crew from mines; designing the floor to permit heavy sandbagging would be of some value.
2. Better protection for the radiator. A regimental reconnaissance company had twelve cars knocked out of action due to shell fragments through the radiators during a period of less than one month.
3. Add a .30 Cal. MG bow-gun to be operated by the assistant driver.

Interesting variants of the M8 that never went into production included the multiple-gun motor carriage T69. It consisted of the chassis of an M8 mounting an open-topped powered turret fitted with four air-cooled .50 caliber machine guns. Testing of the vehicle led the Antiaircraft Artillery Board to conclude that existing antiaircraft weapon platforms were superior and there no was no need to continue development of the vehicle.

Another version of the M8 that was proposed was called the armored chemical car T30. It had provisions for mounting ten 7-inch rocket launchers, five on either side

Taking part in a World War II reenactment is this light armored car M8. The vehicle is 16 feet 5 inches long and 8 feet 4 inches wide. It is 7 feet 4.5 inches tall. It had authorized storage for 80 main gun rounds, 1,500 rounds for the coaxial .30 caliber machine gun, and 400 rounds of .50 caliber machine-gun ammunition (if fitted). (Michael Green)

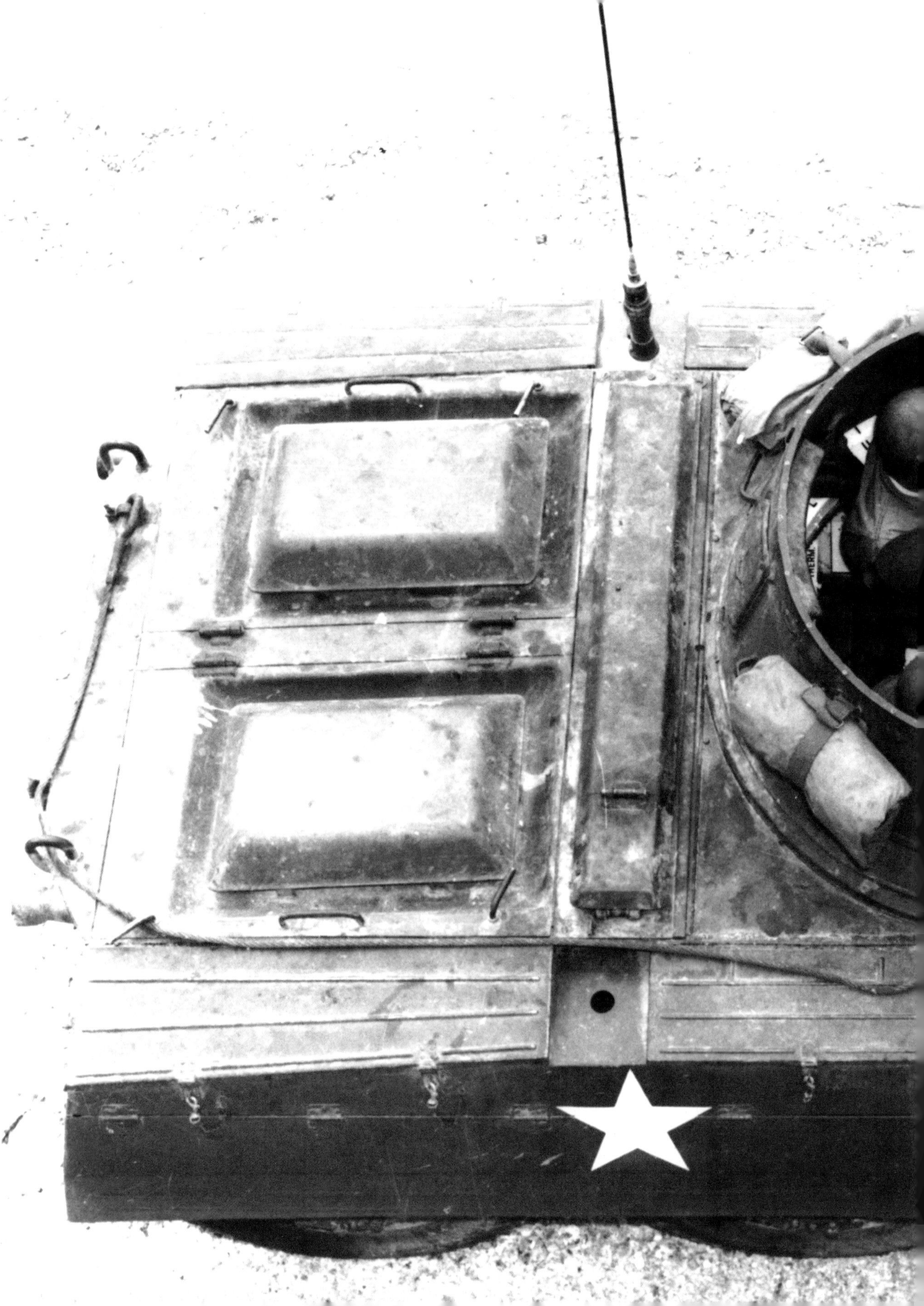

Pictured is an early-production light armored car M8 undergoing testing at Fort Knox, Kentucky. This particular vehicle has no fitting on the turret for a .50 caliber antiaircraft machine gun. There was no stabilizer system on the M8, and the traverse and elevation of the turret was manual. (Patton Museum)

of the existing M8 turret. No prototypes or pilots were built of the proposed vehicle before the project was canceled in November 1943.

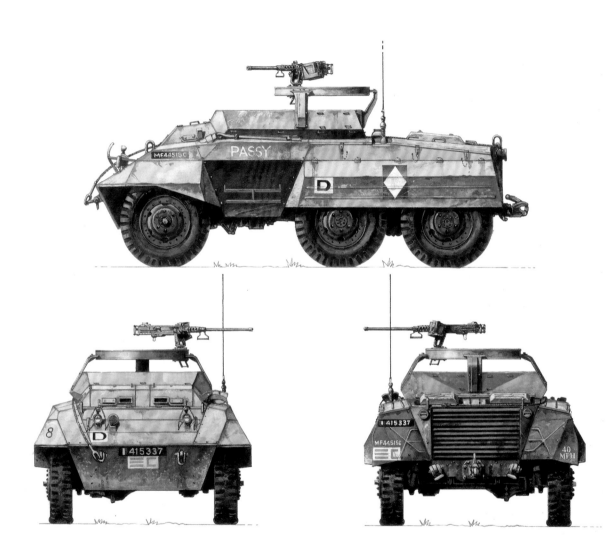

A derivate of the M8 light armored car was the open-topped armored utility car T26 that was standardized as the armored utility car M20 on May 6, 1943. As with the M8 light armored car, many were supplied under Lend-Lease to the Free French Army during World War II. The M20 shown is serving with the 2e Regiment des Dragons, of the French First Army, in the Alsace region of France in 1944. As with all the American tanks and AFVs supplied under Lend-Lease it has the base olive drab. The Free French have added their national insignia on the hull and various other tactical markings. The French military registration number is in white on a black background. (© Osprey Publishing Ltd.)

PROPOSED M8 REPLACEMENT

As they had with the T17 and T17E1 armored cars, the Ordnance Department investigated replacing the unsatisfactory truck suspension system components on the M8 with an independently sprung suspension system, as developed for the armored cars T13 and T19. The result of that research led to the construction of two pilot vehicles, one by the Chevrolet Division of General Motors and the other by the Studebaker Corporation.

The Studebaker contender was an (8x6) vehicle, designated the light armored car T27. The Chevrolet submission was a (6x6) vehicle, referred to as the light armored

The light armored car M8 pictured could handle a 60 percent grade and cross a trench 1.5 feet in width. It could climb a vertical wall 12 inches tall and had a fording depth of 24 inches. The minimum turning circle of the M8 was 56 feet. (Christophe Vallier)

ABOVE
On display at APG was the light armored car T28 that was tested as a possible replacement for the light armored car M8. The big advantage of the T28 over the M8 was its independently sprung suspension system that provided much better off-road mobility than the truck-type suspension system on its predecessor. (TACOM Historical Office)

RIGHT
Restored by a European collector is this armored utility car M20. As with the light armored car M8, the M20 was powered by a Hercules JXD six-cylinder, liquid-cooled, gasoline engine. It could develop 110hp and provided the vehicle a maximum speed on level roads of 55mph. (Pierre-Olivier Buan)

car T28. Both four-man vehicles weighed in at about 7 tons and featured an open-topped cast armor turret armed with the 37mm gun M6 and a coaxial .30 caliber machine gun. The hulls were a combination of cast and welded armor.

Testing of the T27 and T28 led the Cavalry Board to conclude that both vehicles were superior in off-road mobility to the M8. Of the vehicles tested, however, the T28 was clearly superior off-road, resulting in the cancellation of the T27. The T28 was suggested for standardization as the light armored car M38 in February 1945, once necessary modifications were made to the vehicle's design. The British Army even went ahead and named the M38 the "Wolfhound" in anticipation of receiving the vehicle under Lend-Lease. The end of World War II resulted in the M38 program being canceled before series production began.

CHAPTER SEVEN

ARMORED HALF-TRACK

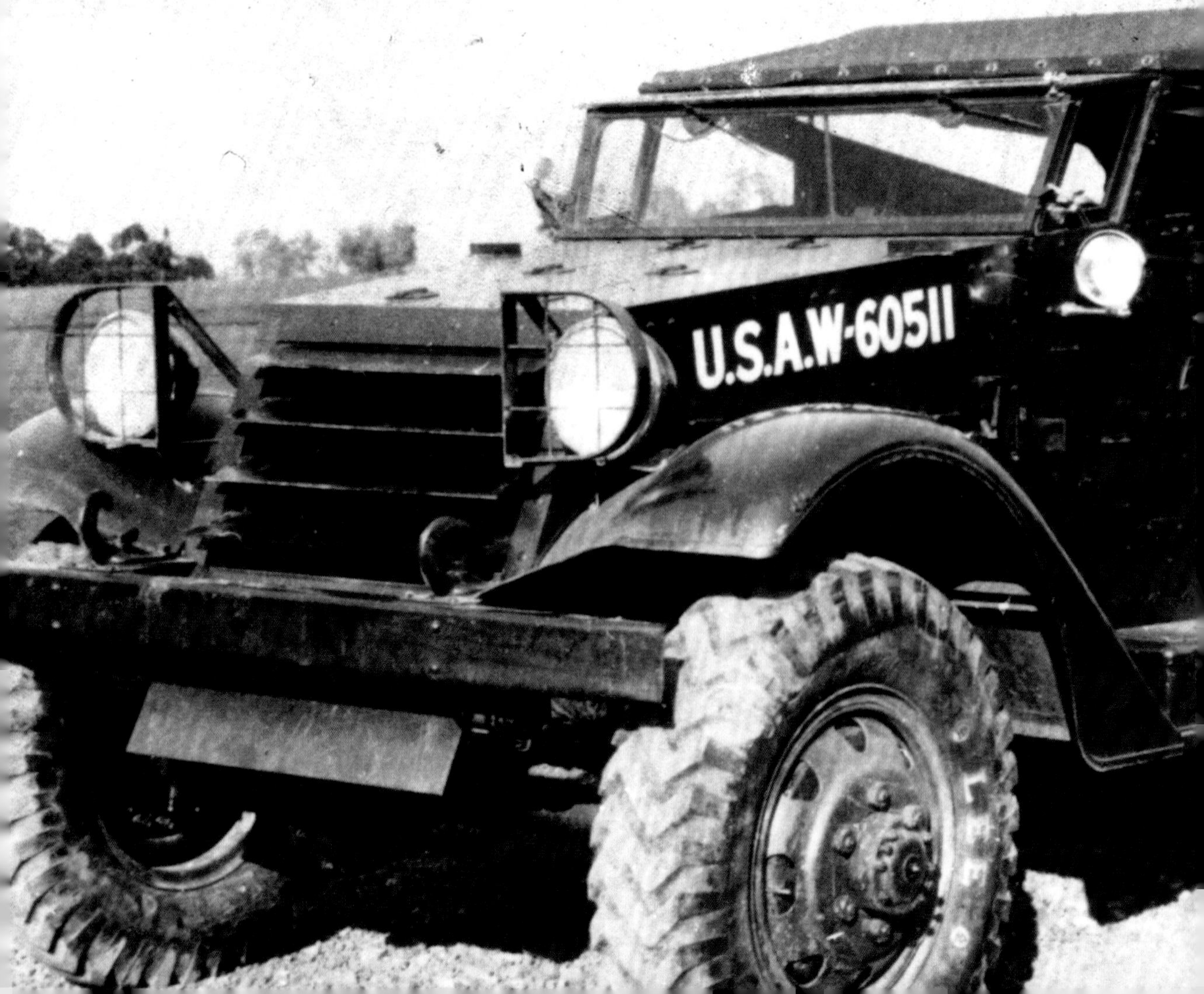
U.S.A.W-60511

As the U.S. Army flirted with mechanization in the cash-strapped 1930s, it was clear a vehicle was needed to allow infantry to cross shell-swept battlefields, while keeping up with fast-moving tanks. The first experiment the Ordnance Department conducted with an open-topped armored half-track took place in 1938. It involved taking a scout car M2A1, and converting it into a half-track by removing its rear wheels and replacing them with a volute spring suspension system and rubber band tracks, similar to that installed on the half-track truck T9, and experimentally on the pilot T1. The conversion work was carried out at the Rock Island Arsenal with the assistance of the White Motor Company. The completed pilot vehicle was designated the half-track personnel carrier T7 and was tested from September to October 1938.

The T7 featured a maximum armor thickness of 0.375 inches (9.5mm) of FHA and was envisioned as serving many roles, from a scout vehicle to an artillery prime mover. Unlike the half-track car M1, the front wheels on the T7 were powered and synchronized as they were on the half-track truck T9. This combination greatly aided the vehicle's off-road mobility. The T7 also appeared with an un-ditching roller in place of its steel front bumper to aid it in climbing in and out of shallow ditches. The biggest drawback of the vehicle was that the Ordnance Department considered the vehicle underpowered. At the conclusion of the test cycle, the vehicle was returned to its original configuration.

The U.S. Army reacted to the beginning of World War II by quickly expanding in size. Among the many items of equipment that were seen as key to the fielding of a modern mechanized army was the armored half-track. In response, the Ordnance Department released in December 1939 the specifications for a 7-ton vehicle designated the half-track scout car T14. It was intended that the vehicle incorporate all the lessons learned from the experimental work done on half-tracks prior to that time.

The half-track personnel carrier T7 pictured here was a converted scout car M3 done as a proof of concept in 1938. It had its rear wheels replaced by a volute spring suspension system and rode on band tracks. The addition of tracks did much to improve the vehicle's off-road mobility. However, the M3's engine proved underpowered for its new role. (Patton Museum)

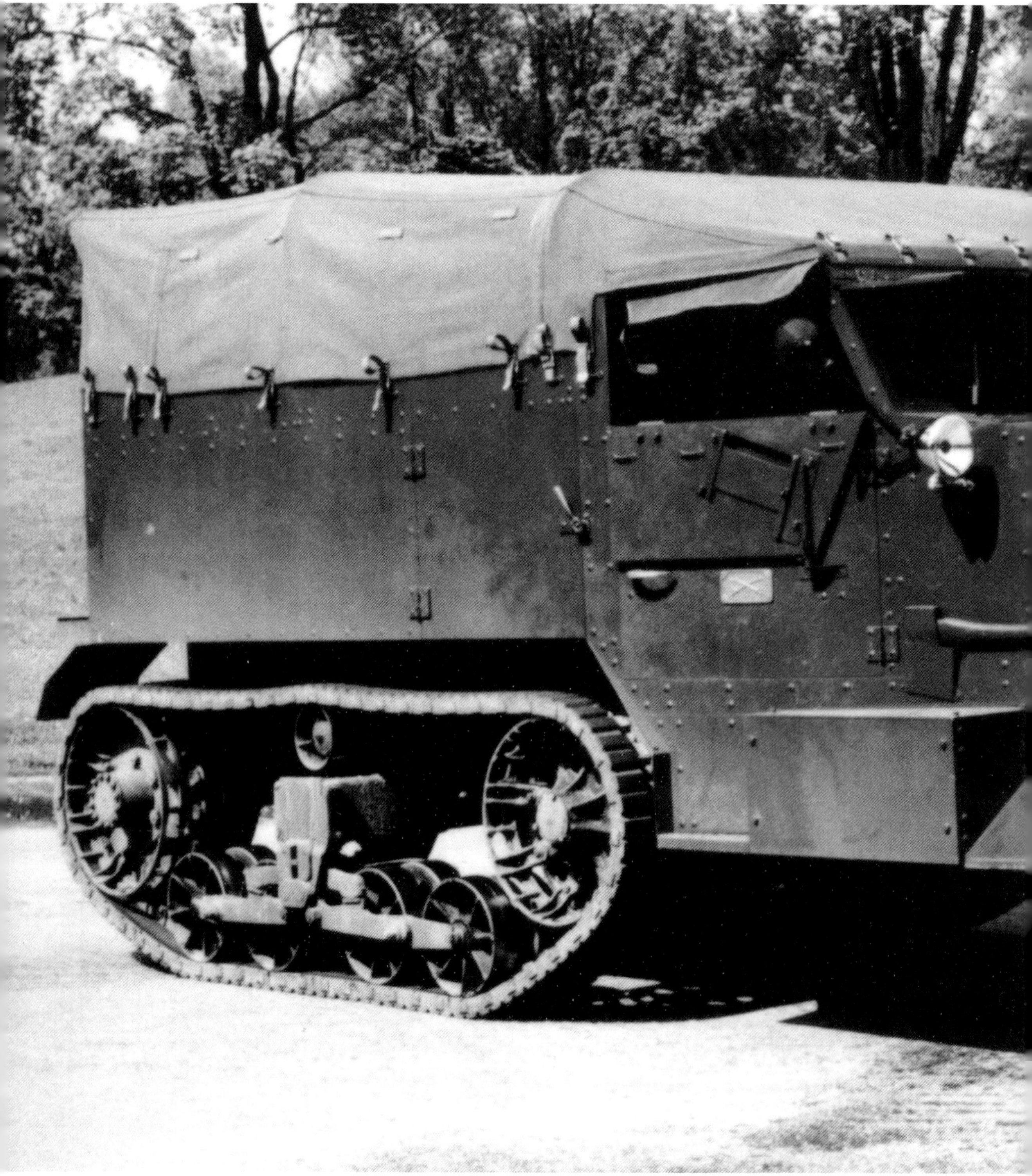

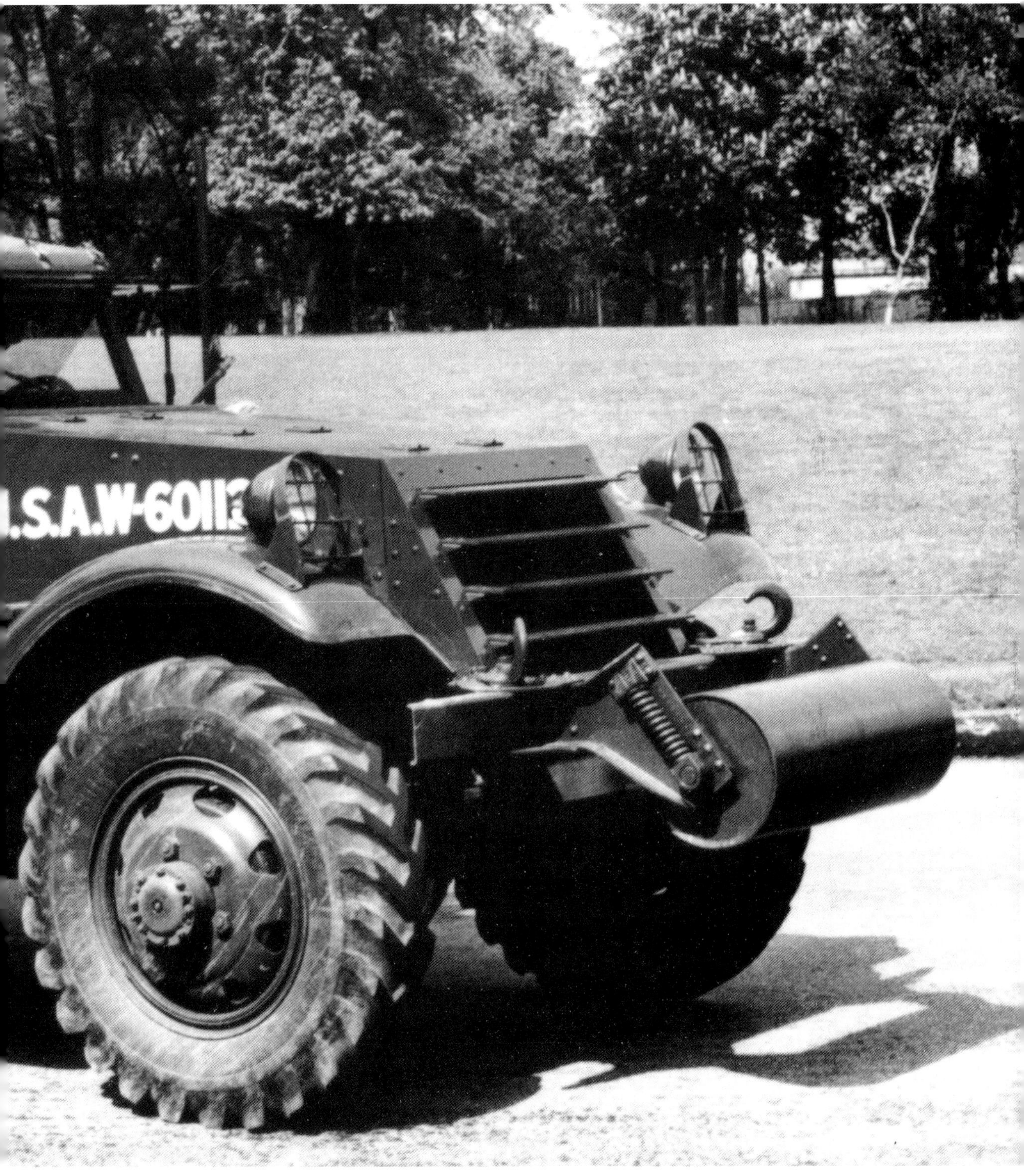
.S.A.W-60113

PREVIOUS PAGE
Following the outbreak of
World War II, the U.S. Army
realized that it would require a
great many armored
half-tracks. Using the
successful testing of the
half-track personnel carrier T7
as a model, in 1940 they
developed an improved pilot
half-track, seen here, with an
uprated engine. It was
designated the T14 half-track
scout car. (Patton Museum)

The T14 was built by the White Motor Company and had a maximum armor thickness of 0.5 inches (13mm) of welded FHA. To speed up the development of the T14, the pilot vehicle was fitted with the tracked suspension system from the T7. The original specifications had called for the drive sprocket on the T14 to be at the front of the vehicle.

In the name of expediency it was decided that for test purposes it was okay to use the rear-mounted drive sprockets from the tracked suspension system on the T7. Instead of the 95hp six-cylinder Hercules JXD gasoline engine that powered the T7, the T14 pilot vehicle was eventually powered by the White Motor Company six-cylinder gasoline engine known as the 160AX that produced 147hp.

ARMORED HALF-TRACKS

In its haste to re-arm, the Ordnance Department decided in September 1940 to recommend that the pilot T14 be standardized even before its test results were in, and to have three versions of the vehicle built, all with welded FHA. These included the 9.6-ton half-track car M2 with seating for ten men that could be utilized as a prime mover for artillery pieces up to 155mm in size, or in a secondary role as a reconnaissance vehicle.

The original role of the half-track scout car T14 was envisioned by the U.S. Army as a prime mover and ammunition carrier for the 105mm howitzer M2. In this overhead picture of the T14 can be seen the large ammunition storage boxes for the 105mm rounds. The vehicle has no rear-hull door due to the skate rail. (Patton Museum)

Also based on the M2 chassis was an 8.7-ton mortar-carrying version, referred to as the 81mm mortar carrier M4. There was a lengthened version of the half-track car M2 that could transport 13 men, referred to as the half-track personnel carrier M3. The M3 weighed in at 10.2 tons, and was capable of fulfilling many roles, from ambulance to command vehicle.

The M2, M3, and M4 featured the un-ditching roller that first appeared on the T7. Some later production units of the three armored half-tracks were fitted with a 10,000-pound capacity Tulsa Model 18G-winch mounted behind the front bumper.

There were some external differences between the M2 and M3. The M2 had two large ammunition storage compartments, one on either side of the vehicle, just behind the driver's compartment, which reflected its primary role as an artillery prime mover. These ammunition storage compartments do not appear on the M3.

On the M2, the rear compartment of the vehicle did not extend past the tracked suspension system and featured two steps not seen on the rear of the M3. The M3 was 10 inches longer than the M2, with its rear compartment extending past the vehicle's tracked suspension system. Reflecting the vehicle's primary role as an armored personnel carrier (APC), there was a rear door on the M3 that is not seen on the M2.

Because the number of armored half-tracks needed by the U.S. Army was constantly increased, the Ordnance Department quickly concluded that no one company had the industrial capacity to build as many vehicles as might be necessary. So, it was decided in September 1940 that the White Motor Company, the Diamond T Motor Company, and the Autocar Company would collaborate and build identical versions of the same vehicle with interchangeable parts, except for the armor plates that protected the vehicles.

The Ordnance Department recommendation in May 1940 that the three armored half-track versions based on the T14 should be standardized was approved and procurement was soon authorized. That same month, the first production batch of 62 half-track car M2s rolled off the assembly line of the White Motor Company.

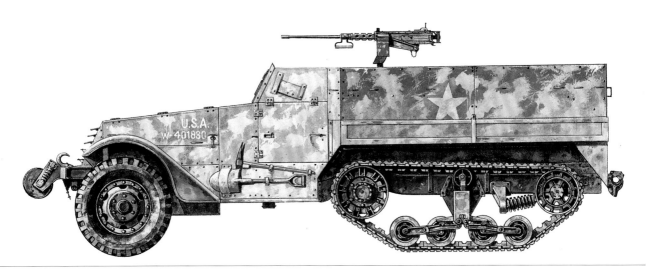

The M3 half-track carrier was not well received by the U.S. Army soldiers who initially served on it during the fighting in North Africa; they felt it was under-armored. However, it was what the Ordnance Department and American industry could come up with in short order, and in the numbers required to meet U.S. Army requirements. The M3 half-track carrier pictured is in the markings of the 1st Armored Division, in Tunisia in February 1943. It sports a national insignia in yellow. The vehicle's registration number is in white rather than the authorized yellow. Because the factory-applied olive drab proved too dark for the light-colored terrain in which it was serving, the crew of the vehicle pictured have applied a wash of local mud to lighten the shade of the half-track's paint. (© Osprey Publishing Ltd.)

Eventually, the White Motor Company, in conjunction with the Autocar Company, built a total of 11,415 units of the M2 by September 1943. The M2 was followed into production by the modified 9.8-ton M2A1 that did away with the skate rail around the interior of the vehicle for mounting .30 caliber or .50 caliber machine guns. In their place was a new armor-protected ring mount, designated the M49, located over the right front of the vehicle, armed with a .50 caliber machine gun. There were also three pintle socket mounts arrayed around the rear compartment of the vehicle for fitting a .30 caliber machine gun. Between October 1943 and March 1944, the White Motor Company and the Autocar Company built 1,643 units of the M2A1.

Between March and October 1942, the White Motor Company assembled 572 units of the six-man half-track 81mm mortar carrier M4. It was originally thought that the mortar carried by the vehicle would be dismounted before firing would commence. Field use showed the need for the mortar to be fired from the vehicle, and it was modified to do so over the rear face of the half-track. This modification resulted in the new vehicle designation M4A1. Production of the 9-ton M4A1 by the White Motor Company began in May 1943 and continued until October of that same year, with 600 units constructed. Both vehicles were reclassified as limited standard in July 1943.

HALF-TRACK PERSONNEL CARRIER M3/M3A1

As with the half-track car M2, the half-track personnel carrier M3 was eventually fitted with the M49 ring mount on the right front of the vehicle, armed with a .50 caliber machine gun. This changed the vehicle's designation to M3A1. The M3 was originally fitted with a tall pedestal mount in the crew compartment that was intended to be fitted with a .30 caliber machine gun. Platoon leader vehicles were supposed to carry a .50 caliber machine gun. In the field, many crews replaced their .30 caliber machine guns with .50 caliber machine guns as they favored its extra punch.

Unlike the M2 and M2A1 which were built only by the White Motor Company and the Autocar Company and not the Diamond T Motor Company, the M3 and M3A1 were built by all three firms, with 12,391 units of the M3 constructed between May 1941 and September 1943. A total of 4,222 units of the 10.2-ton M3A1 were assembled between October 1943 and June 1945, of which a total of 1,360 of them were converted from the 75mm gun motor carriage M3 when that vehicle's battlefield role was taken over by more suitable vehicles.

Among the various models of the half-track scout car T14 standardized in September 1940 was the half-track personnel carrier M3 seen here. It was a lengthened version of the half-track car M2/M2A1 and had a capacity of 13 men. The roller in front of the bumper aided the vehicle in crossing obstacles. (Patton Museum)

The U.S. Army never stopped trying to improve the vehicles it fielded. Pictured at a World War II reenactment event is an improved version of the half-track car M2, designated the half-track car M2A1, which had a capacity of ten men. Notice the armored pulpit for the .50 caliber machine gun and the external rack for M1A1 antitank mines. (Michael Green)

In December 1942 there was some thought of rationalizing production by doing away with the half-track car M2 in favor of a modified version of the half-track personnel carrier M3. However, only a single pilot vehicle – originally designated the T29 and later as the half-track car M3A2 – was ever constructed. In early 1944 the need for additional new half-track cars vanished with the introduction of the light armored car M8, which took over the reconnaissance role once performed by the M2/M2A1, and new, full-tracked unarmored prime movers which took over the artillery towing role once performed by the vehicles.

With the thought that the M3A2 would eventually replace the half-track cars M2/M2A1 and the M3/M3A1 half-track personnel carriers, the Ordnance Department reclassified all of them as limited standard in 1944.

Unhappiness with the rear-facing mortar on the M4A1 caused the Ordnance Department to experiment with mounting a front-facing 81mm mortar on a modified

M3. Tests of a pilot vehicle went well, and the vehicle was later standardized as the half-track 81mm mortar carrier M21 in June 1943. The White Motor Company built a total of 110 units of the 10-ton vehicle between January and March 1944.

There was some development work done on mounting the 4.2-inch mortar used by U.S. Army chemical mortar battalions on the half-track 81mm mortar carrier M4. Tests showed the chassis could not absorb the recoil of the larger weapon. The Ordnance Department then switched to using the modified and reinforced chassis of the half-track personnel carrier M3A1, which proved a little better at absorbing the 4.2-inch mortar's recoil. The vehicle was designated the 4.2-inch mortar carrier T21. When it was ordered that the vehicle have the mortar facing forward, the vehicle was re-designated as the T21E1. The project was later canceled in March 1945 when the U.S. Army's interest turned to a full-track mortar carrier vehicle.

IN SERVICE

Actual combat experience with the M2/M2A1 and the M3/M3A1 in the U.S. Army was mixed. In their first serious action during the fighting in North Africa in early 1943, the vehicles were often sarcastically referred to by those who served on them as "Purple Heart Boxes" in reference to the American medal that was awarded to those wounded in combat. It was the open-topped design of the vehicle that made them so vulnerable to artillery airburst and their thin FHA that did not endear them to the average American soldier or his officers.

Some American infantry officers who saw action in North Africa were so disgusted with the limited protection offered by the armored half-tracks that they suggested they be replaced with unarmored six-wheel-drive 2.5-ton cargo trucks. They felt that the added maintenance requirements of the more complex armored half-tracks did not warrant their usefulness. General Omar Bradley, who commanded U.S. Army units in North Africa, was more enthusiastic about the armored half-track and felt that its poor showing in that theater of conflict "resulted from the inexperience of our troops who attempted to use it for too many things."

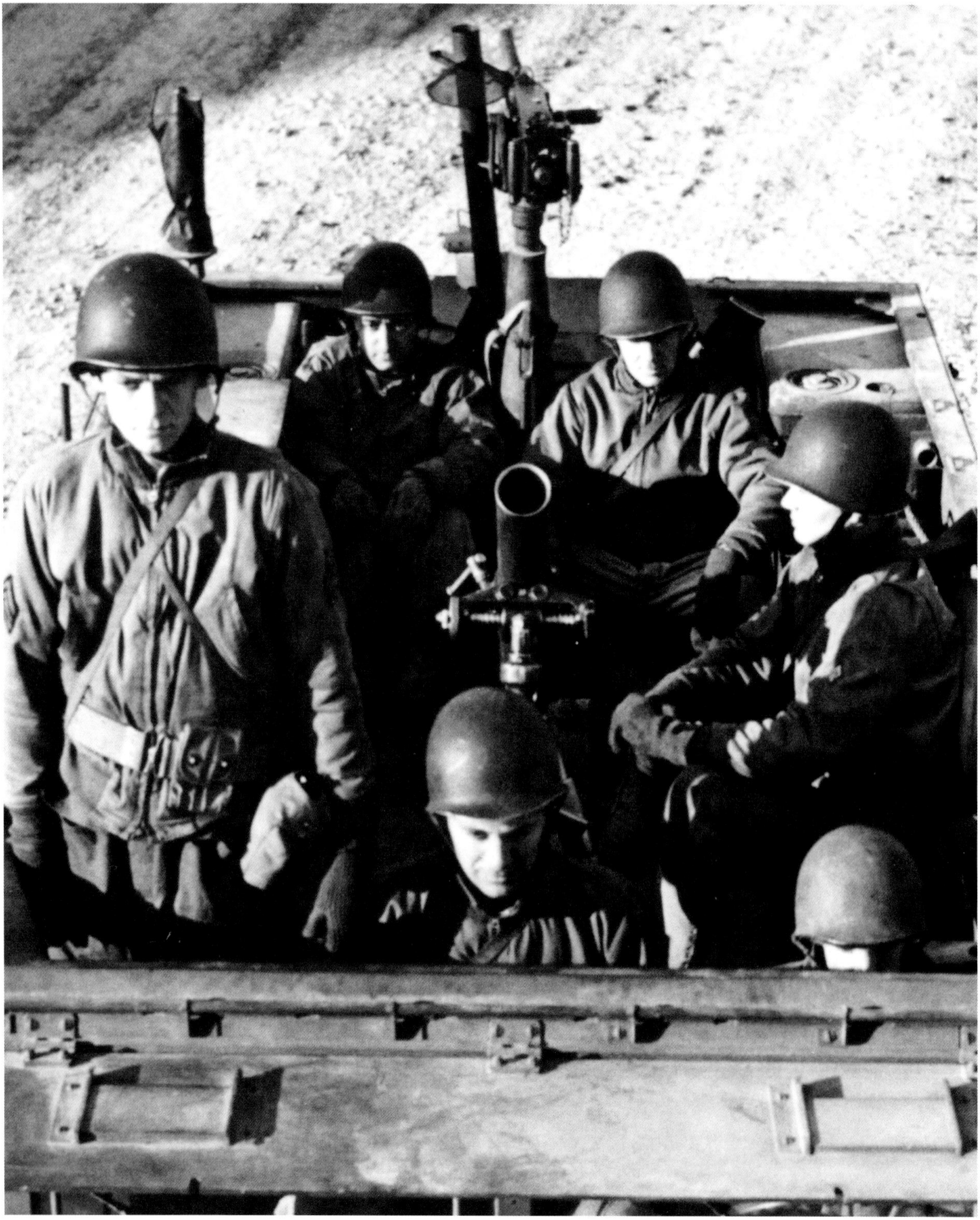

The U.S. armored divisions that took part in the fighting in North Africa had a table of organization and equipment (TO&E) referred to as "heavy" that called for 163 of the half-track car M2/M2A1, and 441 of the half-track personnel carrier M3/M3A1. By the time of the invasion of France in June 1944, 14 of the 16 U.S. Army's armored divisions had a new TO&E that was referred to as "light." These armored divisions contained 448 half-tracks, most of them being the half-track personnel carrier M3/M3A1. Of that number, 216 were divided among three armor infantry battalions that formed a key part of every U.S. Army armored division with the light TO&E.

The half-tracks of the armored infantry in both the heavy and light armored division structure are described in a U.S. Army World War II field manual titled *Armored Infantry Battalion* and labeled as FM17-42:

> The half-track personnel carriers provide protection for the troops against small arms fire up to close range. The armor also gives protection against bomb and shell fragments. Troops are transported as far forward as possible in each situation; terrain, cover, and the type of weapon available to the enemy governing the dismounting. The vehicular weapons are used to protect the attacking troops against air attack. Armor on the half-track does not protect crews against antitank weapons and direct hits by assault guns and light artillery.

Almost always in the forefront of battle, armored infantrymen had an extremely high casualty rate throughout World War II. Private Tom Sator, a tanker with the 4th Armored Division, remembered always feeling sorry for the armored infantrymen that often accompanied his tank into battle as they were so vulnerable to enemy machine-gun and artillery fire. On the other hand, many armored infantrymen who saw the burning deaths suffered by many American tankers would never want to serve on any tanks.

The mobility issues and lack of sufficient armor protection on the armored half-tracks were addressed by Colonel S. R. Hinds of the 2nd Armored Division in *United States vs. German Equipment*: "While our half-track vehicle is far superior to any other similar vehicle, it falls short of the required cross-country mobility. It should be the equal of the tank in that respect. I believe a full track vehicle with slightly more armor on the sides is necessary in order to have the complete support of the armored infantry when most needed."

Ordnance Department efforts to redress the various issues with the armor protection and off-road mobility on the U.S. Army's inventory of armored half-tracks had come to naught, as by the time they had begun testing the pilots of larger, improved, up-armored half-track vehicles (designated the half-track trucks T17 and T19), the user community's interest had turned to full-tracked vehicles.

There was the half-track car T16 based on the chassis of the half-track car M2 that was a bit longer and had a new and improved track suspension system. Something not seen on previous armored half-tracks was an arrangement of 0.25-inch-thick steel armor folding plates on the roof of the T16. Testing of the vehicle showed the improved track suspension system and roof armor unsatisfactory and the project was canceled.

The 81mm mortar carrier M4 and M4A1 were based on the half-track car M2. With both versions, the mortar fired over the rear of the vehicle's hull. Another model, based on the half-track personnel carrier M3 seen here, had the mortar facing forward and was designated the half-track 81mm mortar carrier M21. (Patton Museum)

A NEW MANUFACTURER OF ARMORED HALF-TRACKS

Following America's official entry into World War II, there was a requirement for even more armored half-tracks than originally envisioned. It was concluded that the three existing manufacturers of the vehicles would not be able to meet the increased demand, so a fourth firm was brought in, the International Harvester Company. It built its own modified versions of the half-track car M2 and half-track personnel carrier M3 that were respectively designated as the half-track car M9, and the half-track personnel carrier M5. The 10.2-ton M5 was approved for standardization in June 1942 and the M9 the following month.

The International Harvester Company incorporated into their vehicle's designs all the lessons learned by the U.S. Army in testing and field use with the earlier versions of the armored half-tracks built by the original three manufacturers. This resulted in their vehicles having heavier springs and axles fitted, plus a host of other improvements. Power for these new vehicles would come from an International Harvester Company designed and built six-cylinder gasoline engine that produced 143hp, and was known as the RED-450-B.

Externally, the International Harvester Company vehicles looked very similar to their predecessors from a distance. A closer inspection reveals that the M9 and M5 were no longer protected by FHA plates bolted together. Instead, they were protected

The half-track personnel carrier M3 had a rear hull door to aid ingress and egress of its passengers. The vehicle's thin armor and open-topped design were unpopular features to all those who served on them. The U.S. Army evolved tactics for its mechanized infantry units that minimized the exposure of their personnel carriers to frontline combat situations. (Pierre-Olivier Buan)

Taking part in a World War II reenactment event is a half-track personnel carrier M3A1 that can be identified by the armored pulpit for the .50 caliber machine gun. With a front-mounted winch, the vehicle was 20 feet 9 inches long and had a width of 6 feet 5.25 inches. (Bob Fleming)

The half-track personnel carrier M3A1 shown here was 8 feet 10 inches tall with the pulpit-mounted .50 caliber machine gun fitted. The vehicle had authorized storage for 700 rounds of .50 caliber ammunition and 7,750 rounds of .30 caliber machine gun ammunition. In addition, it could carry 24 M1A1 antitank mines. (Christophe Vallier)

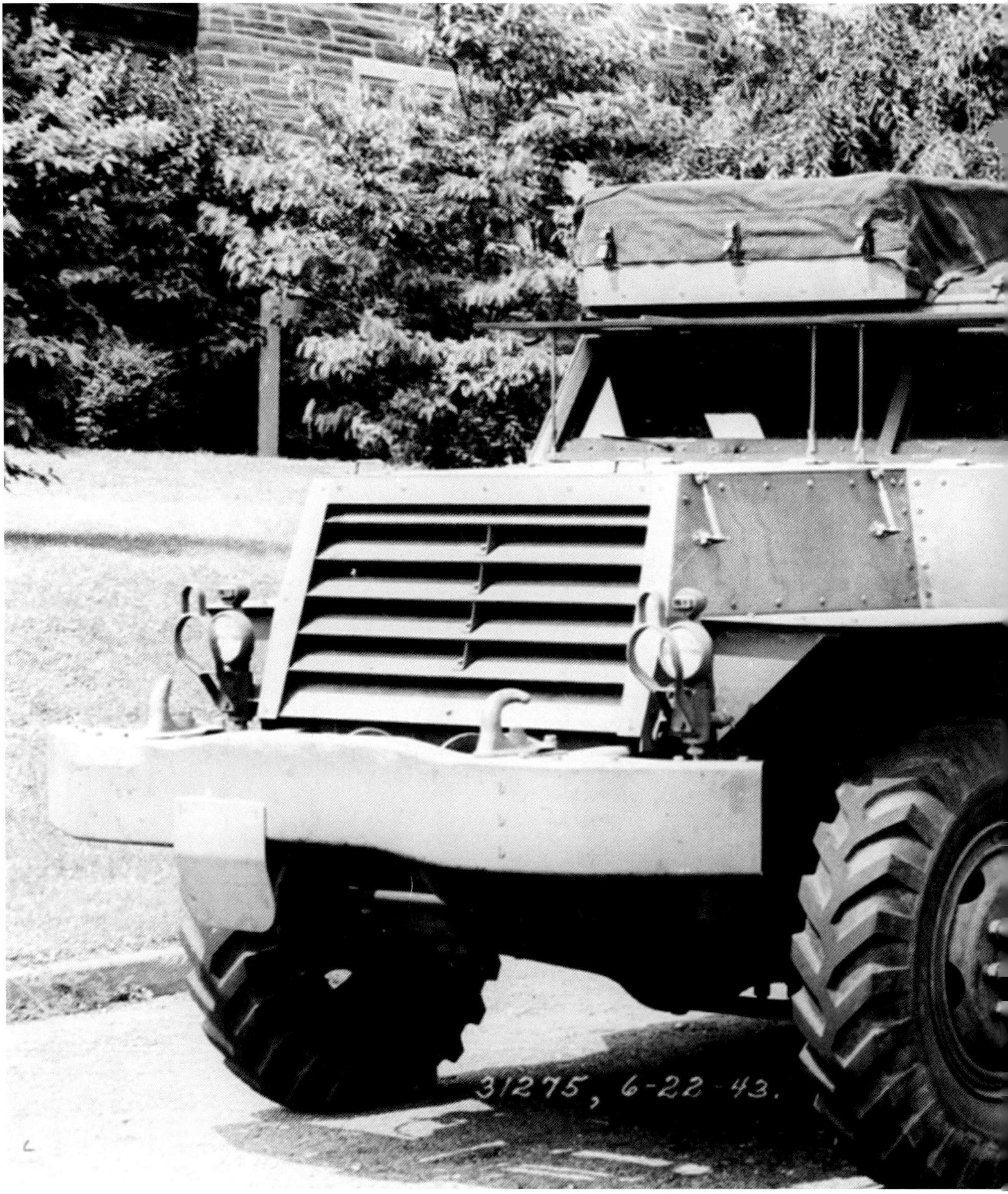
31275, 6-22-43.

In October 1941 a call went out for an improved family of armored half-tracks that were larger, better protected, and had more powerful engines to improve on-road and off-road performance. This resulted in five new half-track designs, including the half-track truck T17 pictured here. In the end, none were standardized for a variety of reasons. (TACOM Historical Office)

by RHA plates welded together. This change in construction method resulted in the smooth rounded corners of the rear compartment of the M9A1 and M5, which was in contrast to the sharp edged corners seen on the armored half-tracks built by the other three companies.

As the RHA plates on the M9 and M5 provided a bit less ballistic protection than the FHA plates on the M2 and M3, it was necessary to make them a bit thicker to compensate. On the M9 and M5 the maximum armor thickness went up to

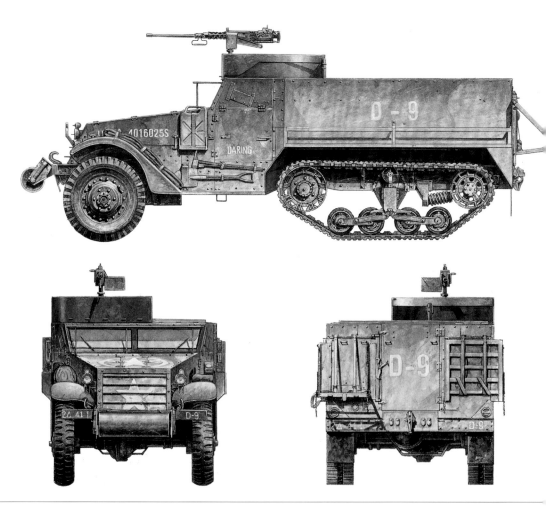

As with almost all U.S. Army tanks and AFVs employed during World War II, the manufacturers and the Ordnance Department attempted to improve them, sometimes based on the user community's feedback. An improved version of the M3 half-track carrier was the M3A1 half-track carrier seen here in the markings of the 41st Armored Infantry, 2nd Armored Division, in Normandy, France, June 1944. The vehicle shown has had its base olive drab painted with a pattern of earth-brown splotches. In the process of over-painting the half-track base olive drab, the national symbol has been covered over on the sides of the vehicle. However, it was retained on the radiator armor louvers without a segmented circle, and on the engine hood with a segmented circle. Tactical markings are in yellow, with the vehicle's nickname and registration number in white. (© Osprey Publishing Ltd.)

0.625 inches (16mm). The slight increase in weight brought on by the thicker armor and heavier automotive components did not detract from the mobility characteristics of the M9A1 or M5. The front fenders on the International Harvester Company vehicles also differed from the ones mounted on the other companies' vehicles and can act as a key identifying feature.

As events turned out, there were more than enough armored half-tracks being built by the three original manufacturers to meet all the U.S. Army's requirements.

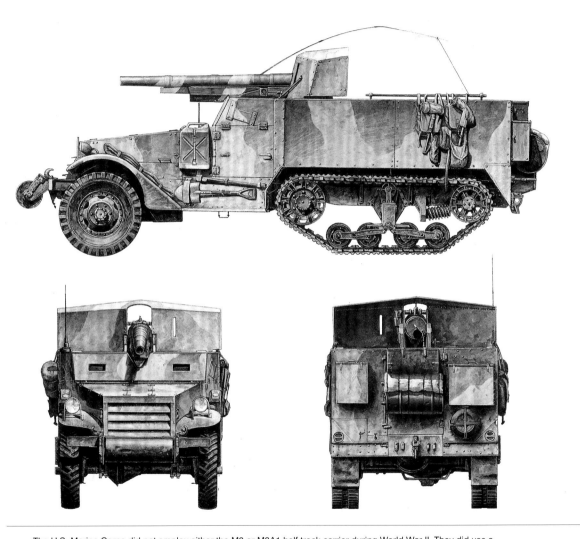

The U.S. Marine Corps did not employ either the M3 or M3A1 half-track carrier during World War II. They did use a small number of M3 half-tracked-based tank destroyers, designated the M3 75mm GMC. They were issued two per infantry battalion, or twelve per divisional special weapon battalion, until early 1945. The M3 75mm GMC pictured is shown in service with the Special Weapons Company, 2nd Marine Division, on the island of Tinian in July 1944. The vehicle shown has had its base olive drab over-painted in areas with earth yellow. There is no national symbol or tactical or registration number on the vehicle. Pictorial evidence shows other Marine Corps M3 75mm GMCs in a variety of camouflage schemes, some with a few markings. (© Osprey Publishing Ltd.)

The International Harvester Company (IHC) version of the half-track personnel carrier M3/M3A1 was the half-track personnel carrier M5A1 shown here. An identifying feature of all IHC vehicles was the rounded rear corners of the rear hull as compared with the squared rear hull corners on all the vehicles built by other firms. (TACOM Historical Office)

As a result, almost all M5/M5A1 and M9A1 armored half-tracks were allocated to Lend-Lease, except for some that were employed in the United States as training vehicles. In total, the International Harvester Company built 4,625 units of the M5, 2,959 units of the M5A1, and 3,433 units of the M9A1.

The biggest recipients of the International Harvester Company's armored half-tracks were the British Army and the Red Army. The British Army eventually passed on some of these same vehicles to the Free French Army.

As with the M2 and M3 armored half-tracks, some thought had been given to rationalizing production by doing away with the half-track car M9A1 in favor of a modified version of the half-track personnel carrier M5. A single pilot version of a modified M5, referred to as the M5A2, was built, but the changing military requirements meant it was never standardized. The Ordnance Department later classified the M5A2 as substitute standard.

VARIANTS

German air superiority during the invasions of Poland in late 1939 and France and the Low Countries in the summer of 1940 had led to heavy losses in vehicles among its opponents. To deal with this future threat, the Ordnance Department set about developing suitable antiaircraft vehicles. As with the early tank destroyers and self-propelled artillery, the half-track personnel carrier M3 was chosen as the expedient mount for such weapon systems.

ATION · ENGINEERING STANDARDS VEHICLE LABORATORY · DETROIT, MICH.   W.A.# E.L. 1

CAR, HALF-TRACK, M9A1 - I.H.C.   ORD.# 2583
w - with .30 and .50 cal. Machine Guns mounted on vehicle.

The first of these new antiaircraft armored half-tracks was designated the multiple gun motor carriage M13. Placed in the rear compartment of the vehicle was an electrically powered, one-man armored turret armed with two air-cooled .50 caliber machine guns. It was designed and built by the W. L. Maxson Corporation and standardized in March 1943 as the twin .50 caliber machine gun mount M33. To allow the gun mount M33 to fire at minus 10-degree elevation the upper portions of the vehicle's rear compartment were hinged to fold downwards if the need arose.

The White Motor Company built a total of 1,103 units of the 18,500-pound M13 between January and May 1943. Of that number, only 139 made it overseas with the U.S. Army. A large number of the remaining vehicles were converted into an improved version known as the multiple gun motor carriage M16. The main difference between the five-man M13 and the five-man M16 was the fact that there was a new gun mount, designated the M45, in the M16 that was armed with four air-cooled .50 caliber machine guns, doubling the firepower.

To meet the demand for more armored half-tracks, the International Harvester Company (IHC) was brought into the building program. Their standardized counterpart to the half-track car M2 was designated the half-track car M9A1. A spotting feature of all the IHC-built vehicles was a different front fender arrangement. (TACOM Historical Office)

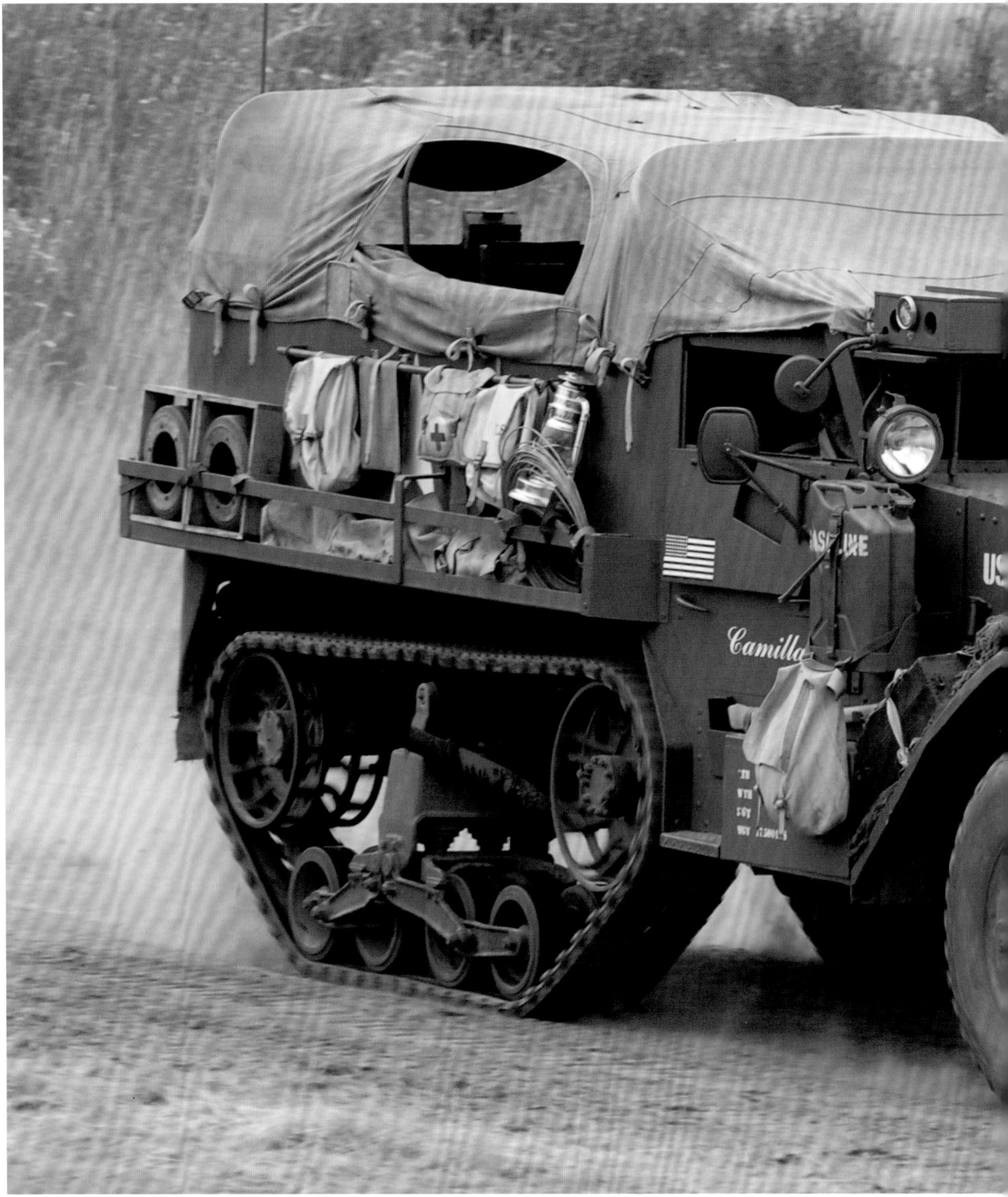
Camilla

On display at "Tankfest," held every year at the Tank Museum, Bovington, is this restored half-track personnel carrier M5A1 being driven around by its owners for the viewing pleasure of the assembled guests. This particular vehicle has its overhead canvas foul weather gear fitted. (Christophe Vallier)

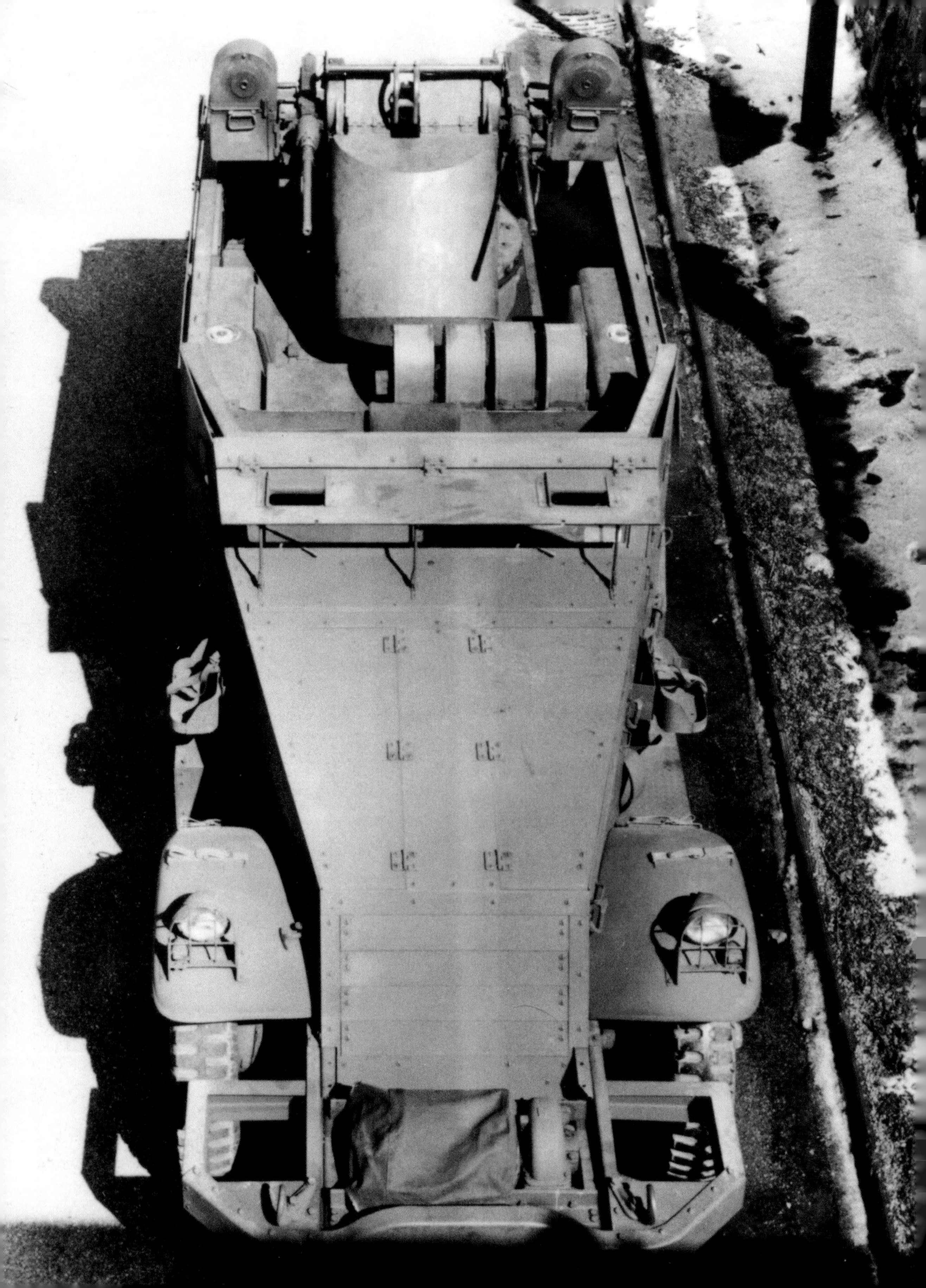

The M16 was approved for standardization in December 1942. Series production of the 19,000-pound vehicle by the White Motor Company began in May 1943 and concluded in March 1944 with 2,877 units being built. Counting additional half-track personnel carrier M3s converted to the M16 configuration from other roles, there were 3,614 units of the vehicle assembled. The M13 was reclassified as substitute standard when replaced by the M16.

Prior to the invasion of France in the summer of 1944, the Chief Ordnance Officer of the First U.S. Army felt that there were an insufficient number of M16s in the inventory. What there was in the inventory in adequate numbers was a towed, four-wheel trailer fitted with the gun mount M45 armed with four air-cooled .50 caliber machine guns. He therefore ordered that these gun mounts be stripped from their trailers and mounted on surplus half-track personnel carrier M3s or the half-track car M2. First Army documents report that 332 such conversions took place. The converted vehicles were unofficially called the M16B and saw service with the U.S. Army until the war in Europe concluded, and then up through the Korean War.

The M16B differed from the M13 and M16 as the sides and rear of the vehicle's rear compartment did not fold down to allow the gun mount M45 to engage targets at minus 10 degrees. To compensate for this, the gun mount M45 was raised 12 inches off the floor of the vehicle.

The twin gun mount M33 was also fitted to the International Harvester Company's half-track personnel carrier M5, which resulted in the designation multiple gun motor carriage M14. There was a total production run of 1,605 units of the 9.6-ton

OPPOSITE

In the early stages of World War II, the U.S. Army recognized a need for an antiaircraft vehicle. The answer was the mounting of a power-operated turret armed with two .50 caliber machine guns on the chassis of the half-track personnel M3. The vehicle seen here was designated the multiple gun motor carriage M13. (TACOM Historical Office)

No doubt somebody who looked at the multiple gun motor carriage M13 quickly figured out that if they added two more .50 caliber machine guns the vehicle might increase its effectiveness. This resulted in the fielding of the multiple gun motor carriage M16 shown here. (Michael Green)

M14 between December 1942 and December 1943, with all being allocated for Lend-Lease. When the quad gun mount M45 was fitted to the International Harvester built half-track personnel carrier M5, it was re-designated as the multiple gun motor carriage M17. The firm built 1,000 units of the 9.85-ton M17 between December 1943 and March 1944, all of which were supplied to the Red Army under Lend-Lease.

COMBINATION ANTIAIRCRAFT GUN HALF-TRACKS

Based on a request from the Coast Artillery Branch of the U.S. Army, the Ordnance Department came up with another antiaircraft vehicle, this time based on a modified half-track car M2. It was fitted with a manually operated gun mount that contained a single 37mm gun M1A2 and two water-cooled .50 caliber machine guns. The pilot versions of the vehicle were referred to as the multiple gun motor carriage T28. Based on testing of the pilot vehicles, the Coast Artillery Branch rejected the vehicle as being unstable and hence inaccurate. The Ordnance Department suggested to the Coast Artillery Branch that the weapon system be mounted on the larger chassis of the half-track personnel carrier M3, but this idea was rejected and the project was ended in April 1942.

In the summer of 1942, the U.S. Army identified a need for a dual-purpose weapon that could double as both an antiaircraft and antitank weapon. The Ordnance Department thought that a modified version of the multiple gun motor carriage T28, this time based on the chassis of the half-track personnel carrier M3, would be the perfect candidate for that role. The newest version of this weapon system was labeled the multiple gun motor carriage T28E1.

Eighty units of the T28E1 were built by the Autocar Company in time to take part in the American military invasion of North Africa in November 1942, codenamed Operation *Torch*. The U.S. Army Air Force did not enjoy air superiority in the North African Theater of Operations, and the T28E1 proved a most useful vehicle in warding off attacks by the German Luftwaffe. In the AAR *Operations of the 1st Armored in Tunisia*, Major-General E. N. Harmon stated: "The combination antiaircraft weapon of two .50 caliber machine guns and one 37mm gun was the most efficient antiaircraft weapon in the division."

Based on a British Army requirement for a vehicle mounting a weapon able to perform the role of an antiaircraft and antitank gun, the U.S. Army came up with the multiple gun motor carriage T28E1 seen here. It was armed with two water-cooled .50 caliber machine guns and one 37mm automatic cannon. (Patton Museum)

The U.S. Army was generally pleased with the multiple gun motor carriage T28E1 and continued refining the concept. The end result was the fielding of the multiple gun motor carriage M15 shown here. It retained the 37mm automatic cannon, but the water-cooled .50 caliber machine guns were replaced with air-cooled versions. (Patton Museum)

The Antiaircraft Artillery Board of the U.S. Army liked what they saw with the stopgap T28E1, and decided an improved version with two air-cooled .50 caliber machine guns in place of the two water-cooled .50 caliber machine guns would make more sense. That improved vehicle was referred to as the multiple gun motor carriage M15. Unlike the T28E1, the M15 had an open-topped armored shield around the front and sides of the weapons to protect the gun crew.

The first series production unit of the 10-ton M15 rolled off the assembly line of the Autocar Company in February 1943. Production of the seven-man M15 ended in April 1943 with a total of 600 units built. It remained in U.S. Army service until the end of the war in Europe when it was reclassified as limited standard.

There was also an improved version of the M15 constructed by the Autocar Company, referred to as the M15A1. The key external spotting feature of the 10-ton M15A1 was the mounting of the two air-cooled .50 caliber machine guns above the 37mm gun. On the M15, the two .50 caliber machine guns were mounted below the 37mm gun. The Autocar Company built 1,652 units of the M15A1 between October 1943 and February 1944. One hundred units of the M15A1 were supplied to the Red Army under Lend-Lease. The vehicle was reclassified as substitute standard when the war in Europe concluded.

Following the fielding of the multiple gun motor carriage M15, a slightly different version appeared in service designated the M15A1. It had a different gun mount with the two air-cooled .50 caliber machine guns installed below the 37mm automatic cannon, as is seen in this photograph. (Patton Museum)

During World War II, an Ordnance Depot located in Australia modified a small number of half-track personnel carrier M3s to mount a 40mm automatic gun in an open-topped turret as seen here. These improvised vehicles are sometimes referred to as the "M15 Special," despite the fact they were not based on the multiple gun motor carriage M15. (Patton Museum)

CHAPTER EIGHT

SELF-PROPEL
ARTILLERY

U.S.A.
W-403610

CATHY

In the years before America's entry into World War II, senior Army officers in charge of the field artillery showed little interest in the development of self-propelled guns or howitzers, preferring towed artillery. They believed that towed artillery was more dependable than a complex, self-propelled tracked chassis, which was more prone to mechanical problems. It was also felt that towed artillery pieces would be less conspicuous on the battlefield than a larger, self-propelled tracked vehicle. Most important, with the dearth of funding during the interwar period, it was the cost that pushed the Field Artillery Branch to stick with towed artillery, as it was more affordable than self-propelled artillery.

The Ordnance Department continually pushed field artillery officers to explore the possibility of adopting self-propelled artillery. The field artillery leadership finally had a change of heart with the outbreak of World War II and the expansion of the Armored Force in 1941. The first tangible example of the field artillery's new-found interest in self-propelled artillery was the half-track 105mm howitzer motor carriage (HMC) T19, which entered service in January 1942. It remained in production for only four months, with 324 units built by April 1942. The 10-ton HMC T19 was based on the chassis of the half-track personnel carrier M3 and mounted the 105mm howitzer M2A1 in a fixed forward-firing position with limited traverse and elevation. The vehicle had a crew of between six and seven men.

From the beginning, a full-tracked chassis was preferred, and the half-track configuration was chosen only in the name of expedience until such time that a full-tracked chassis could be made available to mount a 105mm howitzer. The HMC T19 saw service during the fighting in North Africa, and some lasted in service with the U.S. Army until August 1944 during the invasion of Southern France. The vehicle was declared obsolete in July 1945.

The successful mounting of the 105mm howitzer M2A1 led to the subsequent development and fielding of an

In a rush to field a self-propelled howitzer, in fall 1941 the U.S. Army pushed into service the half-track 105mm howitzer mortar carriage T19. It combined the strengthened frame of a half-track personnel carrier M3 mated with the 105mm howitzer M2A1. The pilot of the T19 is shown here. (Patton Museum)

In October, 1941, the U.S. Army authorized the development of what became known as the 75mm howitzer motor carriage T30, seen here. It consisted of the half-track personnel carrier M3 mounting the 75mm howitzer MA1. The howitzer had entered service with the U.S. Army in 1927 and had a maximum range of 9,760 yards. (Patton Museum)

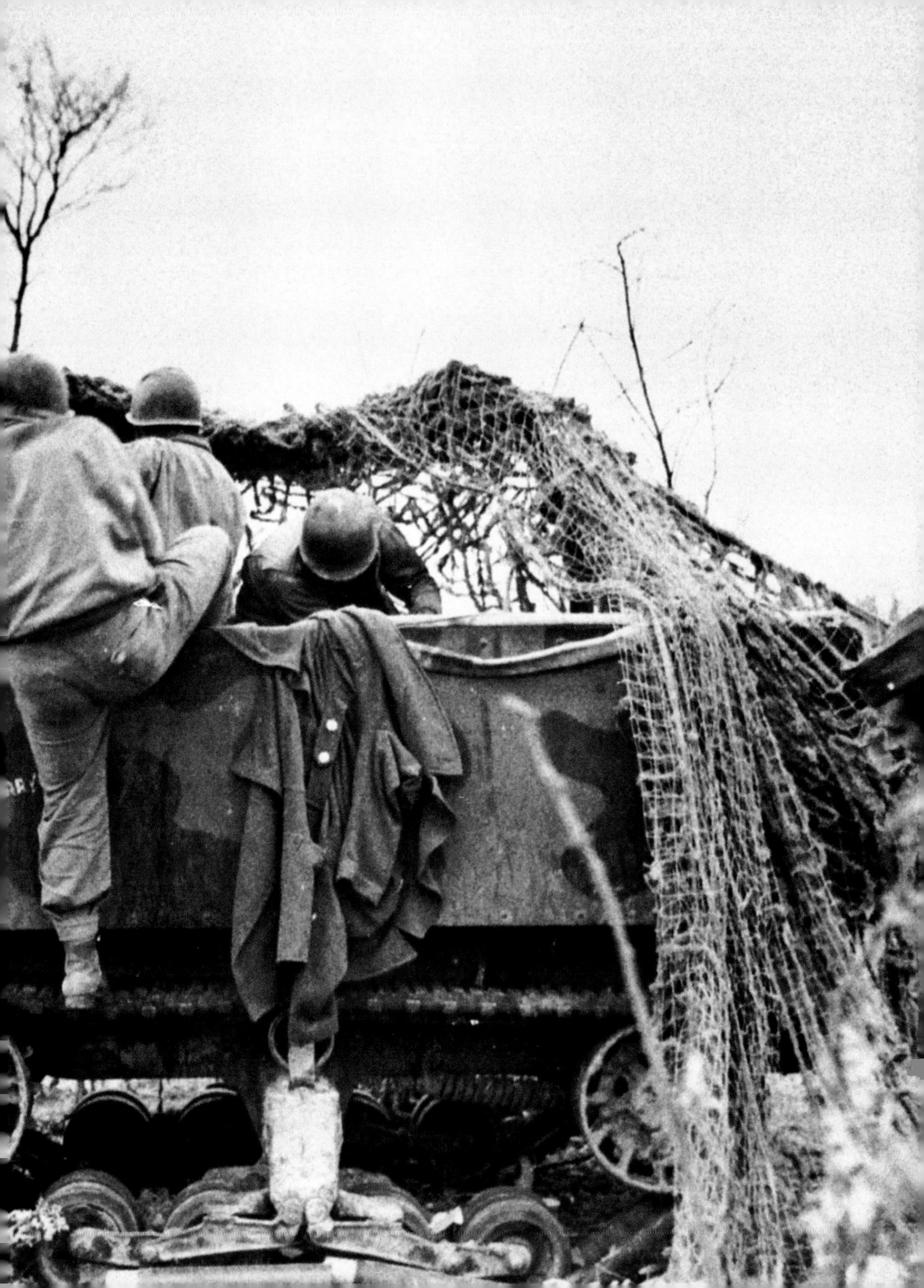

open-topped armored half-track mounting the 75mm howitzer M1A1. This combination was assigned the designation 75mm HMC T30 and had a five-man crew. The White Motor Company began series production of the 10.25-ton vehicle in February 1942. Production continued through November 1943 with a total of 500 units completed.

During the fighting in North Africa, the 75mm HMC T30 performed the role of assault gun in armored reconnaissance battalions with great success. It provided both indirect and direct fire when called upon and proved to be one of the few weapons, and the only one organic to these units, that could deal with German medium tanks. In the *Cavalry Journal* article "Reconnaissance Lesson from Tunisia," Lieutenant-Colonel Charles Hoy, commander of the 81st Armored Reconnaissance Battalion (ARB) stated: "We are sold on the assault gun. It gives us poise and confidence."

Thought was given to mounting the 105mm howitzer M3 on a modified half-track personnel carrier M3. In this configuration the vehicle would have been designated the 105mm HMC T38. However, the vehicle never progressed past the concept stage.

Unlike the various tank destroyers the U.S. Army employed during the fighting in North Africa that senior officers criticized so harshly for their poor performance, the M7 105mm HMC was an overwhelming success story based on its combat debut in North Africa. The M7 pictured is in the markings of the Cannon Company, 34th Infantry Division, at Rabat, Morocco, in December 1942. The base olive drab has been over-painted in areas with a sand color. There is a large national symbol on the hull side as well as the vehicle's nickname, both in white. However, no registration number is visible. (© Osprey Publishing Ltd.)

FULL-TRACKED SELF-PROPELLED ARTILLERY

The replacement of the HMC T19 was standardized in April 1942 as the 105mm HMC M7, which consisted of the 105mm MA1 howitzer based on the opened-topped chassis of the medium tank M3. The maximum armor thickness on the vehicle was 0.5 inches. Series production of the 25.3-ton vehicle began in April 1942 and continued until August 1943 with 2,814 units completed. The seven-man vehicle first saw combat with the British and U.S. Armies in North Africa and was named the "Priest" by the crews due to the pulpit-like appearance of the .50 caliber machine gun mount located at the right front of the vehicle's hull.

Lieutenant-Colonel Douglas G. Dwyre, in "The Field Artillery Puts on Armor," an article in the December 1943 issue of *The Field Artillery Journal*, described the effect of fielding the brand new M7:

> The Armored Field Artillery Battalion rolls into action, with big track laying howitzer carriages crashing through brush and timber, spanning gullies, swinging into position close behind the first wave of tanks, ready in a matter of minutes to lay crushing fire on the enemy vehicles and

The U.S. Army's replacement for the half-track-based howitzers was the 105mm howitzer motor carriage M7, pictured here on display at Aberdeen Proving Ground. Early-production vehicles were based on the chassis of the M3 medium tank powered by an air-cooled radial engine. Later production units incorporated components from the M4 series of medium tanks. (Christophe Vallier)

installations visible ahead. Only a few months ago this same battalion hitched its 105s to 2½ ton trucks, dug them in well behind the lines, and delivered fire at 6,000 yard range over the heads of advancing infantry. This change from an infantry to an armored battalion involved reorganization of the firing battalion, adoption of new equipment, a fundamental revision of tactics, and the retraining of all personnel. The end result is a battalion with a more powerful punch and greater maneuverability, ideal support for the tanks and armored infantry in combat.

As time went on the HMC M7 was improved by incorporating components from the M4 series tanks. These included differential and final drive housings, as well as improved suspension system components. A total of 3,490 of these improved HMC M7s were built between March and October 1944. At the same time, another 826 units of the HMC M7 powered by the Ford GAA engine from the M4A3 were built. Reflecting the change in the engine, these vehicles were designated the M7B1 and were in production up to the end of World War II. The M7/M7B1 were reclassified as substitute standard in January 1945.

M4 SERIES SELF-PROPELLED HOWITZER

One of the original characteristics of the first-generation M4 series tanks was the removable front plate of the turret, constituting the gun mount, which could accommodate any one of five different combinations of weapons. One of the suggested combinations was that of a 105mm howitzer with a coaxial .30 caliber machine gun. Work on standardizing this configuration began in early 1942.

The final version of the 105mm howitzer motor carriage M7 is pictured here and was designated the M7B1. It employed the chassis and power plant of the M4A3 medium tank. External differences between the late-production units of the M7 and the M7B1 are few except for the grill work on top of the rear engine deck and the angle of the upper rear hull. (TACOM Historical Office)

In November 1942, two first-generation M4A4s were modified to permit installation of the M2A1 105mm howitzer and were designated M4A4E1 by Ordnance Committee action in December 1942. The mount itself was designated the T70. Modifications to the vehicle included substitution of 105mm ammunition containers in place of the smaller 75mm main gun racks, installation of an elevation-only gyrostabilizer designed for the 105mm howitzer mount and minor changes in internal stowage. Inclusion of the stabilizer clearly reflected the intention to use these vehicles as direct-fire assault guns, rather than as indirect-fire artillery as was the case of the M7, whose fire direction method precluded it from firing on the move.

During a conference held at Fort Knox, Kentucky, in February 1943, the many new modifications to the proposed 105mm-howitzer-armed version of the M4 series tanks received approval for further development. Other decisions made at this same conference included using a partial turret basket, omission of the power traverse and elevation-only gyrostabilizer, allowance for a partial recoil guard for the howitzer,

The howitzer on this M7 is in full recoil as the weapon has just been fired. Early-production units carried 57 main gun rounds for the onboard M2A1 105mm howitzer. This was later increased to 69 rounds in later production units by reducing the number of folding seats within the vehicle. (Patton Museum)

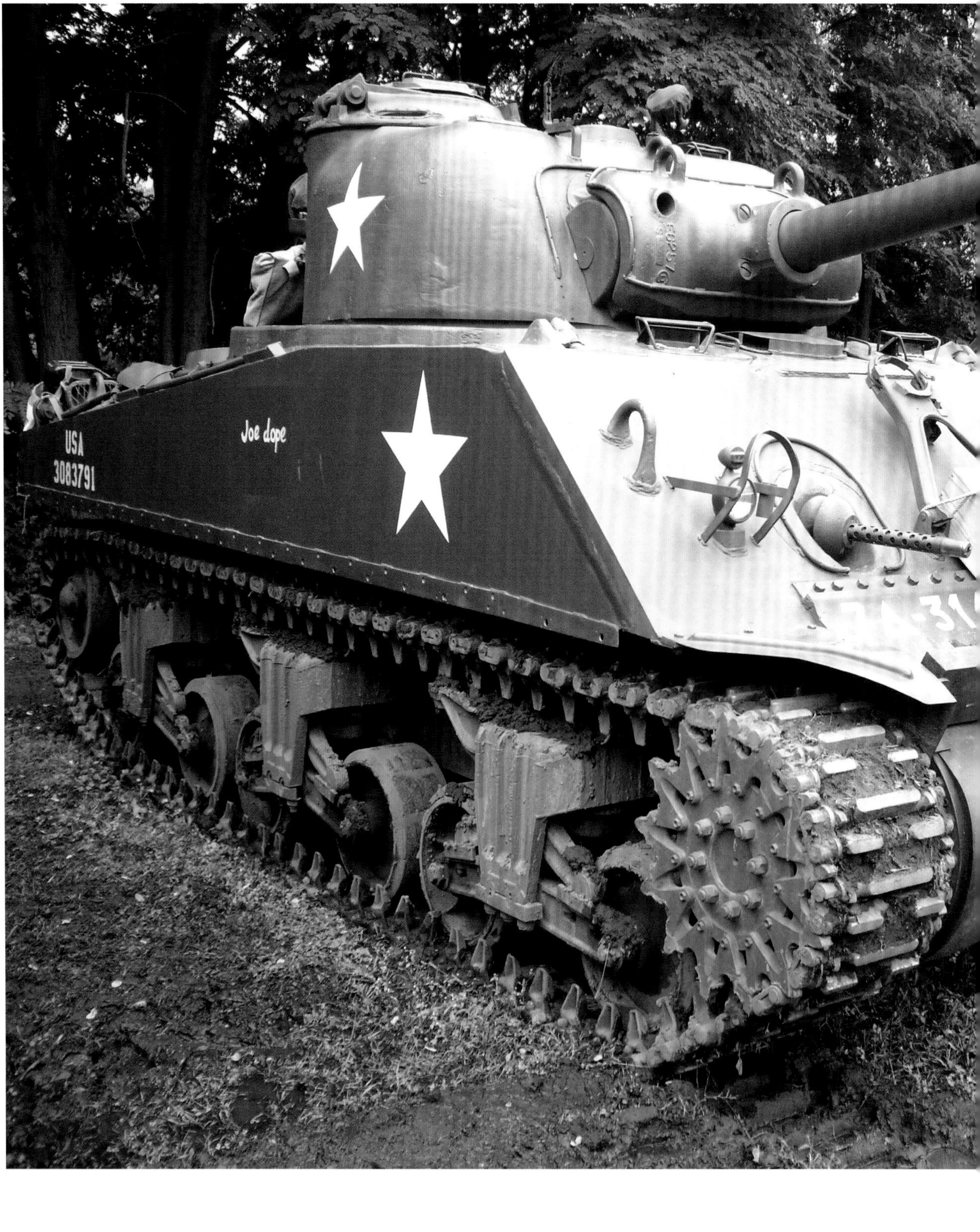
USA
3083791
Joe dope

and redesigning the interior to stow more 105mm howitzer ammunition. The increased ammunition load was a mix of HE and WP, with a few HEAT (High-Explosive Antitank) rounds added for good measure.

Two pilot models with a modified 105mm howitzer installation soon appeared and received the designation M4E5. Testing of the pilot models proved satisfactory, and the Armored Force concluded that the M4E5 was ready for fielding after the inclusion of a ventilating fan and an improved T93 direct sight telescope into the turret.

The production versions of the M4E5 used second-generation M4 or M4A3 tanks built by Chrysler, minus the wet stowage system. Designations for the vehicles included the M4(105) and the M4A3(105). Approximately 4,680 units came off the assembly lines between February 1944 and June 1945, with 1,641 being the M4(105) and the remaining 3,039 being the M4A3(105). Most rode on the HVSS system; however, some that saw service during World War II also rode into combat on the VVSS. An M4A3(105) that rode on an HVSS system weighed approximately 36 tons.

What American tankers thought of howitzer-armed M4 series tanks can be seen in a quote by Staff Sergeant Clarence A. Kreitzer, which appeared in the March 1943 U.S. Army report *United States vs. German Equipment*: "In all known respects our 105mm howitzer mounted in M4A3 tanks is superior to any comparable German assault gun. Certainly it is superior to the German 105mm assault gun, mounted on a half-track, in speed, maneuverability, traverse, armor-plating. It is superior to the German 105mm assault gun mounted on a Mark IV chassis in traversability [sic] (this tank has no turret). It is believed that our tank is more maneuverable and faster than the German tank."

As the 105mm howitzer motor carriages M7/M7A1 were open-topped, they were extremely vulnerable to a wide range of battlefield threats. To address this issue, the U.S. Army decided to field a turret-mounted 105mm howitzer M2A1 based on a modified M4 or M4A3 medium tank. Pictured is a medium tank M4(105). (Pierre-Olivier Buan)

A dozer-equipped medium tank M4A3(105). The vehicle had authorized storage space for 66 rounds of main gun ammunition. Spotting features for the M4A3(105) when compared with the standard second-generation M4A3 armed with a 75mm main gun include a thicker barrel and a different gun shield. (Mathieu George)

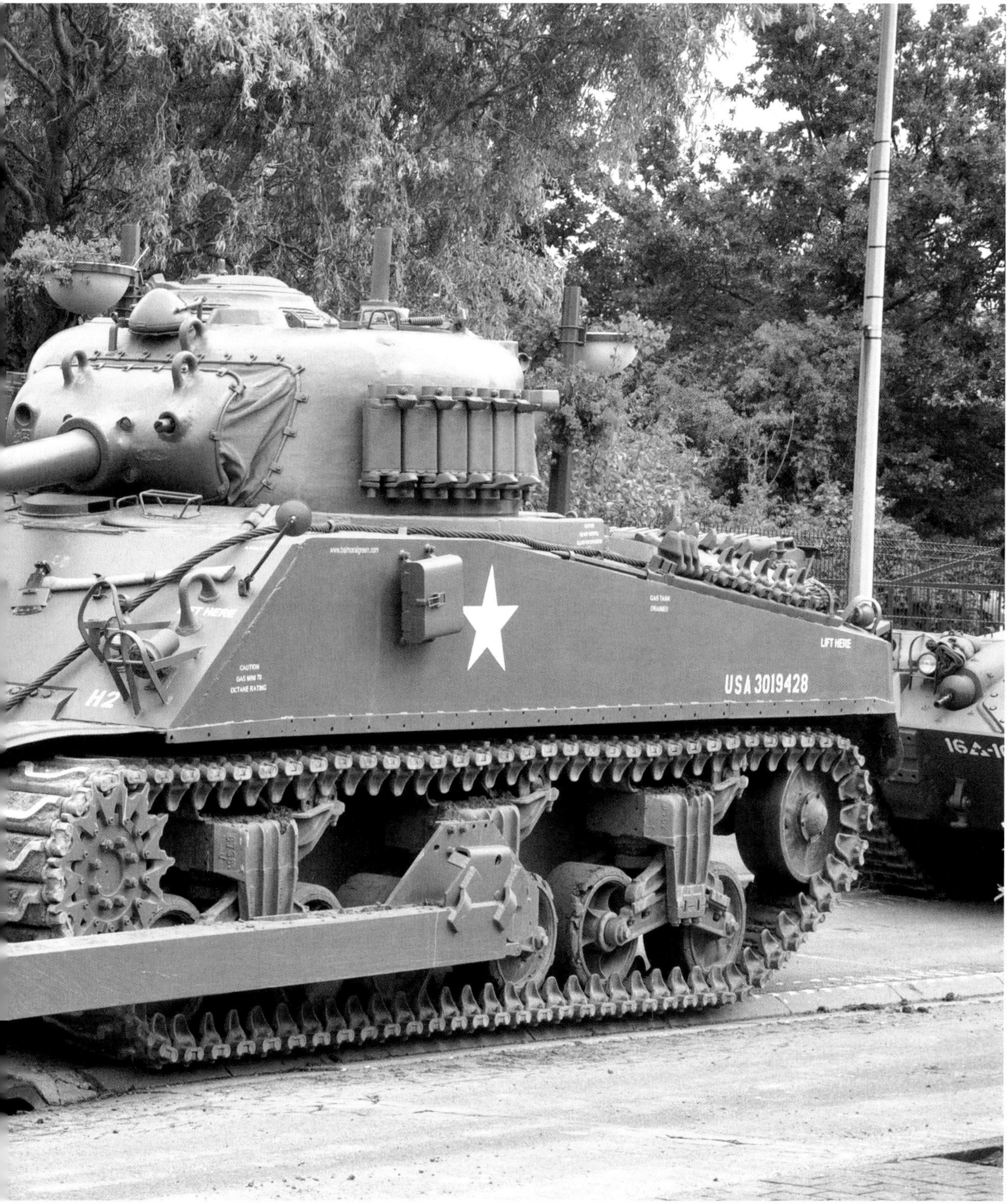
LIFT HERE
USA 3019428
H2

The M4(105) shown here and the M4A3(105) were the intended replacement for the 75mm Howitzer Motor Carriage M8 in the battalion headquarters companies of U.S. Army medium tank battalions. The 105mm howitzer mounted in the M4(105) and M4A3(105) was a redesign of the U.S. Army's standard 105mm towed howitzer M2A1, with a cut-down breechblock to better fit within the small confines of the existed M4 series medium tank turret. The towed version of the M2A1 howitzer began entering the U.S. Army inventory in 1940 and proved the workhorse of the American field artillery during World War II, with 8,536 units built. (Patton Museum)

From the same report were these additional comments on the M4(105) or M4A3(105), the first from Captain John A. McNary, 2nd Battalion 66th Armored Regiment, and the second by Sergeant Harold A. Rovang, assault gun platoon commander:

1. The most frequent complaint of the assault gun, 105 How, M4A3, is there is no power traverse. The men feel that if the vehicle is brought forward to be used as a direct fire weapon it should be able to traverse quickly.
2. I have seen HEAT fired from a 105mm howitzer at a Mark VI [Tiger] at 400 yards. The track was hit and damaged and a direct hit exploded on the turret which only chipped the paint.

Other comments about the use of the M4(105) or M4A3(105) were more positive when used against German late-war tanks. According to the S-3 Journal of the 3rd Armored Group, the 105mm HEAT rounds were "very effective" against Tiger tanks. A 1945 AAR by the 736th Tank Battalion stated than when 105mm HEAT rounds were fired at a captured Panther tank, they reliably penetrated the glacis at a range of

1161   5-9-44   ORDNANCE OPERATION - ENGINEERING STANDARDS VEHICLE LABORATORY - DETROIT, MICH.   W.A. # E.L. 1
TANK, MEDIUM, 105-MM HOWITZER, M4 - CHRYSLER. ORD. #56940
Right side view

350 yards. The downside of the test firing was the fact that the flight path of the HEAT rounds was very erratic, which would have made it very difficult to achieve any degree of accuracy at other than short range.

Each U.S. Army tank battalion featured six of the 105mm-armed M4 tank series howitzer tanks, three in the headquarters company and one in each of the three medium tank companies, although sometimes they were all grouped together under the command of the headquarters company. Private Tom Sator of the 4th Armored Division does not recall any of his battalion's howitzer tanks being a normal part of his medium tank company in the ETO.

The second-generation M4 and M4A3 tanks armed with the 105mm howitzer replaced the open-topped M7 self-propelled howitzer, also armed with a 105mm howitzer, in U.S. Army armored divisions. They provided indirect-fire support to the various platoons and companies of the battalion with HE and could supply a smoke screen at short notice with their WP round.

Taking part in an historical military event is this M4A3(105). The howitzer was designated the M4 and was 8 feet 4 inches in length and weighed 1,140lbs. The muzzle velocity of the standard high-explosive round was 1,550 fps. (Chun-Lun Hsu)

ROCKET ARTILLERY

No doubt influenced by the German use of towed rocket launchers, the Ordnance Department came up with the rocket launcher T34, nicknamed the "Calliope," mounted on the turret of first-generation M4 series tanks. It originally consisted of 60 molded plywood tubes that were 90 inches long stacked on top of each other in two separate banks. The top bank contained 36 tubes in two levels of 18 each, and the bottom bank was made up of two levels with 12 launching tubes each. Each launching tube was loaded with an M8 4.5-inch fin-stabilized HE rocket that packed the same explosive punch as a 105mm howitzer round. Maximum range on the M8 rocket was 4,100 yards.

Fully loaded, the T34 weighed 1,840 pounds and was originally elevated and depressed by a metal elevation arm that clamped onto a vehicle's main gun tube. As the tankers could not use their main gun without removing the T34 first, they soon began modifying the device so that the metal elevation arm that raised and

The usefulness of rocket launchers to provide a high density of fire ahead of assault troops soon attracted the attention of the U.S. Army. The first such system to be mounted on a tank was the rocket launcher T34, seen here on a heavily sandbagged first-generation M4A1 medium tank. It fired a fin-stabilized 4.5-inch rocket, M8 series. (Patton Museum)

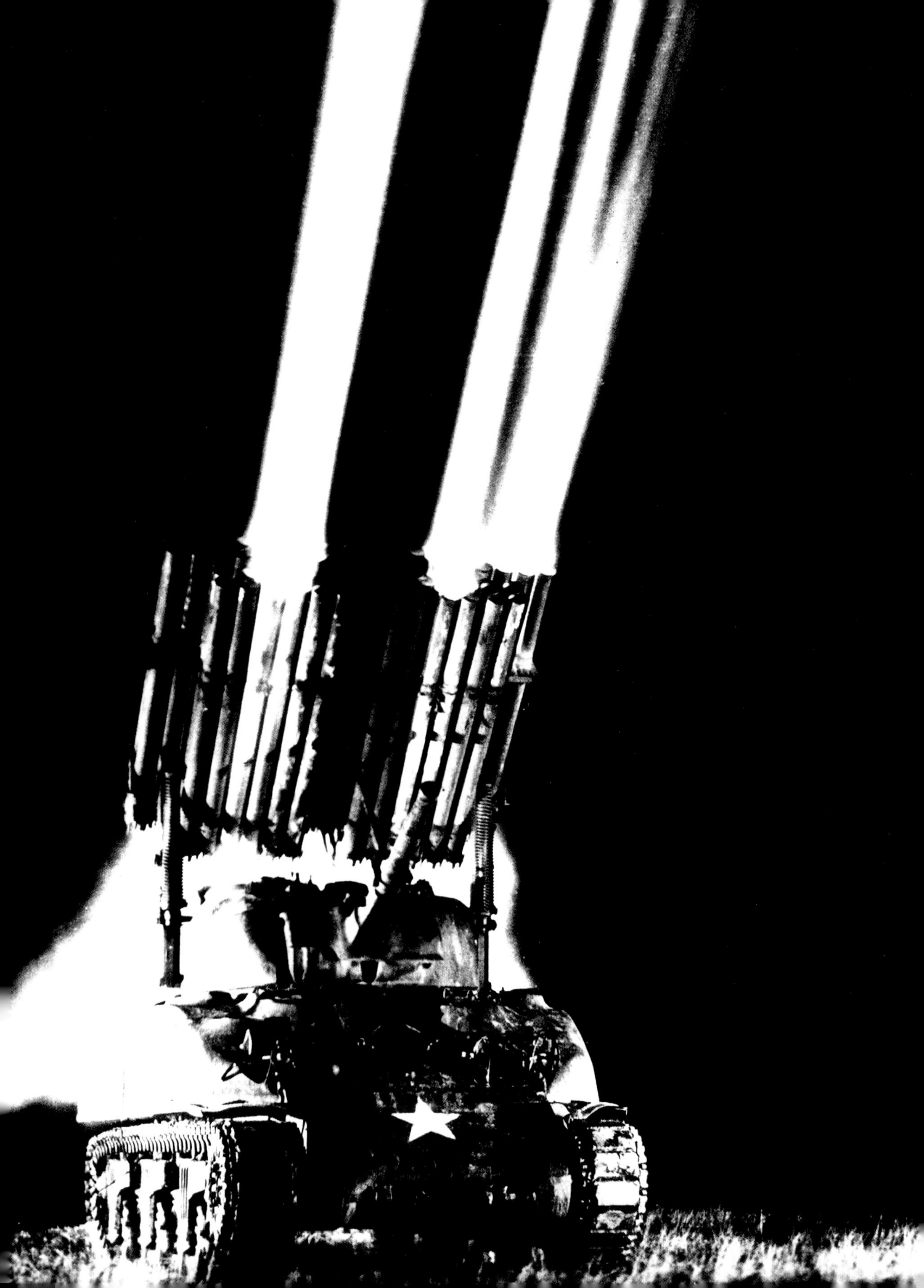

lowered the launcher assembly was attached to the top of the vehicle's gun shield, allowing for use of the tank's main gun. The Ordnance Department soon came up with an improved rocket launcher assembly designated the T34A1 that had the metal elevation arm attached to extensions from the gun shield, providing a full range of motion and allowing the vehicle's crew to use their main gun.

Because rockets have much larger dispersion than conventional tube artillery, the T34 was strictly an indirect-fire weapon. There are a number of reasons for this wide dispersion with World War II rocket weapons. Rockets travel at a low rate of speed when launched, which makes them more prone to drift deviation in contrast to a standard artillery projectile that travels at a much higher speed. Powder charges, particularly in that era, were not always uniform, leading to variations in thrust, affecting the trajectory.

The rockets launched from the T34 could be fired electrically from within the safety of the tank. The 60 rockets were fired at the rate of one round per half second with all being launched in 30 seconds. If needed, the entire T34 assembly could be either reloaded or jettisoned, so as to not hamper the tank crew as they continued their mission. The final version of the rocket launcher assembly used magnesium tubes instead of molded plywood tubes and was designated the T34E2.

Most tankers were not thrilled with the addition of the T34 and felt that was a job best left to the Artillery Branch. A reflection on this thought appears in a postwar U.S. Army report titled *Armored Special Equipment*, put together by the General Board United States Forces, European Theater:

> Users felt that the addition of the T34 rocket launcher to the normal tank armament unnecessarily complicated an already very complicated vehicle and detracted from its primary role of assault by direct fire. It was also felt that the launcher could be used more effectively when mounted on a truck or on the ground, protection being afforded by remote control firing. Specific deficiencies noted were: 1) Frequent misfires. 2) Excessive range dispersion, precluding overhead fire. 3) Jettison feature did not always work.

ROCKET SIDELINE

The Ordnance Department considered tank-mounted rocket launchers for use on engineer-armored vehicles to deal with enemy beach defensive obstacles during the planned Allied invasion of France. One of these concepts resulted in the development of the T40 rocket launcher, which fired twenty 7.2-inch rockets and was mounted on the roof of an M4 series tank. Nicknamed the "Whiz Bang," the mounting of the T40 was similar to that of the T34 as well as the manner in which it was fired. It also had a jettison feature like the T34.

Testing of T40s belonging to the 743rd Tank Battalion quickly led to the conclusion that it was hazardous not only to the crew of the vehicle upon which it was mounted, but to anybody in the general area. It was therefore not employed during the invasion of France on June 6, 1944, although pictorial evidence shows that the M4 series tanks that landed in France were still fitted with the mounting brackets for the T40.

Eventually the U.S. Army planned to mount 30 of the T40s on first-generation M4 series tanks belonging to the 743rd Tank Battalion for an operation planned for December 1944 in Belgium. In that operation, the T40-equipped tanks would be employed in support of an attack by the U.S. Army's 30th Infantry Division. The massive German offensive in the Ardennes that began on December 16, 1944 resulted in the cancellation of the American attack and the removal of all the T40s to a safe location. They were not reinstalled on the tanks of the 743rd Tank Battalion. Eventually, eight of the tank-mounted T40s were sent to a U.S. Army tank battalion fighting in Italy where it was intended that they be employed in artillery fire support missions.

One of the unusual vehicles that came out of the Ordnance Department's research on a suitable engineer-armored vehicle was referred to as the demolition tank T31, based on the M4 series tank. It consisted of a mass ive turret armed with a T94 7.2-inch rocket launcher mounted in a blister on each side of the vehicle's turret.

The M4 series of medium tanks was also employed to mount the 7.2-inch rocket launcher M17, an example of which is seen here fitted on a first-generation medium tank M4A1 somewhere in Italy. Unlike the rocket launcher T34, the rocket launcher M17 was armored reflecting its intended close-range employment in destroying enemy-emplaced defensive obstacles via direct fire. (Patton Museum)

The strangest looking vehicle developed during World War II by the U.S. Army was the demolition tank T31, with a specially designed turret armed with two 7.2-inch rocket launchers. As tests showed the design to be flawed, only the single pilot example of the vehicle seen here was built. (TACOM Historical Office)

The T31 turret itself had a centrally mounted dummy gun tube and two ball-mounted .30 caliber machine guns. This awkward-looking turret was mounted on the chassis of an M4A3(W) tank riding on the HVSS system. Inside the vehicle's turret were two revolving drums loaded with five rockets each, from which each rocket launcher was automatically loaded. Inside the vehicle's hull was storage space for an additional 30 rockets. To complement the rocket launcher there were provisions made to mount a flame gun. Work on this project was ended in January 1946.

LIGHT TANK SELF-PROPELLED GUNS/ HOWITZERS

The most numerous variant of the U.S. Army light tanks employed during World War II was the 75mm HMC M8 that was standardized in May 1942. This vehicle was based on the modified chassis of the light tank M5 fitted with a large, open-topped turret containing a 75mm howitzer M2 or M3 in a mount designated the M7. A two-man crew serviced the piece with the gunner on the left of the turret, and the loader on the right side of the turret. For indirect fire the gunner used the M12A5 panoramic telescope. For direct fire, the gunner originally used the M56 direct sight telescope until this was replaced with the M70C direct sight telescope.

The 75mm howitzer M2 and M3 mounted in the M8 were both derived from the towed pack howitzer M1A1, which was first fielded with the U.S. Army in the 1930s. The weapon could fire two types of HE rounds out to a maximum range of about 5 miles. For antitank purposes, the weapon was provided with a shaped-charge round, designated the HEAT M66 shell. Also carried on the M8 was a WP round. Secondary armament on the vehicle consisted of a .50 caliber machine gun on a ring mount at the rear of the turret.

The overhead hatches for the driver and assistant driver/bow gunner on the M5 were done away with on the M8, and ingress and egress for these personnel were provided through the open-topped turret of the vehicle. There was no Browning .30 caliber bow machine gun mounted in the glacis of the M8 as there was in the M5. Instead of the 360-degree periscopes provided to the driver and bow gunner on the M5 in their overhead hatches, the driver and assistant driver on the M8 each got a large hatch that

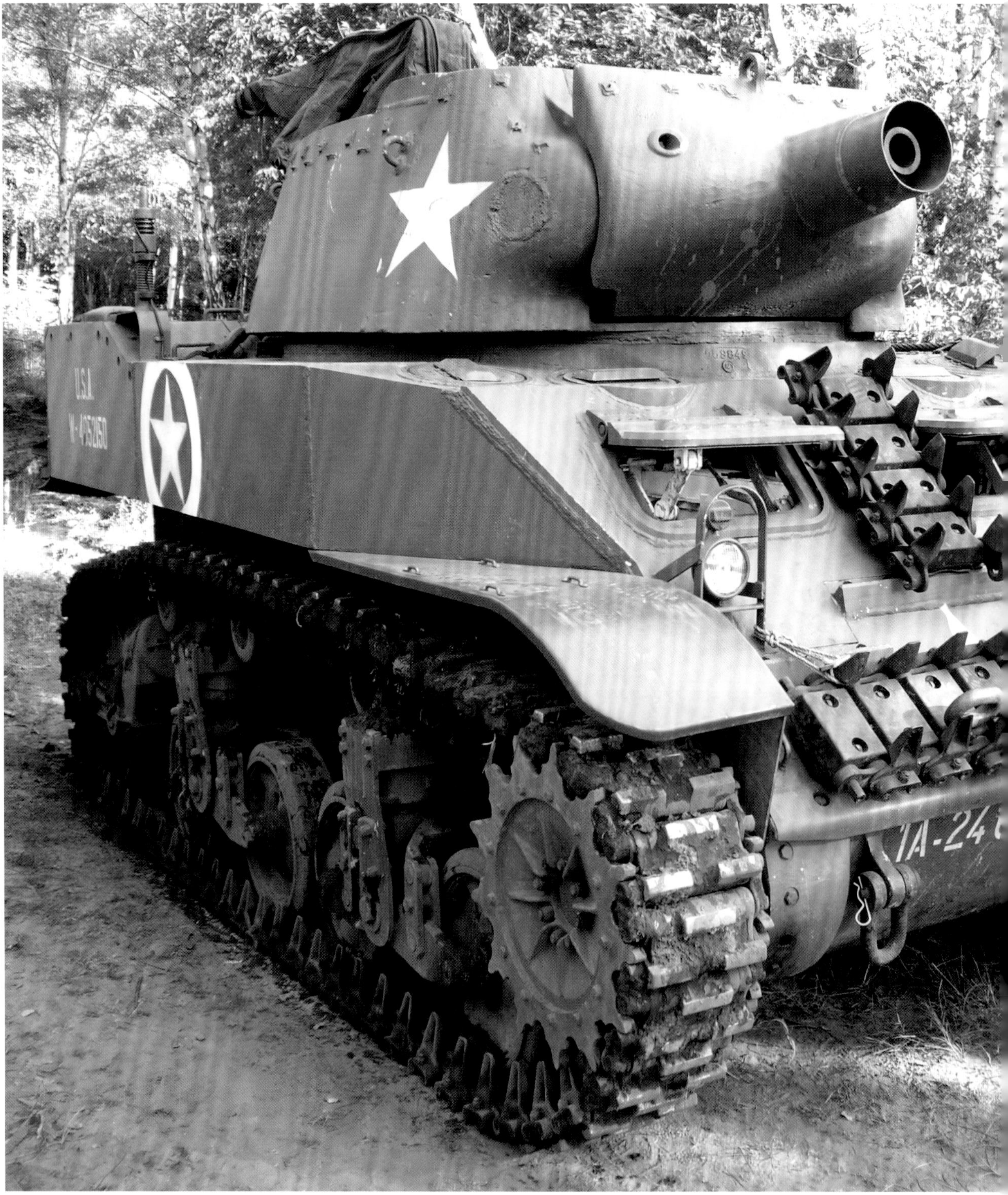
U.S.A.
W-4152160

could be opened for direct vision. When the need arose to close these large vision doors, they each had access to two M9 periscopes in the front hull roof. Maximum armor thickness on the vehicle was 1.75 inches (44mm).

Production of the 17.3-ton M8 began at the Cadillac Motor Car Division of the General Motors Corporation in September 1942, and continued until January 1944, with a total run of 1,778 vehicles. One hundred and seventy-four units of the M8 were provided under Lend-Lease to the Free French Army during World War II.

In the U.S. Army, the M8 was typically assigned to armored cavalry squadrons in order to provide close-in fire support when needed against enemy defensive positions or unarmored targets. A February 24, 1944 War Department manual titled *Cavalry Reconnaissance Troop (Mechanized)* gives this description of the role of the M8:

> Assault guns are used to supplement other available weapons in the base of fire. Their fire is coordinated with that of mortars, 37mm guns, and automatic weapons, to avoid duplication of targets. Both HE and smoke ammunition are available. The former is employed to destroy antitank gun and machine gun crews and tanks, and to pin other enemy personnel to the ground, providing freedom of maneuver for other elements. Smoke is employed to mask the action of maneuvering elements and to prevent observation by enemy gunners.

In Captain Michael J. Reagor's article, "The Guns of the Cavalry," published in the November–December 1991 issue of *Armor* magazine, Lieutenant-Colonel Preston Utterback, former commander of the 43rd Squadron, 3rd Cavalry Group, described his unit's use of the M8:

> We were committed to combat in August 1944. I shall never forget our first encounter with dug-in German 88 guns and tanks. We countered with frontal fire, called for indirect fire from our assault gun troop, and sent a recon troop around the left flank. They dropped those 75mm HE shells with pinpoint accuracy and destroyed enemy equipment and personnel far beyond our expectations. We used this strategy time and time again in the next nine months we spent in combat.

A useful addition to the inventory of U.S. Army self-propelled howitzers during World War II was the 75mm howitzer motor carriage M8, based on the modified chassis of the M5 light tank. The driver and assistant driver lacked overhead hatches on the M8 and entered and left the vehicle via the open-topped turret. (Pierre-Olivier Buan)

During a pause in between fire missions, a member of the crew of a 75mm howitzer motor carriage M8 uses a swab to clean the barrel of the vehicle's main weapon. The vehicle had authorized storage space for 46 main gun rounds and mounted either the 75mm howitzer M2 or M3. (Patton Museum)

The M8 took over the job once done by the 75mm HMC T30. The M8 also served in U.S. Army tank battalions until replaced by other self-propelled howitzers, such as the M7 or the M4(105) or M4A3(105). It remained in use with some U.S. Army tank battalions in the PTO until the end of the war.

There was also a 105mm-howitzer-armed version of the M24 developed during the latter stages of World War II that was designated the 105mm HMC M37. Looking much like a smaller version of the 105mm HMC M7, the seven-man vehicle weighed 16 tons and carried an ammunition load of 58 rounds. Series production of this open-topped vehicle had been approved in January 1945. However, the end of the world-wide conflict resulted in only 150 vehicles being completed, with none being shipped overseas to see combat.

M26 HOWITZER TANK

A variant of the M26 tank that entered into production too late to see service in World War II was a 105mm howitzer-equipped version, eventually designated heavy tank M45. Series production of the vehicle began in July 1945 and ended with the conclusion of the conflict, with only 185 units being completed by the end of 1945. The M45 came about because of the success of the M4(105) and M4A3(105) and the desire by the U.S. Army to have as many weapons mounted on a common chassis as possible to ease the logistical burden and the training of maintenance personnel.

Since the 105mm howitzer mounted in the M45 was much lighter in weight than the normal 90mm gun M3, it meant the turret gun shield on the

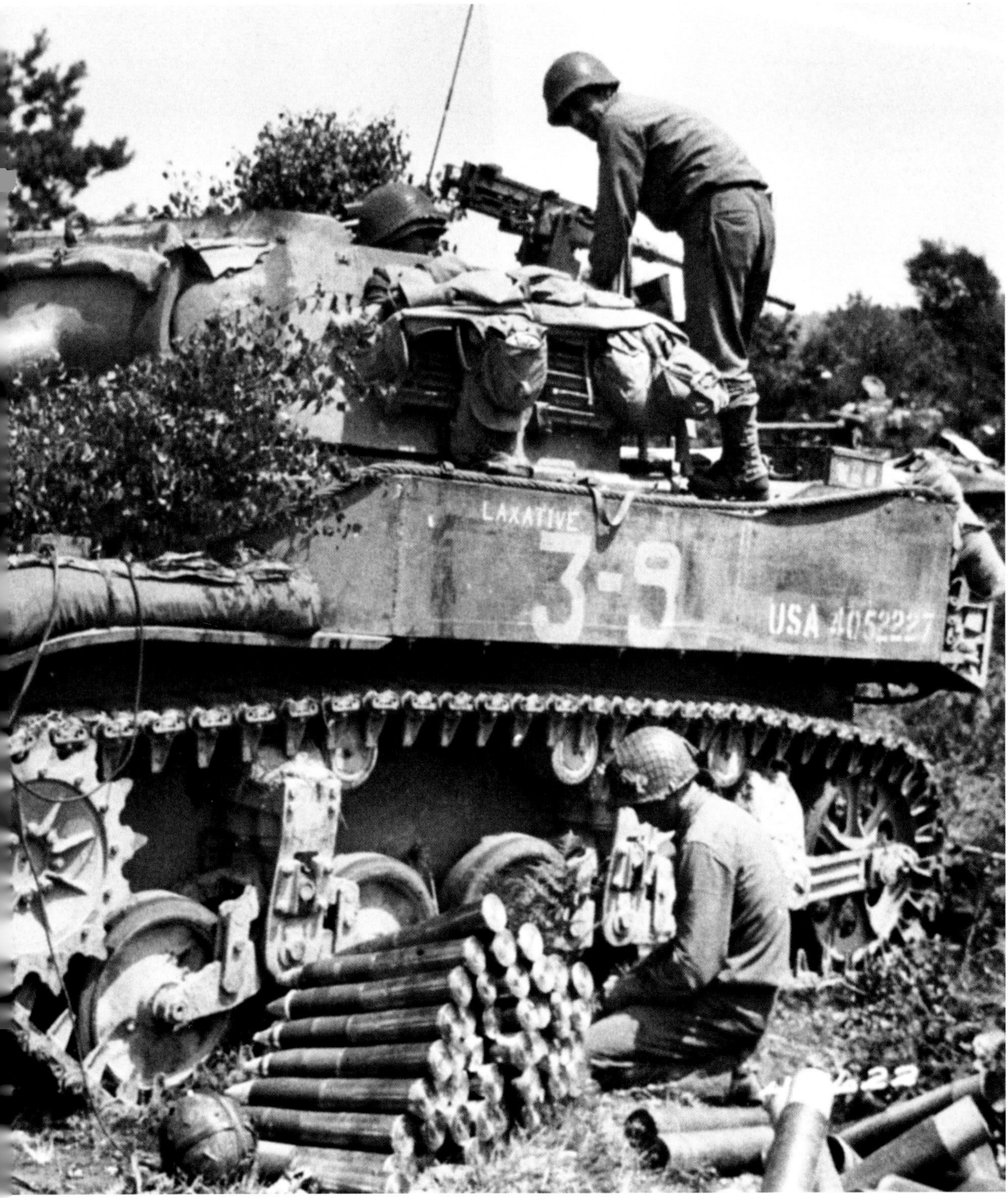
LAXATIVE
3-9
USA 4052227

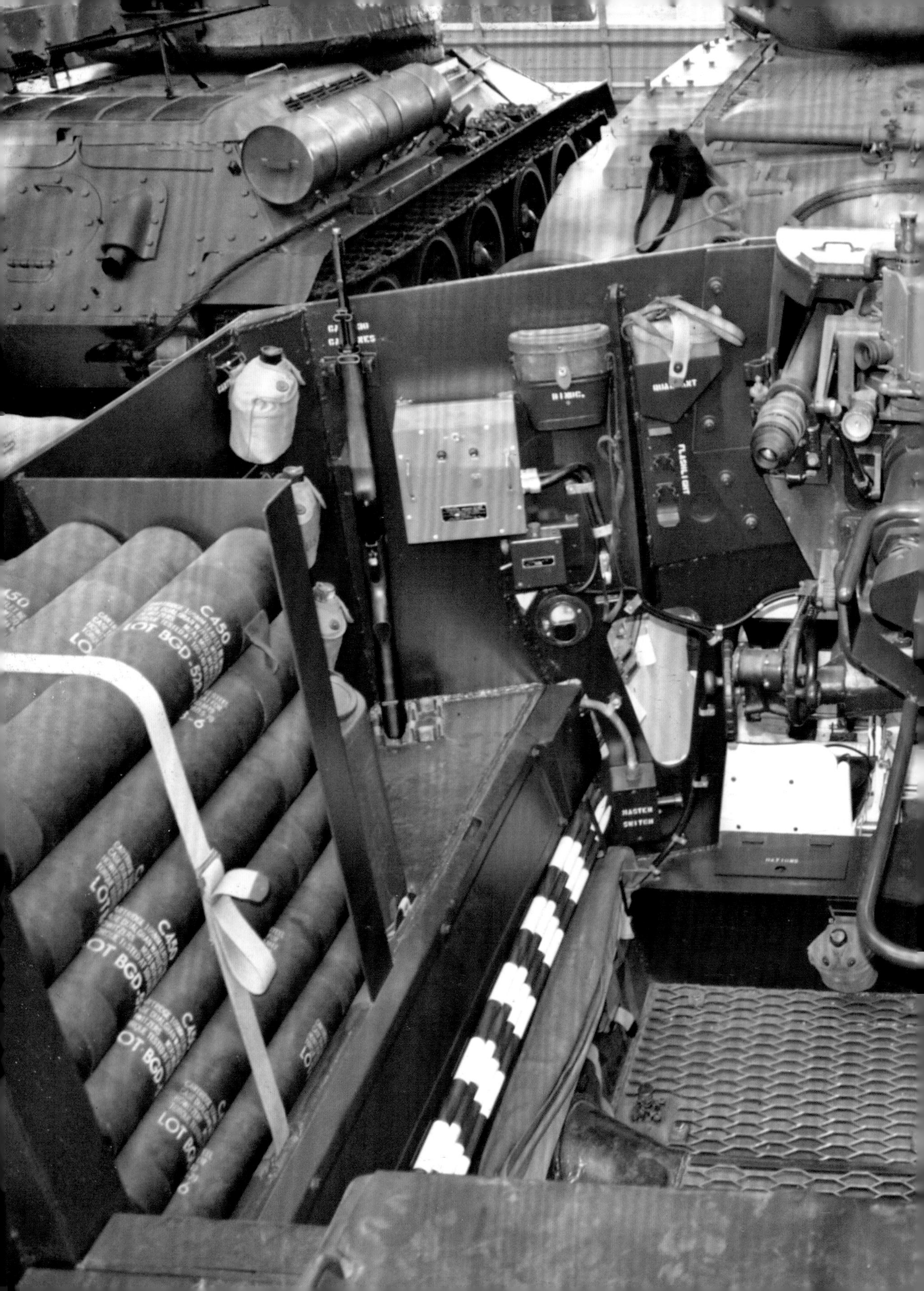
CA50
LOT BGD
QUADRANT
FLASHLIGHT
MASTER
SWITCH

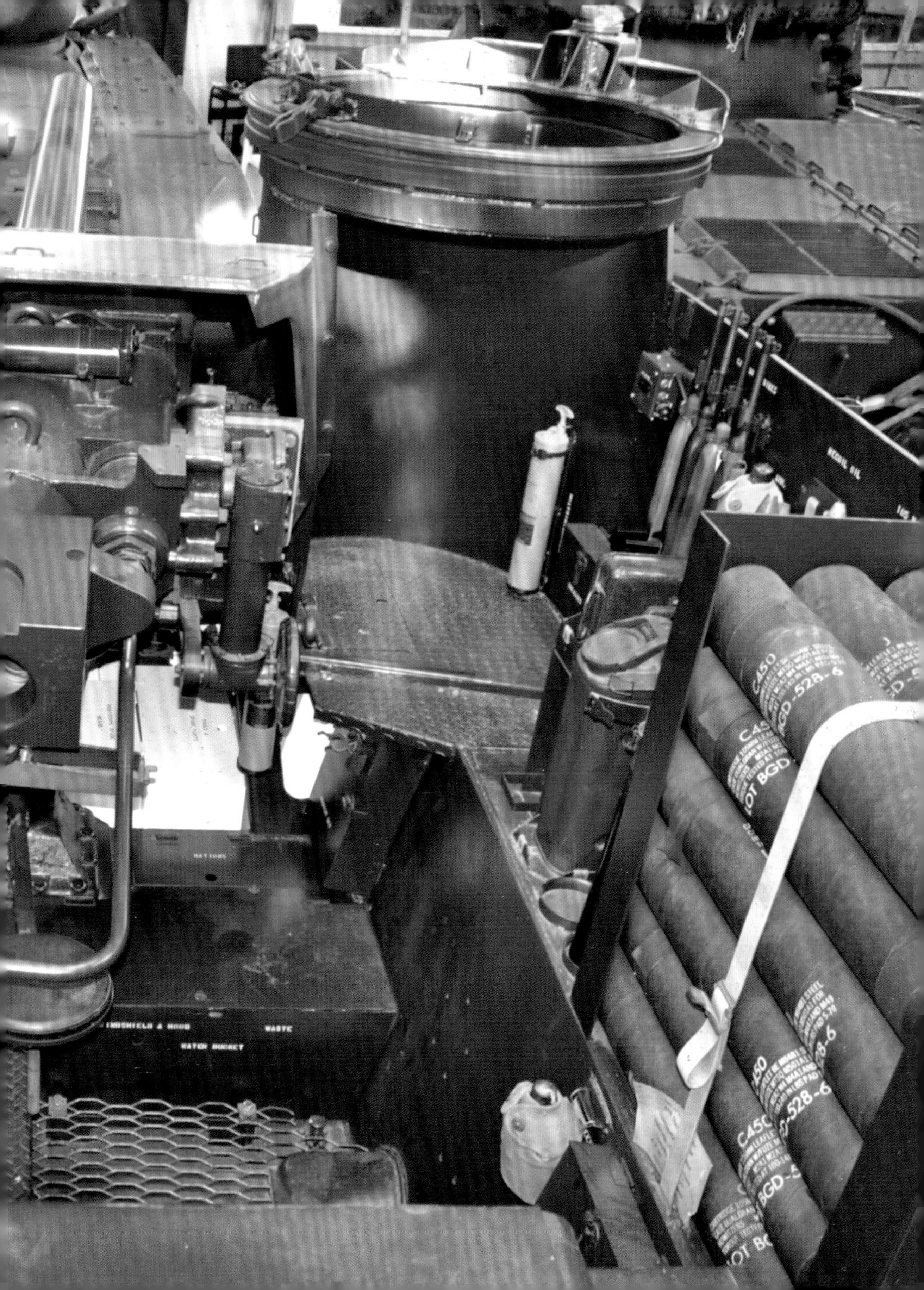

C450
LOT BGD-
BGD-528-6
WASTE
WATER BUCKET

PREVIOUS PAGE
Looking into the fighting
compartment of a 105mm
howitzer motor carriage M37
belonging to the Military
Vehicle Technology
Foundation, one can see the
tight confines for the crew to
load and fire the howitzer.
Notice the recoil guard around
the rear of the breech ring. The
vehicle had authorized storage
space for 126 main gun rounds
stored horizontally on either
side of the hull. (Chris Hughes)

In June 1941, the U.S. Army
authorized development of a
self-propelled gun that joined
a World War I-era 155mm gun
and the open-topped chassis
of the medium tank M3. That
vehicle became the 155mm
gun motor carriage M12, seen
here. The maximum range of
the weapon firing an HE round
was 25,715 yards (14.6 miles).
(Patton Museum)

vehicle could be thickened to 8 inches (203mm) and the front of the turret on either side of the gun shield made 5 inches (127mm) thick. Unlike the M4(105) and the M4A3(105), the turret on the M45 was power operated and was also stabilized in elevation. There was storage onboard the M45 for 74 rounds of 105mm ammunition. These vehicles saw service with the U.S. Army during the early stages of the Korean War (1950–53).

155MM SELF-PROPELLED ARTILLERY PIECES

The Ordnance Department began studying the possibility of mounting a 155mm gun on the chassis of the medium tank M3 in June 1941. After having reached an agreement with the Artillery Branch, the Ordnance Department began construction of a pilot vehicle at Rock Island Arsenal that would be designated the 155mm gun motor carriage (GMC) T6. Testing of the vehicle began in February 1942. The weapon mounted on the open-topped T6 was referred to as the 155mm gun M1918M1, and was an American license-built copy of a French-designed weapon, employed by both the French and American Armies during World War I.

The only major design shortcoming of the T6 proved to be the hydraulically actuated spade located at the rear of the vehicle's chassis that assisted in absorbing the weapon's recoil. After redesigning the spade and coming up with a dust shield for the gun mount traversing mechanism, the vehicle was sent out for additional testing. The testing process dramatically demonstrated to the Field Artillery Board the superiority of a self-propelled vehicle mounting a 155mm gun over its towed counterpart, when supporting fast-moving motorized formations. With some minor

stowage changes, this led to the standardization of the T6 as the 155mm GMC M12 with an initial order for 50 units that was soon bumped up to 100 units in August 1942.

The first production unit of the 29.5-ton M12 drove off the assembly line in September 1942, with the last vehicle delivered in March 1943. Along with the M12, a full-tracked personnel and ammunition carrier built on the chassis of a medium tank M3 – called the cargo carrier M30 – accompanied the vehicle in the field. As the buildup for the planned American invasion of France was taking place, it was decided in December 1943 to put 74 units of the M12 and the accompanying M30 through a modernization process to prepare them to see service on the European Continent. This work was completed between February and May 1944.

During the summer 1944 fighting in France, the M12 proved itself a much-desired weapon in the indirect-fire role for fast-moving armored units. When the U.S. Army came up against the German defensive positions –the Siegfried Line – in fall 1944, the M12 proved its worth in the direct-fire role when taking on and destroying enemy bunkers with only one or two rounds. These same bunkers had previously proven themselves impervious to the main gun rounds of the first and second-generation M4 series medium tanks.

Shown during a fire mission in Northwest Europe is a U.S. Army M12 GMC. The first of these self-propelled guns arrived in France in July 1944 and were attached to Bradley's First Army and Patton's Third Army. The 155mm gun mounted on the M12 was located at the rear of the vehicle's hull on the M4 pedestal mount. To provide room at the rear of the vehicle for installing the gun mount, the engine was moved forward to just behind the driver's and assistant driver's compartment. On-board ammunition storage was limited to only ten of the 155mm shells, located on the carriage floor. (Patton Museum)

NEXT-GENERATION SELF-PROPELLED GUNS AND HOWITZERS

So successful was the M12 that the field artillery requested additional vehicles. However, the supply of French- and American-built 155mm M1918M1 guns was exhausted. Although the U.S. Army placed into service a new and improved replacement 155mm gun, designated the 155mm gun M1, studies showed the medium tank M3 could not absorb the recoil from the more powerful, longer-ranged weapon.

To come up with a suitable full-tracked chassis that could successfully mount the 155mm gun M1 and withstand the weapon's powerful recoil, it was decided to use various components from second-generation M4 series medium tanks, including the HVSS system, to come up with an open-topped chassis powered by a gasoline-powered, air-cooled radial engine. This vehicle concept was approved in March 1944

The intended replacement for the 155mm gun motor carriage M12 was the 155mm gun motor carriage M40, seen here on display at APG. Instead of using World War I-era 155mm guns, the vehicle was fitted with the new and more powerful 155mm gun M1, popularly nicknamed the "Long Tom." (Christophe Vallier)

with five pilot vehicles being authorized. Testing of the vehicles went extremely well and the vehicles were assigned the designation 155mm GMC T83.

The T83 was later designated the 155mm GMC M40 in May 1945 when it was standardized. It began coming off the assembly line in February 1945, with a total of 418 units being completed by the end of the year. That same year, the U.S. Army had 24 units of the M40 chassis changed to mount the 8-inch howitzer M1. In this configuration, the vehicle became known as the 8-inch HMC T89. The T89 chassis was considered by the U.S. Army as a universal chassis that could mount either the 155mm gun M1 or the 8-inch howitzer M1. In November 1945, the T89 was standardized as the 8-inch HMC M43.

Entering into production late in World War II, only a single example of the 40.5-ton M40 and the 40-ton M43 made it to Northwest Europe in time to fire any shots in anger. Both vehicle types saw more widespread employment during the Korean War.

OVERLEAF
Shown at the moment of firing is a 155mm gun motor carriage M12 with the gun tube in full recoil. Notice the front end of the vehicle has been driven onto a pile of timber to increase elevation and, in turn, the range of the weapon. The M12 only had onboard storage for ten main gun rounds. (Christopher Vallier)

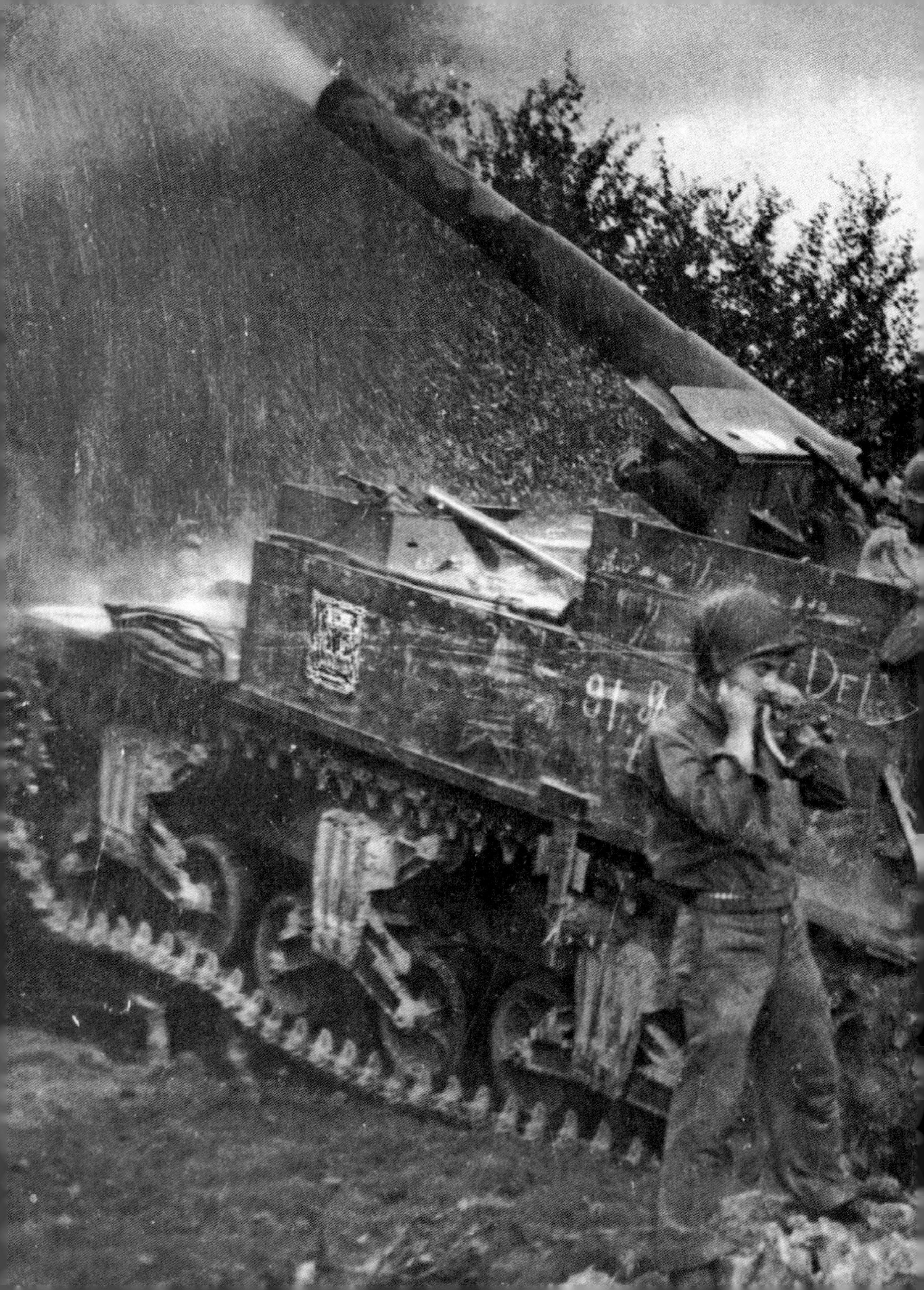

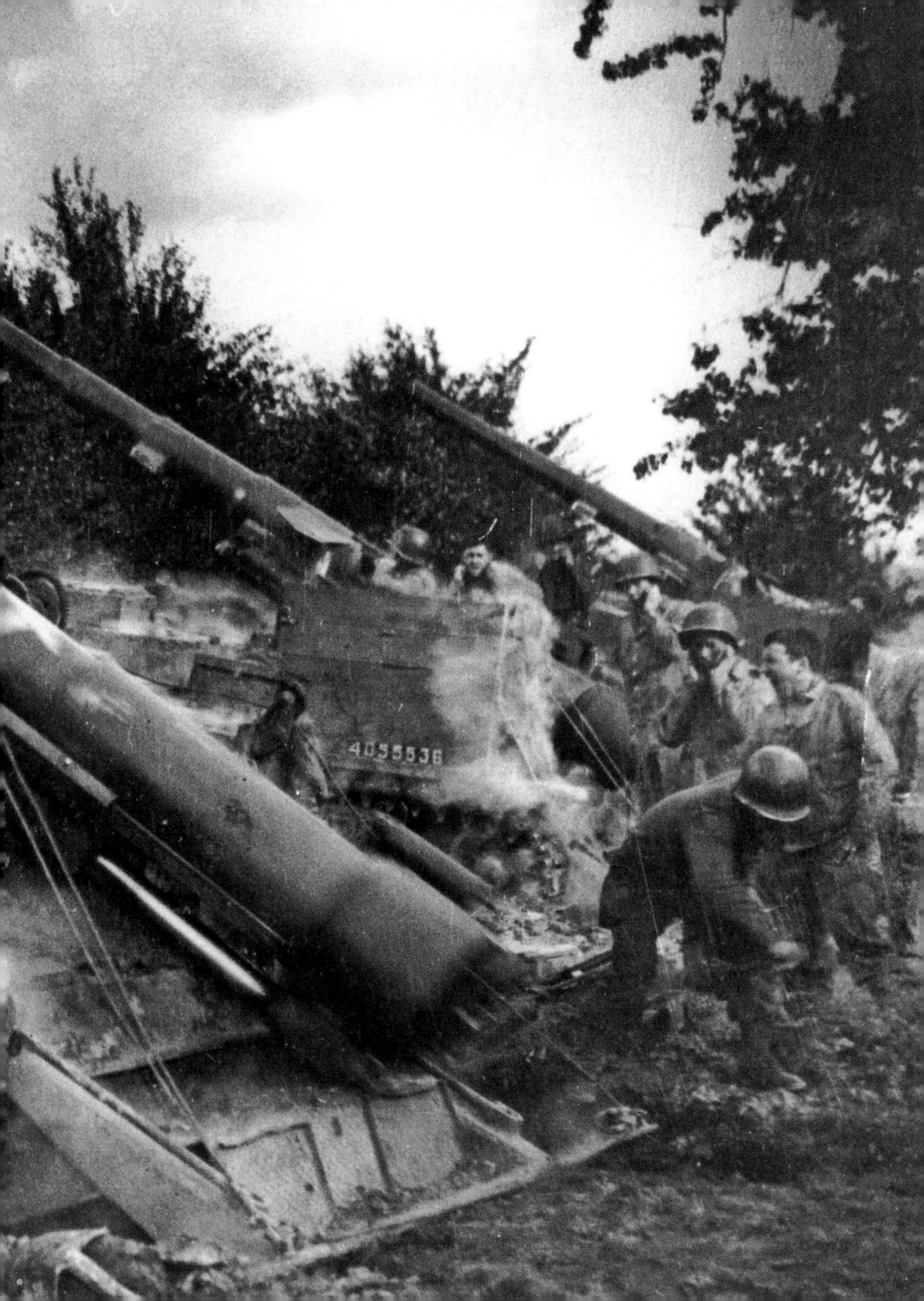

The narrow M3 medium tank chassis proved unsuitable for mounting the 155mm gun M1, and so a new, wider, full-tracked chassis seen here was placed into production. It consisted of second-generation M4 series medium tank components. The vehicle pictured is the pilot model and was designated the 155mm gun motor carriage T83. (Patton Museum)

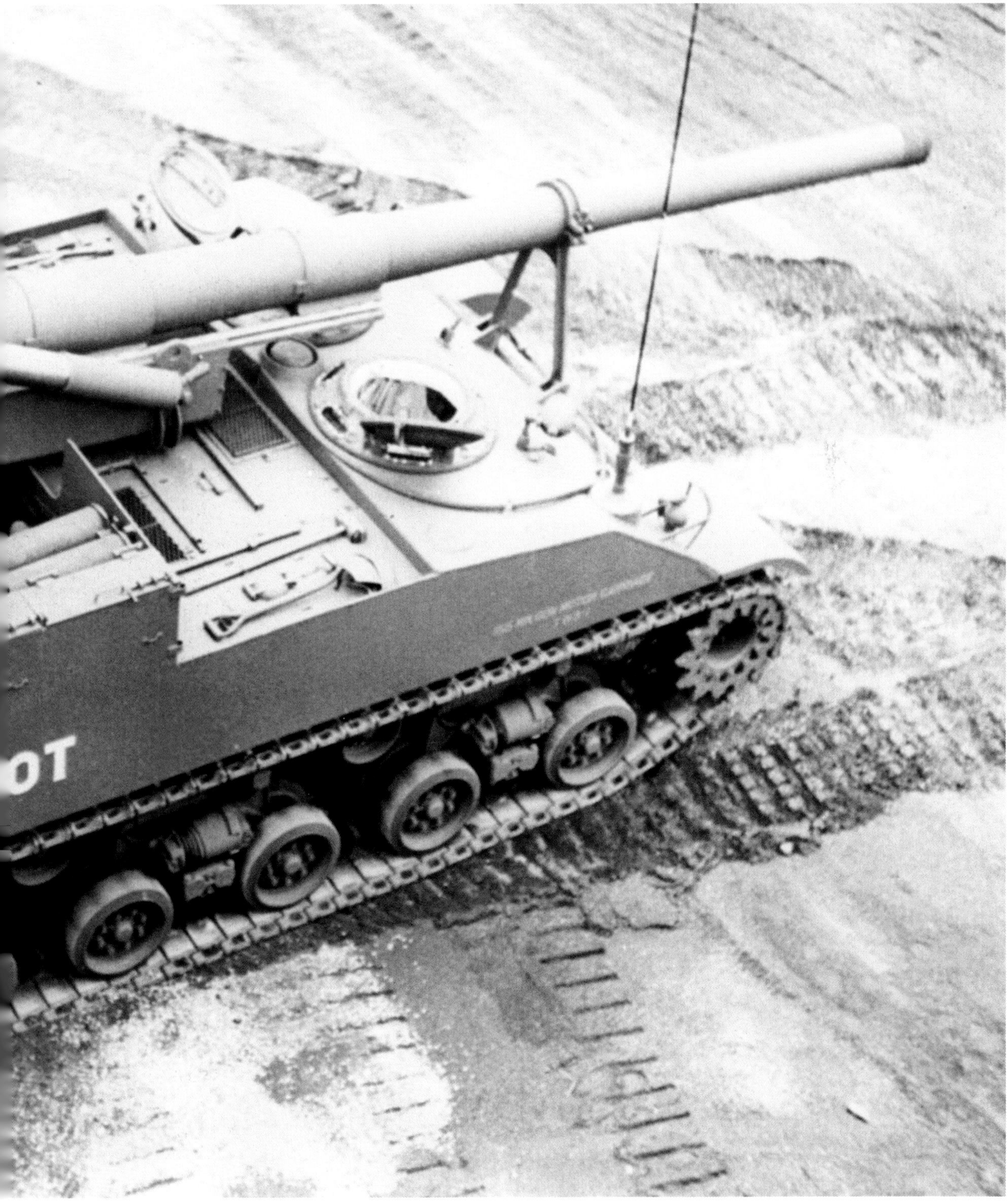
OT

TOO LATE FOR SERVICE

The success of the M12 in combat encouraged the Army Ground Forces to approve the development of an 8-inch howitzer mounted on the much modified chassis of the medium tank T26E1. The vehicle was designated as the 8-inch HMC T84. By the time the first pilot was completed, the Ordnance Department had switched to a new chassis using a great many components of the medium tank T26E3, which later became the heavy tank M26. The T84 weighed an estimated 41 tons. To carry the bulk of the T84's ammunition, the cargo carrier T31 was also built on the same chassis. The end of World War II brought the project to a close without the vehicles being standardized and going into production.

The Ordnance Department also tried to come up with even heavier self-propelled artillery pieces – based on components from the medium tank T26E3 – that would mount the 240mm howitzer M1 or the 8-inch gun M1. Pilot vehicles were built to test the concept, referred to as the 240mm HMC T92 or the 8-inch GMC T93. The T92 came in at 63.7 tons and the T93 weighed 66.3 tons. Both vehicles were propelled by the same gasoline-powered, liquid-cooled Ford GAF engine mounted in the heavy tank M26, which meant they were underpowered for their weight. With the Japanese surrender in August 1945, the need for such vehicles quickly vanished and they were eventually all scrapped.

A companion vehicle to the 155mm gun motor carriage M40 was the 8-inch howitzer motor carriage M43, seen here on display at Fort Hood, Texas. It was based on the same chassis as the M40. Only one production pilot each of the M40 and M43 made it to the ETO before the war ended in that part of the world. (Michael Green)

Using components from the M26 heavy tank, a new self-propelled mount was developed that could carry either the 240mm howitzer M1 or the 8-inch gun M1. These vehicles were designated the 240mm howitzer motor carriage T92, and the 8-inch gun motor carriage T93 shown here. Neither vehicle was standardized as it was felt they were underpowered. (Patton Museum)

CHAPTER NINE

LANDING VEHICLES, TRACKED

E ven before World War I, the U.S. Navy was making contingency plans for what many considered an inevitable war with the Empire of Japan. These plans were formalized in the early 1920s. Part of these contingency plans called upon the U.S. Marine Corps to seize a number of Japanese-occupied islands in the Pacific Ocean in a leap-frog arrangement that would eventually bring land-based American airpower within operational range of the Japanese home islands. To prepare for this role, the Marine Corps investigated a number of methods of getting their men and equipment from ship to shore and the supplies needed to keep them going inland.

A possible solution to this problem appeared in the form of a gasoline-engine-powered, tracked amphibious vehicle, constructed by American civilian inventor Donald Roebling in 1935. His vehicle, nicknamed the "Alligator," was made out of aluminum to save weight and fitted with balsa floats for added buoyancy. It was propelled through the water at a maximum speed of 2.3mph by cleats on its tracks. Roebling saw his invention as a rescue vehicle in swamps and flooded areas after a hurricane.

The U.S. Marine Corps quickly grasped the military potential of Roebling's Alligator and was enthralled by a heavily redesigned version of his vehicle that debuted in 1937. It had a maximum water speed of 8.6mph. At the behest of Brigadier-General Emile Moses, president of the Marine Corps Equipment Board, Roebling came up with a militarized prototype of his newest amphibious tractor and submitted it for testing by the Corps in October 1940. Unfortunately, the U.S. Navy, who funded Marine Corps acquisitions, was unimpressed with the water-handling characteristics of this newest version of Roebling's amphibious vehicle, sometimes referred to as the "Crocodile," and nothing much would happen until the outbreak of war in Europe.

The sudden influx of funding from Congress that came with the beginning of World War II loosened the U.S. Navy's purse strings, and they approached Roebling about building a second prototype of his Crocodile with a more powerful gasoline-powered engine. Roebling agreed and

The LVT-1 pictured was the first military version of inventor Donald Roebling's tracked rescue amphibian, propelled in the water by the cleats or grousers on its tracks. Unlike its civilian predecessors that were constructed of aluminum alloy, the LVT-1 was built out of light sheet steel. Maximum speed in the water was 7mph. (Patton Museum)

the second prototype impressed both the U.S. Navy and U.S. Marine Corps observers during fleet exercises conducted between January and February 1941. In spite of the impressive showing of the Crocodile, the U.S. Navy believed that its aluminum construction lacked the durability to survive the rigors of military service and did not provide any protection against small arms fire. The U.S. Navy felt that mild steel would be a better choice for the vehicle along with an even bigger engine. Roebling acquiesced to U.S. Navy requirements and redesigned his vehicle.

Since Roebling did not have any manufacturing capability, he subcontracted with the Food Machinery Corporation (FMC) to build the vehicle that would meet U.S. Navy requirements. Roebling already had a prior working relationship with FMC as they had built some of the components for the earlier versions of his amphibious tractors. The U.S. Navy would go on to order 100 units of a vehicle that would become known as the Landing Vehicle, Tracked, Mk I (LVT-1). The transporting of cargo was originally seen as the main purpose of the LVT-1. For this reason, the U.S. Navy saw no need for an armored version of the vehicle.

The first production unit of the LVT-1 rolled off the FMC assembly line in August 1941. By the time production concluded in 1943, a total of 1,225 units of the LVT-1

Coming ashore is a machine-gun-armed LVT-1. It was 21 feet long and 9 feet 10 inches wide. To the top of the driver's compartment, the LVT-1 had a height of 8 feet 1 inch. On land, the maximum speed of the vehicle was 18mph with an operational range of approximately 150 miles. (Patton Museum)

were provided to the Marine Corps. To many in the U.S. Navy and Marine Corps, it was known by Roebling's original "Alligator" nickname, while others would refer to them as "Amtracs" or just as tractors. Eventually, the Marine Corps transferred 485 of its LVT-1s to the U.S. Army, which referred to them as "Water Buffalos." Another 200 LVT-1s were supplied to the British military under Lend-Lease. The British military called them "Buffalos."

The LVT-1 weighed 10.9 tons and could carry 2.25 tons of cargo or 24 passengers, not counting its three-man crew. Power for the vehicle was supplied by a Hercules WXLC-3-9A water-cooled engine (the Continental engine was on the LVT-4). Propelled through the water by W-shaped grousers bolted to the vehicle's steel tracks, the vehicle had a maximum water speed of 7mph and a maximum road speed of 18mph. LVT-1s would participate in the fighting for Guadalcanal (August 1942–February 1943) in their designed cargo role.

Prior to the November 1943 invasion of Betio Island, part of the Tarawa Atoll located in the Central Pacific, the Marine Corps became concerned that the very shallow coral reefs surrounding the island might prevent the U.S. Navy's landing craft from reaching the island itself. This would in turn force the attacking Marine

infantrymen to disembark from their landing craft and wade long distances across a shallow lagoon in the face of heavy enemy fire. To deal with this terrain issue, the Marine Corps decided that their existing inventory of LVT-1s could double as amphibious assault vehicles. With their tracked suspension systems, they could easily climb over the shallow coral reefs surrounding Betio Island and deliver the combat troops onto their chosen landing beaches.

To better prepare their LVT-1s for the upcoming operation, the Marine Corps welded on 0.35-inch-thick (9mm) armor plates to the cabs of their vehicles and provided their crews with machine guns to suppress enemy defensive positions as they approached the landing beaches. The usefulness of this last-minute addition of thin armor plates to the LVT-1s during the assault on the island of Betio was questioned by some as it did not protect the vehicle's crews or passengers from Japanese heavy machine-gun fire, which easily penetrated the improvised armor added to the vehicles.

A description of the initial Marine Corps assault wave that went onto Betio appeared in this passage from the history pamphlet "Across the Reef: The Marine Assault of Tarawa," written by Colonel Joseph H. Alexander and published by the History and Museums Division, Headquarters, U.S. Marine Corps:

> For wave one, the final two hundred yards to the beach were the roughest, especially for those LVTs approaching Red Beaches One and Two. The vehicles were hammered by well-aimed fire from heavy and light machine guns and 40mm anti-boat guns. The Marines fired back, expending 100,000 rounds from

A Marine Corps manned LVT-1 is shown here having been fitted with two machine gun for self defense. It lacks the improvised armor added to the vehicles that took part on the invasion of the island of Betio, in the Tarawa atoll, in November 1943. It was Major David M. Shoup, the 2nd Marine Division operation officer that conceived the idea of armoring the existing LVT-1s to better protect the troops that would be in the first assault wave on the island. It fell to Major-General Holland M. Smith to overcomes the objections of the senior U.S. Navy officers in the employment of the LVT-1s at Betio. He also managed to convince the U.S. Navy to supply the attacking Marine Corps units with 100 brand-new LVT-1s. (Marine Corps)

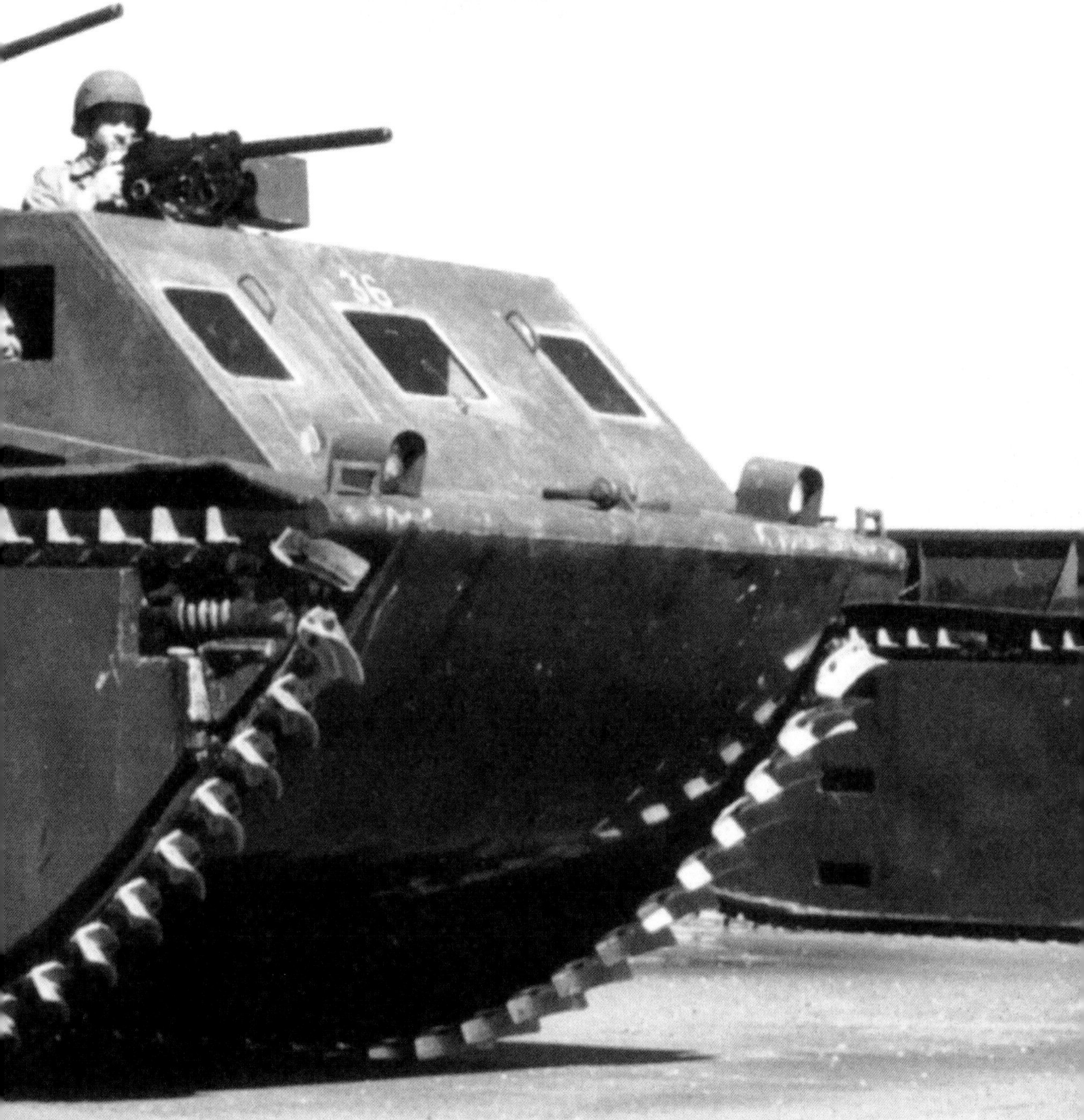
36

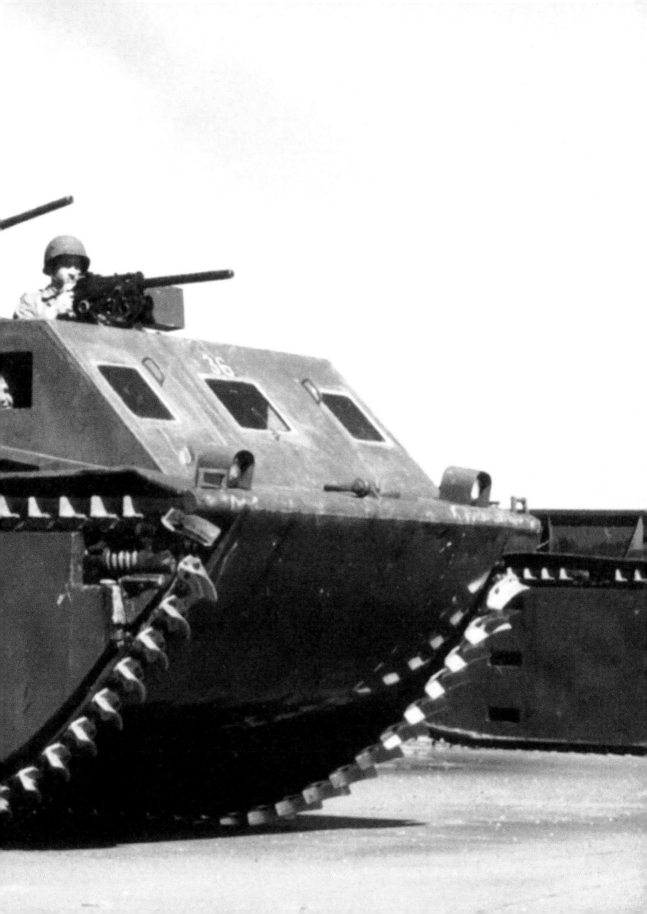

The LVT-1 had a number of design problems centered mainly on its fragile suspension system, which proved too vulnerable to the corrosive effects of saltwater, and as a result had a very short lifespan. To resolve this issue, a new, much improved amphibious tracked vehicle, designated the LVT-2, seen here, was placed into service. (Patton Museum)

the .50 caliber machine guns mounted forward on each LVT-1. But the exposed gunners were easy targets, and dozens were cut down. Major Drewes, the LVT battalion commander who had worked so hard to make this assault possible, took one machine gun from a fallen crewman and was immediately killed by a bullet through the brain. Captain Fenlon A. Durand, one of Drewes' company officers, saw a Japanese officer standing defiantly on the sea wall waving a pistol, "just daring us to come ashore."

Despite the heavy losses sustained by the Marine Corps in the capture of Betio, they came away convinced that if it had not been for the use of LVTs as amphibious assault vehicles the operation would have been a complete failure. As expected by the Marine Corps, those Marines who were transported to the island in subsequent waves by U.S. Navy landing craft were unable to cross the coral reefs fringing the island. They were forced to wade ashore through the offshore lagoon and were massacred by Japanese machine-gun fire. From this point on, the Marine Corps became a firm believer in the use of amphibious assault vehicles. Any objections by the U.S. Navy in the deployment of LVTs in future operations or the Marine Corps acquiring more of them evaporated in the wake of the slaughter at Betio. The U.S. Army, fighting in the Southwest Pacific, also took notice of this new role for the LVT.

AN IMPROVED VERSION

Service use of the LVT-1 led to a degree of displeasure by the U.S. Navy and Marine Corps, both with the vehicle's short mechanical life span as well as its water and land speed. To address these issues, the U.S. Navy approached FMC to see whether they could design and build a superior vehicle. Eventually, what rolled off the FMC assembly line in summer 1942 was referred to as the LVT-2. Other companies that were contracted by the U.S. Navy to build the LVT-2 included Graham-Paige, the Borg-Warner Corporation, and the St. Louis Car Company. By the time production of the vehicle ended in 1945, a total of 2,963 units had been built. The U.S. Navy later transferred 1,507 units of the LVT-2 to the U.S. Army and another 100 went to the British military under Lend-Lease.

The unarmored LVT-2 was larger than its predecessor and heavier at 15 tons. It carried 3.5 tons of cargo or 30 passengers, not counting its six-man crew. The LVT-2 used the proven powertrain from the light tank M3. With a more powerful, gasoline-powered, air-cooled radial Continental engine, the vehicle had a maximum water speed of 7.5mph and could attain 20mph on level roads. Using rubber-encased components, the LVT-2's newly designed suspension system prevented the corrosive effects of saltwater that had shortened the service life of the LVT-1's all-metal suspension system.

The rubber-encased suspension system was a much more durable system than on the LVT-1, which consisted of metal roller-track running over a rail. Like the LVT-1, the LVT-2 had an enclosed crew cab in the front of the vehicle's hull. While the LVT-1 cab had three large windows for the crew to look out, the LVT-2 only had two, which hinged downward and could be opened for ventilation or used as escape windows.

As with the LVT-1, the center of the LVT-2 hull was taken up with an open-topped cargo compartment. Supplies could only be loaded and unloaded from the cargo compartment with hoisting equipment. Passengers had to embark or disembark from the vehicle by climbing over the sides of the vehicle's hull. The engine compartment was located at the rear of the hull. The bottom of the cargo compartment was bisected by the enclosed engine drive shaft that transmitted power from the rear hull-mounted engine to the front hull-mounted transmission.

The prescribed armament of the LVT-2 consisted of four machine guns, a single .50 caliber machine gun and up to three .30 caliber machine guns. The weapons were mounted on M35 skate mounts that attached to two gun tracks located on the inside walls of the vehicle's cargo compartment. The LVT-2 first saw use at the invasion of the island of Betio. Of the 50 that took part in the operation, 30 were lost, mostly to enemy fire. Like the LVT-1s that took part in the operation, the LVT-2s were fitted with improvised armor prior to the assault on Betio.

In conjunction with production of the unarmored version of the LVT-2, there were also 450 units built of an armored version of the vehicle, designated the Landing Vehicle, Tracked (Armored), Mk II or the LVT(A)-2 for short. The maximum armor on the vehicle was 0.5 inch thick (13mm) RHA with the two cab front windows being replaced by a single armored front panel for the driver who could prop it open in a non-combat situation. When in combat, the driver would use a periscope mounted in his overhead hatch. The vehicle's radioman/assistant driver, who sat to the right of the driver, was also provided with a periscope mounted in his overhead hatch.

Armament on the LVT(A)-2 was the same as that of the LVT-2. Reflecting the addition of armor to the vehicle, the LVT(A)-2 was both heavier than its unarmored counterpart and had less cargo-carrying capacity. It first saw action during the Marine Corps assault at Cape Gloucester in December 1943, which was the opening of the New Britain Campaign that lasted until August 1945.

AMPHIBIOUS TANKS

The next step in the evolution of the LVT-2 series was the development and fielding of the Landing Vehicle, Tracked (Armored), Mk I, LVT(A)-1 for short. The vehicle consisted of the modified chassis of the LVT(A)-2 in which an overhead armored superstructure extended from the rear of the crew cab back over the cargo compartment. Centrally mounted upon the roof of this armored superstructure was a two-man welded-armor turret loosely based on that of the light tank M5A1, armed with the 37mm gun M6 and a coaxial .30 caliber machine gun in the gun mount M44.

OPPOSITE
This particular example of the LVT-2 is armed with several machine guns. The vehicle was 26 feet 2 inches long, 10 feet 8 inches wide, and 8 feet 2 inches high. In the water, the vehicle had a maximum speed of 7.5mph and on land it could reach 20mph on a level surface. (Patton Museum)

Armor thickness on the LVT(A)-2 seen here was half an inch on the front crew compartment and down to a quarter of an inch on the less exposed parts, such as the rear engine compartment. The armored cab has roof hatches and overhead periscopes. There was also an armored front panel that could be raised by the driver in non-combat situations. (Patton Museum)

Because the American military inventory of amphibious tracked vehicles were being called upon to assault defended enemy shores, the call went out for an armored version of the LVT-2, seen here, and designated the LVT(A)-2. Reflecting the extra weight carried, it was both slower in the water and on land than its non-armored counterpart. (Patton Museum)

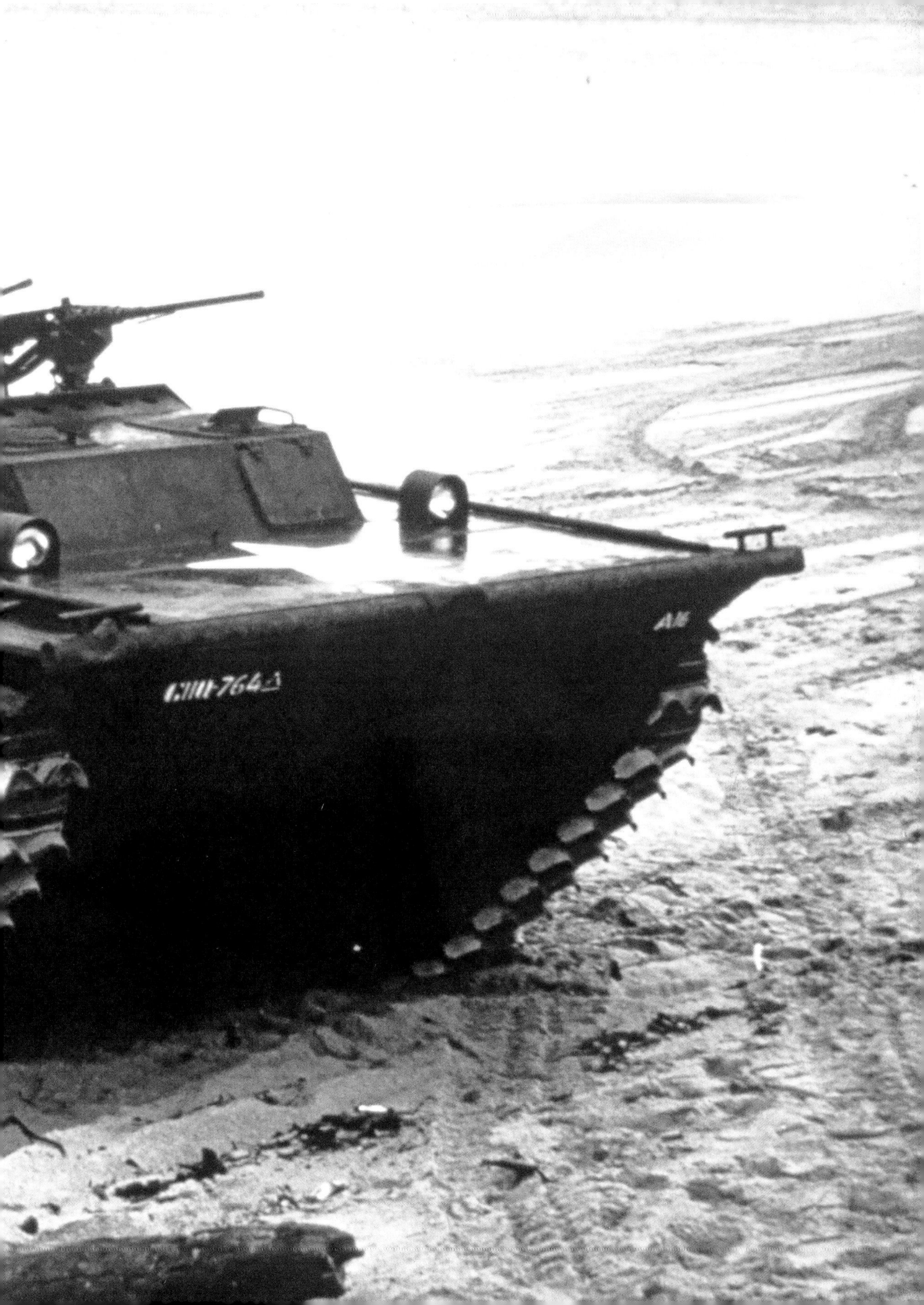

The maximum armor thickness on the 16.4-ton LVT(A)-1 was 1.5 inches (38mm) on its cast-armor gun shield. Besides the turret-mounted weapons, the LVT(A)-1 featured two shield-protected .30 caliber machine guns in open-topped Mark 21 rotating mounts located directly behind the vehicle's turret. The Food Machinery Corporation built 509 units of the LVT(A)-1 between 1942 and 1944 for the U.S. Navy. Of that number, 328 were eventually transferred to the U.S. Army.

Later production units of the LVT(A)-1 appeared with a bow .30 caliber machine gun on the right side of the armored cab. They also appeared with a more elaborate armored shield for the two crewmen operating the machine guns located behind the vehicle's turret.

To distinguish the LVT(A)-1 from the cargo/troop carrier LVTs that were often called Amtracs or tractors, the LVT(A)-1 was sometimes referred to as a "tank," an "amtank," or "amphtank." It was originally envisioned that the LVT(A)-1 would serve as a substitute tank during the initial assault on enemy-held islands until such time that M4 series tanks could be landed. However, the vehicle's thin armor protection and the lack of a main gun that was capable of dealing with well-protected Japanese bunkers rendered it somewhat ineffectual in combat.

One idea that was considered to improve the firepower of the LVT(A)-1 was implemented in June 1944. It involved replacing the existing turret of the vehicle with the turret of the light tank M24 armed with the 75mm gun M6. Tests showed the concept was impractical as the weapon could not be fired when in the water. With the end of World War II, work on the project was discontinued.

The LVT(A)-1 shown here was no longer just a cargo carrier but an amphibious tank based on the chassis of the LVT(A)-2. A superstructure extending from the front crew cab over the vehicle's centrally mounted cargo compartment was topped off by a 37mm main-gun-armed turret, seen here, with 360 degrees of traverse. (Patton Museum)

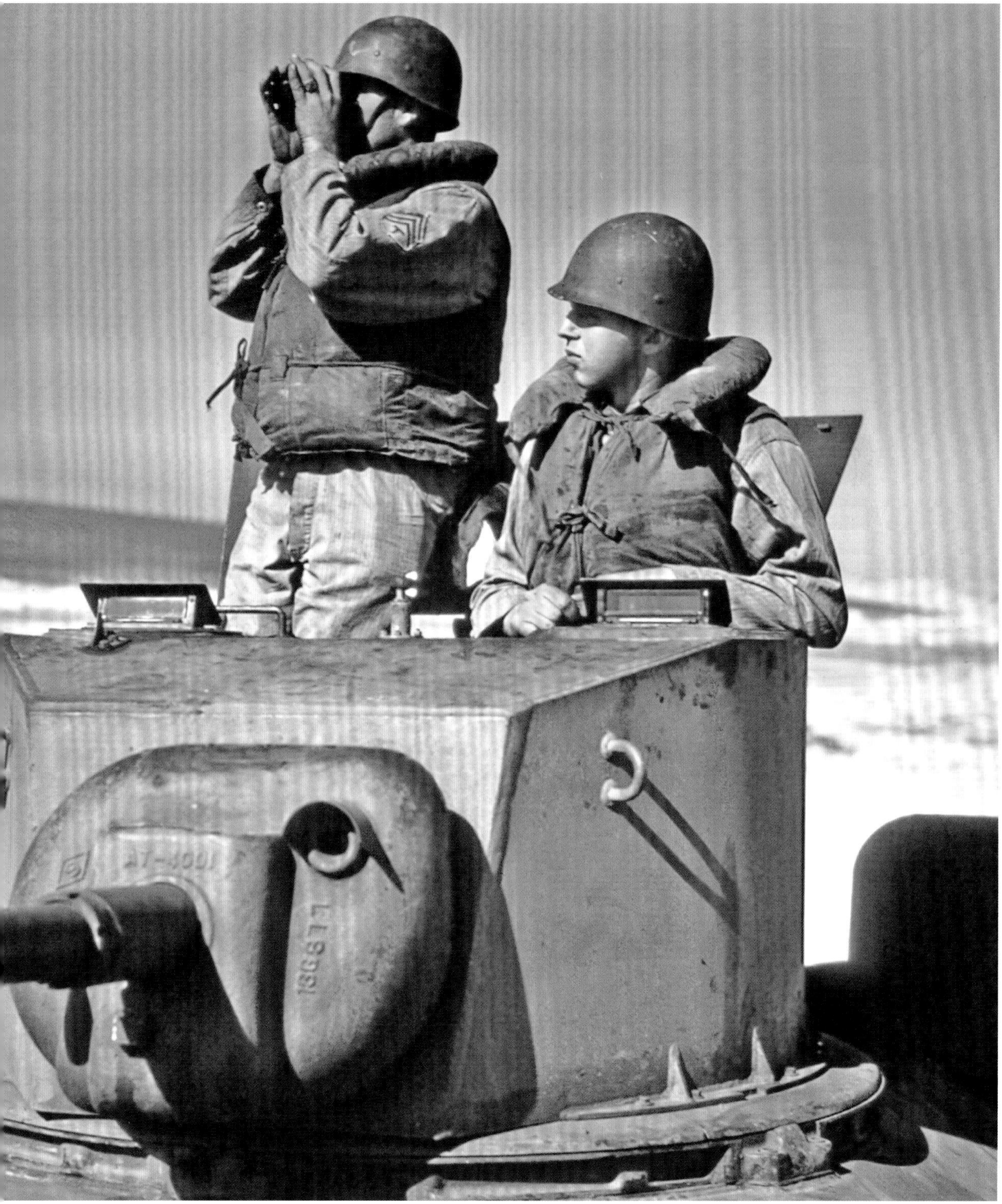

In this dramatically-composed image we see an LVT(A)-1 coming up an American beach for the photographer. Reflecting the fact that the vehicle needed to float, the armor protection was extremely limited. The front hull cab featured armor that was 13mm thick, with the armor on the remainder of the hull being only 6mm thick. The pontoons that extended out from either side of the vehicle's hull were made of 12 gauge mild steel. Also made of 12 gauge mild steel was the hull floor of the vehicle. (Patton Museum)

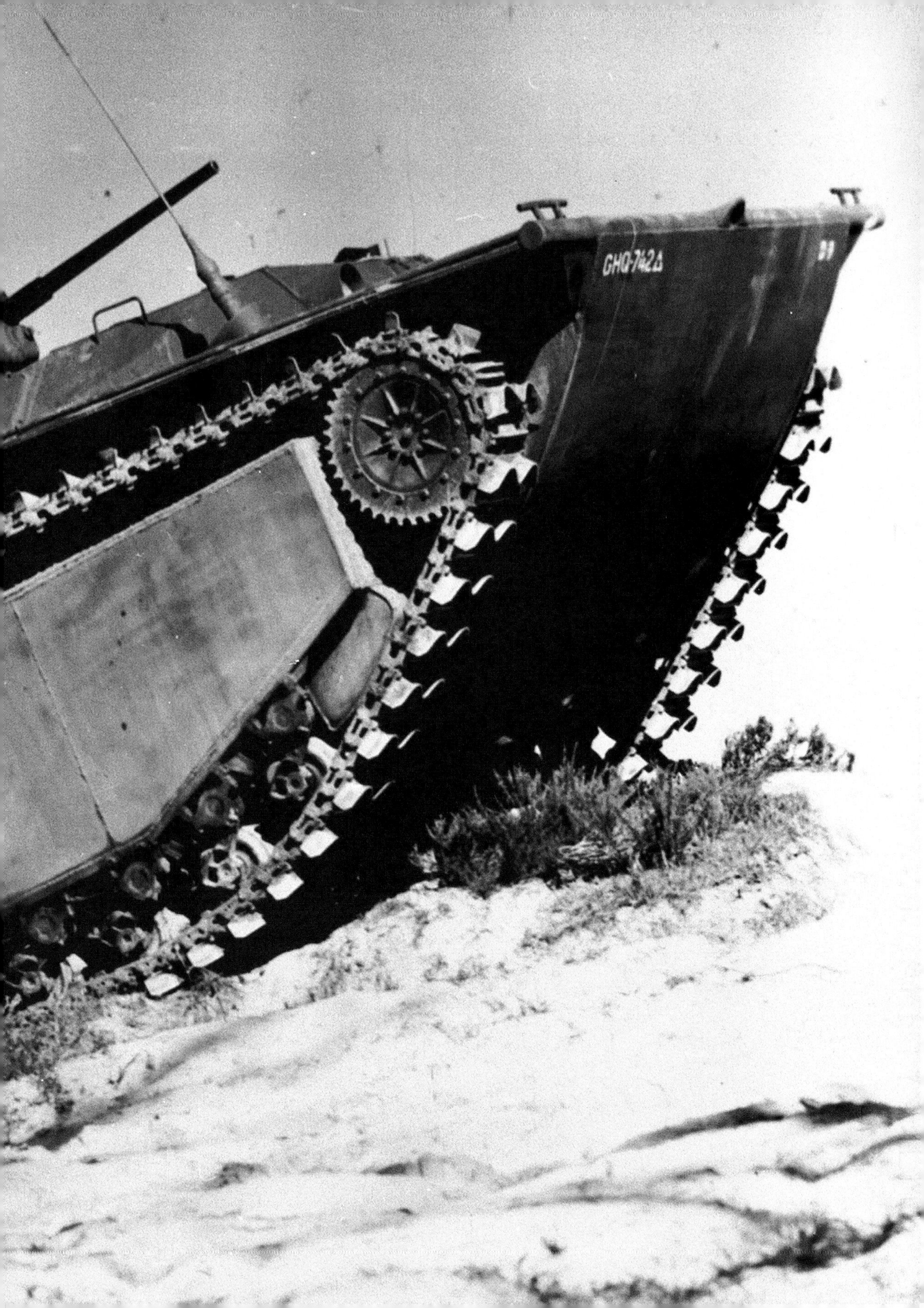

U.S. Army Lieutenant Colonel James L. Rogers, in an article titled *Amphibian Tank Battalion in Combat*, from the 1945 March/April issue of the *Cavalry Journal*, described the crew arrangement of the LVT-2, LVT(A)-2, and LVT(A)-1:

The crew of the LVT(A)-2 was made up of seven men, including a driver, an assistant driver, four machine gunners and a tractor commander. The LVT-2 crew was made up of three men – a driver, an assistant driver, and a vehicle commander. Since the weapons on the LVT-2s were not to be fired except at point targets and on direct orders from an officer, the infantrymen in the vehicle were to man the guns. The crew of the LVT(A)-1 was composed of six men – a driver, an assistant driver, two machine gunners, a 37mm gunner, and an extra man who would serve as an ammunition passer or a replacement for any crew member who became a casualty.

To address the lack of suitable firepower on the LVT(A)-1, an up-gunned version of the vehicle was introduced into service in March 1944. It was designated the Landing Vehicle, Tracked (Armored), Mark IV, LVT(A)-4 for short. It featured a new, larger, open-topped manually operated welded-armor turret taken from the 75mm HMC M8 armed with the 75mm howitzer M3. Early-production units of the vehicle had a .50 caliber machine gun fitted to a ring mount on top of the vehicle's turret.

Combat experience quickly demonstrated that the location of the .50 caliber machine gun on the LVT(A)-4 meant whoever attempted to operate the weapon was exposed to enemy fire. To correct this issue, later production units of the LVT(A)-4 did away with the ring-mounted .50 caliber machine gun and replaced it with two shield-protected .30 caliber machine guns fitted on either side of the turret. Later production units of the LVT(A)-4 also appeared with a bow-mounted .30 caliber

Located directly behind the 37mm main-gun-armed turret of the LVT(A)-1 were two open-topped mounts for the fitting of .30 caliber machine guns. The only protection for the machine gunners was the fitting of a forward-facing gun shield. The 37mm gun-armed turret also had a coaxial mounted .30 caliber machine gun. (Patton Museum)

machine gun in the right front of the armored cab. Vision blocks also began appearing on the turrets of later production units of the LVT(A)-4.

Eventually, a power traverse system for the turret and elevation stabilizer were provided for the LVT(A)-4, which resulted in a designation change to LVT(A)-5. The Food Machinery Corporation built 1,890 units of the LVT(A)-4 for the U.S. Navy between 1944 and 1945, of which 1,307 were eventually transferred to the U.S. Army, and another 50 went to the British military under Lend-Lease. Only 269 units of the LVT(A)-5 were built by FMC in 1945 for the U.S. Navy, of which 141 went to the U.S. Army.

The only use of LVTs in the ETO by the U.S. Army was described in this extract from *Armored Special Equipment*, a postwar U.S. Army report put together by the General Board United States Forces, European Theater:

> The 747th Tank Battalion (Ninth Army), on March 16, 1945 stored its standard tanks and was re-equipped with LVTs, each of its four tank companies receiving seventeen LVT-2s and

eight LVT-4s. Training in driving and maintenance was conducted for approximately ten days. After the Rhine [river] crossing the LVTs were turned in, their standard tanks taken out of storage and the battalion resumed its normal role. There were no other U.S. [Army] units similarly equipped and trained in the European Theater.

The British Army employed American-supplied LVT-2s in their assault on the German-occupied island of Walcheren, which began on October 31, 1944, in an attempt to clear the path for supply ships to the Belgium port of Antwerp. Of the 169 LVTs employed in the operation, which lasted until early November 1944, the majority were eventually rendered unfit for service due to the extremely arduous conditions. Some of the LVT-2s were lost to enemy mines and artillery fire. The British Army also employed LVT-2s during their nighttime crossing of the Rhine River on the night of March 23, 1945, referred to as Operation *Plunder*.

An early-production unit of the LVT(A)-1. The vehicle was 26 feet 1 inch long and 10 feet 8 inches wide. To the top of its turret, the LVT(A)-1 had a height of 10 feet 1 inch. There was authorized storage for 104 rounds of 37mm main gun ammunition onboard the vehicle. (Patton Museum)

LESSONS LEARNED

As with the LVT(A)-1, the thin armor on the LVT(A)-4 meant that it could not absorb the same amount of punishment from enemy fire as the more heavily armored M4 series tanks. The U.S. Army and Marine Corps eventually redirected the LVT(A)-4 away from being a direct-fire platform into an indirect-fire artillery platform, where its lack of sufficient armor protection was less of an issue. This change in doctrine was reflected in an article titled *Amtanks*, by U.S. Army Major John T. Collier, which appeared in the May/June 1945 issue of the *Cavalry Journal*. Here Collier writes regarding their use in the invasion of the Philippines in October 1944:

In the water, the LVT(A)-1 seen here could attain a maximum water speed of 7mph and range of 75 miles. On a level surface on land the vehicle could reach a maximum speed of 25mph and an approximate range of up to 125 miles. (Patton Museum)

When the veterans of the 7th Infantry Division stormed the flaming beaches of Leyte, their assault waves were led by these amtanks, which bombarded the beaches with accurate area fire delivered from the water (the knack of which is a trade secret developed by the battalion). Succeeding waves had the battalion's artillery support when they needed it – and that support was delivered from as many guns as there are in the entire division artillery. Forward observers and liaison officers of the division artillery fired the amtank batteries with great effect, and one early fire mission helped to break up enemy preparations for a counterattack. The artillery support was, in itself, an important innovation in amphibious warfare.

Once the U.S. Army divisions were safety established ashore in the Philippines, the LVT(A)-4s were used for raids behind enemy lines to weaken their defensive ability. Collier also describes this new role:

> The actions on the west coast of Leyte were cut to order for the amtanks to make water envelopments that would land the fighters behind Jap lines, without friendly infantry support. Cutting loose with thousands of rounds of 75mm HE and WP, they blasted supply dumps and troop concentrations on reverse slopes which could not be reached by our artillery. Unable to hold ground for long, the amtanks would do their damage and dash away – usually pulling out to sea under ineffective Jap mortar and gunfire – leaving the enemy too demoralized to pursue, even if they had any means of pursuit.

THE NEXT LOGICAL STEP

Having to employ hoisting equipment to load or unload the early LVTs, or requiring their crews and passengers to embark or disembark over the hull side, especially in combat, was seen as a design shortcoming of the early vehicles. To correct this problem, FMC developed a new version of the LVT-2 with a rear loading ramp.

Belonging to the former Patton Museum of Armor and Cavalry is this LVT(A)-1. The vehicle could tackle a 60 percent grade and cross a trench up to 5 feet wide. When required, the LVT(A)-1 could cross over a vertical wall 36 inches high. Fire protection for the vehicle was provided by a number of onboard carbon dioxide fire extinguishers. (Henry Penn)

Combat experience showed the LVT(A)-1 to be both under-armored and under-gunned. Adding additional armor to the vehicle and the resulting weight penalty would have impacted the sea-handling characteristics of the LVT(A)-1 so that option was discarded. However, it did prove possible to up-gun the vehicle with a 75mm howitzer as seen here, resulting in the designation LVT(A)-4. (Patton Museum)

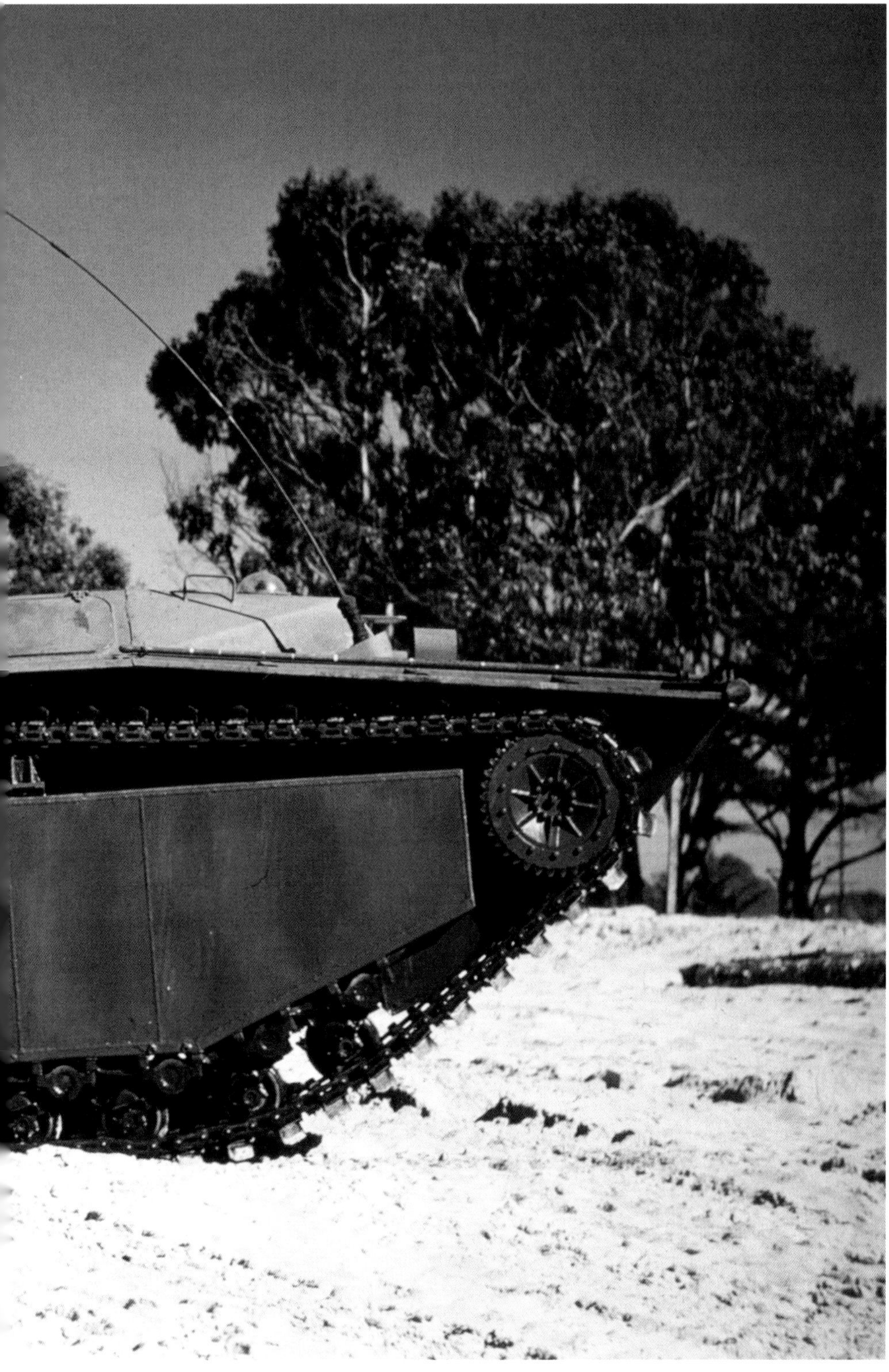

The redesigned vehicle was designated the Landing Vehicle, Tracked (Unarmored), Mk IV, or LVT-4 for short. To provide the LVT-4 with a rear loading ramp, FMC moved the rear hull engine compartment seen on the earlier LVTs to a position behind the vehicle's cab. This arrangement also did away with the enclosed engine drive shaft that had bisected the bottom of the cargo compartment on the earlier LVTs.

The LVT-4 weighed 18.2 tons, and with the rear loading ramp, could transport small vehicles such as jeeps, and artillery pieces. It could also carry 30 passengers not counting its six-man crew. Graham-Paige, the St. Louis Car Company, and FMC built 8,348 units of the LVT-4 between 1943 and 1945. A total of 6,083 were transferred by the U.S. Navy to the U.S. Army. Approximately 500 went to the British military under Lend-Lease.

The LVT-4 first saw use in summer 1944 during the invasion of the Japanese-occupied island of Saipan, part of the Marianas Island chain. They saw service in Italy and Northwest Europe with both the American and British Army. In March 1945, the Royal Marines were supplied with both the LVT-4 and LVT(A)-4 for possible use

The 75mm howitzer fitted on the LVT(A)-4 was mounted in an open-topped turret similar to that on the gun motor carriage M8. In addition to the howitzer, the early-production units of the vehicle carried a .50 caliber machine gun as shown here. There was authorized storage onboard the vehicle for 100 rounds of main gun ammunition. (Patton Museum)

in the invasion of Japan. Some of the LVT-4s were modified by the Royal Marines to mount flame throwers, while others were fitted with rocket launchers in their cargo hold. The Japanese surrender in August 1945 meant these vehicles were never used in battle.

The base model of the LVT-4 was unarmored. However, it was designed from the beginning to accept steel appliqué armor up to 0.5 inches (13mm) thick if the need arose. The unarmored version of the LVT-4 had a cab with two large windows. There were also several versions of an armored cab that could be fitted to the vehicle with a maximum armor thickness of up to 0.5 inches (13mm) thick.

Early versions of the armored cab for the LVT-4 had a front-cab-mounted armored hatch and a periscope in the overhead hatch for the driver to look out of. The radioman/assistant driver also had an overhead hatch fitted with a periscope as well as a bow-mounted .30 caliber machine gun. Later production units of the vehicle did away with the driver's front-cab-mounted armored hatch and overhead periscope. In their place, the driver now depended on a number of vision blocks on the front of the cab and on its side for vision.

Besides the bow-mounted machine gun, the LVT-4 could be fitted with two .50 caliber machine guns mounted on the engine compartment bulkhead. There were also provisions for two .30 caliber machine guns on either side of the vehicle's cargo compartment.

The lack of a rear loading ramp on the LVT-1, LVT-2, and LVT(A)-2 and the high sides of the vehicles made it difficult to load and unload cargo without a crane or hoist. It also made it very hard for the crews and passengers to enter or leave the vehicles quickly. This dilemma resulted in the design and building of the LVT-4, seen here, which had a rear loading ramp. (Patton Museum)

Late-production units of the LVT(A)-4 did away with the single turret-mounted .50 caliber machine gun and replaced it with two .30 caliber machine guns. These were mounted on either side of the open-topped turret as shown here. There was authorized storage for 6,000 rounds of .30 caliber ammunition on the LVT(A)-4. (Patton Museum)

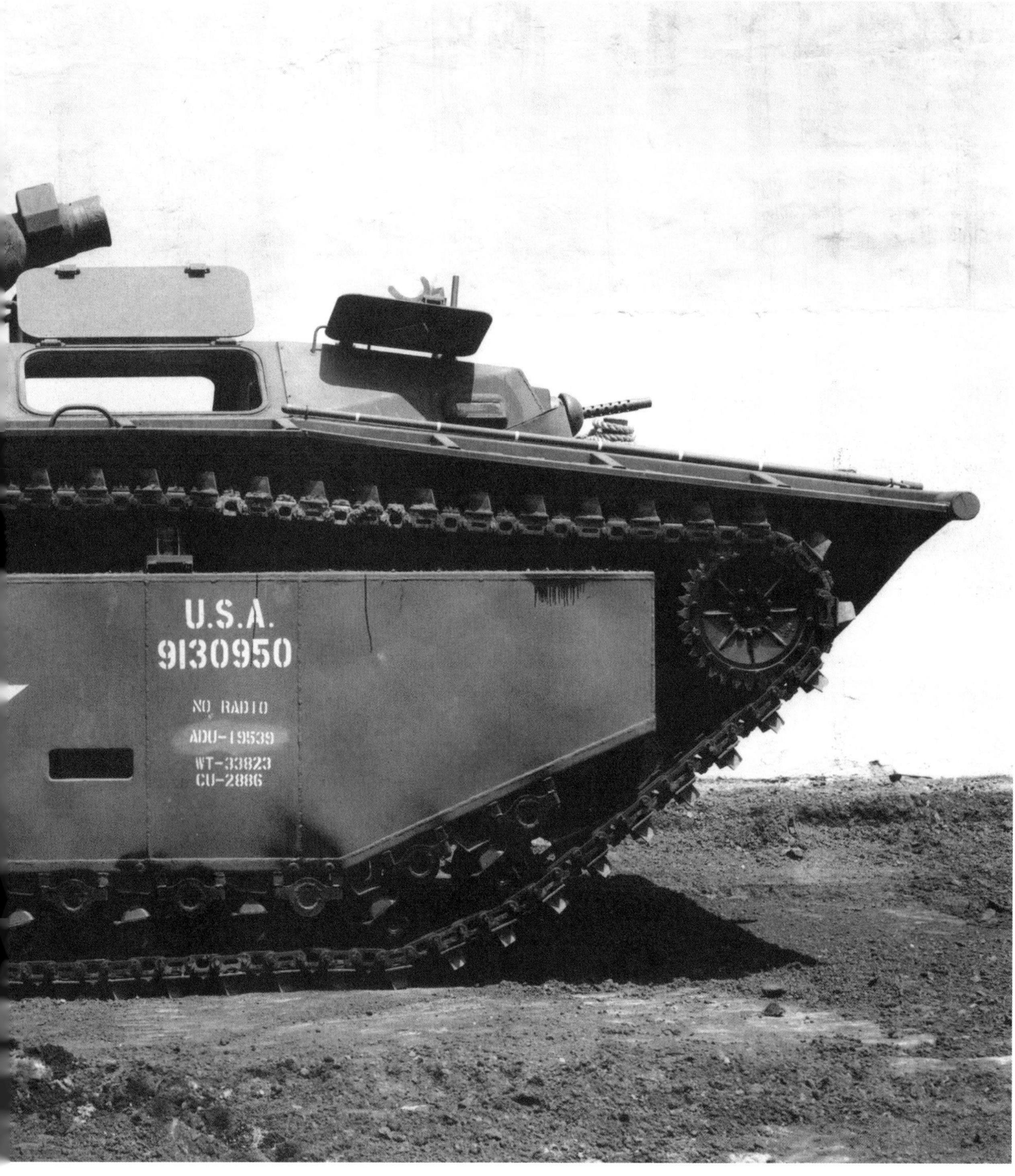
U.S.A.
9130950
NO RADIO
ADU-19539
WT-33823
CU-2886

The 75mm howitzer in the LVT(A)-4s pictured heading towards an enemy-held shore was designated the M3 and weighed 421 pounds. It could fire an almost 19-pound HE projectile out to a range of 9,620 yards. The maximum rate of fire for the M3 was eight rounds per minute. There was also a smoke round available for the weapon. (Patton Museum)

The Borg-Warner Corporation also came up with an LVT with a rear loading ramp in 1944. They managed this task by placing a liquid-cooled, gasoline-powered 110hp Cadillac car engine on either side of the cargo compartment. The new vehicle was designated as the Landing Vehicle, Tracked, Mark III, or LVT-3 for short. The company nickname for the vehicle was the "Bushmaster." Borg-Warner built 2,962 units of the 13.3-ton LVT-3 between 1944 and 1945 for the U.S. Navy. It did not see service until the American invasion of Okinawa, which lasted from April to June 1945. Other than two experimental versions of the vehicle, none were ever transferred to the U.S. Army.

The LVT-3 can be distinguished from the very similar-looking LVT-4, as its cab sits further forward than that on the LVT-4. Like the LVT-4, the LVT-3 could transport small wheeled vehicles and artillery pieces, or up to 30 passengers not counting its three-man crew. As with the LVT-4, the LVT-3 was designed from the beginning to accept appliqué armor if there was a requirement for it. The vehicle could be armed with a single .50 caliber machine gun. An upgraded LVT-3, designated the LVT-3C, was the only vehicle besides an upgraded LVT(A)-5 that was retained by the Marine Corps following World War II.

The last amphibious tracked vehicle to enter into American military service during World War II was the LVT-3. An improved, postwar version designated the LVT-3C is seen here on display at the Marine Corps Landing Vehicle Tracked Museum, located at Camp Pendleton in southern California. The vehicle was 24 feet 6 inches long, 11 feet 2 inches wide, and 9 feet 11 inches high. (Michael Green)

CONCLUSIO

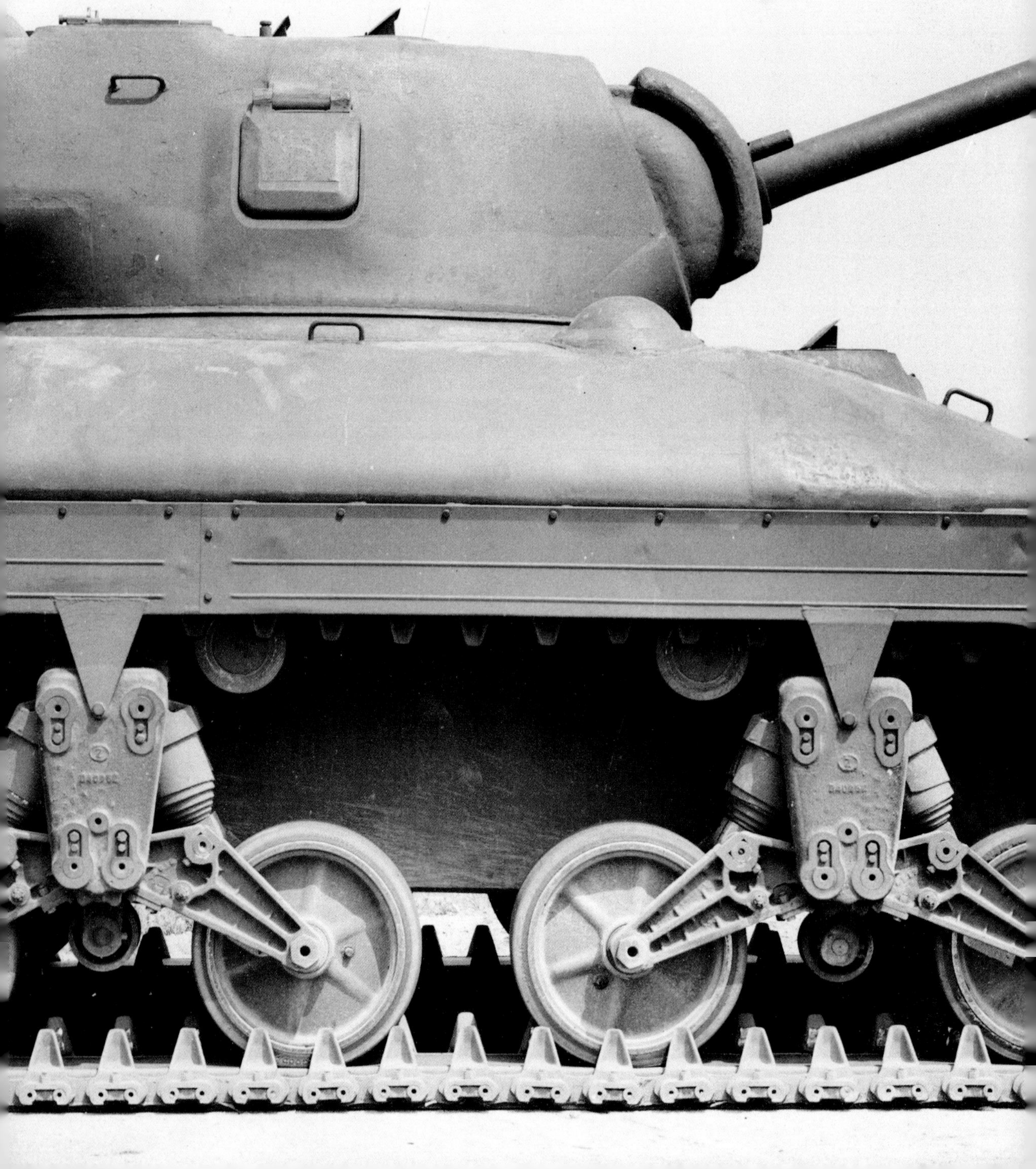

The incredible number of tanks and AFVs built by American industry during World War II is stunning when compared to those of the Axis nations. This is even more amazing considering that shortly after the conclusion of the World War I the U.S. Army's senior leadership had abolished the Tank Corps, which had been created in 1918, and was the only unit in the service that had any experience with the relatively new war machine. The decision on what tanks would be needed for the future then fell to the relatively uninterested Infantry Branch of the service.

U.S. Army interest in tanks would not really be rekindled until the 1930s when the Cavalry Branch of the service began looking at mechanized equipment to supplement its horses. The Cavalry Branch's notice of mechanized equipment soon pushed the Infantry Branch to look at the potential of mechanization. It would fall to the Ordnance Department of the U.S. Army to develop the tanks that would satisfy the user community. Yet, at that time, the Ordnance Department had no specialized tank division, and surprisingly no plans for the wartime procurement of tanks. The office within the Ordnance Department concerned with the design and development of tanks was until 1941 merely a sub-office of the much larger Artillery Division.

One month after the German military invasion of Poland in September 1939 that officially began World War II, the Ordnance Department was allowed to order 329 light tanks. That order went to a civilian firm that built railroad equipment. It would the first tank order placed by the U.S. Army with American industry in 20 years. In the early summer of the following year, when the German military invaded France and the Low Countries, the U.S. Army had only 28 modern medium tanks in service, all hand-built at an Ordnance Department-run arsenal.

As the Axis continued their long string of military victories overseas, additional American railroad equipment companies were provided contracts by the Ordnance Department to build more tanks, both light and medium. Prior to America's official entry into World War II the Ordnance Department had envisioned that America's railroad equipment companies could be depended on to build a sufficient number of tanks to meet the needs of the U.S. Army.

In June 1940, American automobile executive William S. Knudsen reviewed the Ordnance Department's tank production plans. He quickly deduced that they would never come close to meeting the future needs of the U.S. Army.

Knudsen, a recognized expert in the field of mass production had been appointed Chairman of the Office of Production Management in early 1940. He took it upon himself to approach his former automobile executive associates in Detroit, Michigan, then the car production capital of the world, to join in the tank building business. Rather than convert existing automobile factories for the job, Knudsen decided it made more sense to build a brand new specialized tank plant in Detroit and have the management of Chrysler Corporation run it. They could then apply their skills in efficiently mass-producing cars, to the job of mass-producing tanks.

During the first few months of 1941 the production of tanks gradually gained momentum in the United States. However, this improvement in numbers would quickly be rendered meaningless by an outside event, the German military invasion of the Soviet Union in July 1941. President Roosevelt quickly stepped into the picture and directed that the production of tanks be expedited at once, "with the only limiting factor … the ability of American industry to produce tanks." This directive was just a small part of the President's plan to ramp up American industry's production capacity to what he saw as a comprehensive Victory Program aimed at out-producing all "potential enemies." In response, the Ordnance Department finally set up a separate Tank and Combat Vehicle Division during the following two months.

As the Ordnance Department and American industry strived to increase the production of tanks and AFVs, the President dropped another bombshell on them in September 1941. After reviewing the proposed figure of 16,800 tanks to be built in 1942, he now called for that number to be doubled with 33,600 tanks to be built in 1942. Roosevelt wanted enough tanks and AFVs not only for the U.S. Army but sufficient numbers to send to those then fighting the Axis, chiefly Britain and the Soviet Union. To meet these needs an ever wider range of American industry was pulled into the fold, and plans were laid for another specialized tank plant to be built and run by the Fisher Body Division of General Motors.

The month following the Japanese attack at Pearl Harbor, in December 1941, the President now informed the Ordnance Department and American industry that he wanted 45,000 tanks built in 1942 and 75,000 in 1943. Wiser heads would curtail these grandiose projected numbers and when he was later questioned on the industrial practically of mentioning these numbers before the American Congress he is said to have answered, "Oh – the production people can do it if they really try." As events transpired, American industry did construct a respectable total of 88,140 tanks by 1945, with the majority being built between 1942 and 1945. This was in sharp contrast to World War I when American industry proved unequal to the task of outfitting America's armed forces with almost any of the military hardware it needed, including tanks, forcing a reliance on foreign sources, French and British.

Despite the tremendous number of tanks and AFVs built by American industry between 1942 and 1945, there were times that America's ground forces did not have enough tanks and AFVs on hand when needed. This first occurred after the American military invasion of France in the summer of 1944, when fighting in the hedgerows of Normandy. The loss rate of tanks and AFVs would prove much higher than anticipated by the Ordnance Department. It was not that the tanks and AFVs were not being built in sufficient numbers, there was excess capacity during 1943 and early 1944, but the building of specific models of tanks and AFVs could not be increased overnight, or over a period of several months, to meet sudden increases in theater demands.

After breaking out of the terrain restrictions imposed upon it by the hedgerows of Normandy, and sweeping the German Army before it across France, by September 1944, the tank shortages of the summer were forgotten. However, the same issues would crop up again during the heavy fighting in December 1944 and January 1945, known to most Americans as the Battle of the Bulge. With the resulting heavy losses to the advancing German Armies, of both American tanks and AFVs, by January 1945, the call for more tanks was intense and American industry was called upon once again to make an all-out effort. However, as the U.S. Army successfully pushed back the now spent German Armies that took part in the massive German counteroffensive, and the defeat of Germany appeared more imminent, American tank and AFV production was cut back. By November 1945 it had come to almost a complete halt, with the Japanese having surrendered in September 1945 and bringing World War II to a conclusion.

The fact that individual American tanks may not have been the battlefield equal of some of their opponent's vehicles in World War II, a problem that first really arose following the invasion of France in June 1944, when American tankers encountered the German Panther medium tank in large numbers, would prove to be irrelevant in the bigger picture of World War II. This is seen in the maxim (attributed to a great many authors) that "Quantity has a quality all its own." This was a belief that led to the eventual victory of the Red Army's tank fleet in the world-wide conflict, despite having to face many more individually superior German tanks than the U.S. Army ever did.

In a strange turn of events, the postwar U.S. Army would eventually go down the same path with its tanks as the German Army did late in World War II, when facing overwhelming numbers of Soviet tanks. As postwar American industry could never match the production numbers achieved by their Cold War counterparts, it was decided that U.S. Army tanks would have to be built with a qualitative superiority that would allow them to offset the quantitative advantage in tanks possessed by the Soviet Union and the Eastern-Bloc armies How this doctrine would have panned out for the U.S. Army in case of World War III will never be known.

Hindering the effectiveness of U.S. Army tanks on the battlefield during World War II was the fact that they were plagued by a badly thought out prewar doctrine that insisted that they were only weapons of exploitation. Specialized tank destroyers would deal with enemy tanks when encountered. Combat experience would quickly demonstrate to the U.S. Army that depending on specialized tank destroyers to handle enemy tanks was a big mistake, but at that point it was too late to change course and the U.S. Army would have to make the best of the existing organizational structure and doctrine. The fact that U.S. Army tankers managed to prevail in many combat engagements with enemy tanks that were on paper superior in almost every way was a testament to their courage and adaptability in making the best of a bad situation.

What is amazing about the number of American tanks and AFVs built between 1942 and 1945 is placing this achievement in context with all the other military hardware that either rolled out of American factories, or were launched at American shipyards, at that time. An example would be the 197,760 combat aircraft built in American factories during World War II, or the 151 aircraft carriers that joined the good fight. It was only the need to divert critical resources to building these other types of military equipment that prevented President Roosevelt from having even more tanks and AFVs built during World War II. Truly, America earned itself the nickname as the "Arsenal of Democracy" during this time.

SELECTED BIBLIOGRAPHY

Baily, Charles M. *Faint Praise: American Tanks and Tank Destroyers during World War II*. Hamden, CT: Archon Books, 1983

Cameron, Robert S. *Mobility, Shock, and Firepower: The Emergence of the U.S. Army's Armor Branch, 1917–1945*. Washington, DC: Center of Military History, 2008

Cooper, Belton Y. *Death Traps: The Survival of an American Armored Division in World War II*. Novato, CA: Presidio Press, 2000

Estes, Kenneth W. *Marines Under Armor: The Marine Corps and the Armored Fighting Vehicle, 1916–2000*. Annapolis, MD: Naval Institute Press, 2000

Green, Constance McLaughlin. *The Ordnance Department: Planning Munitions for War*. Washington, DC: Office of the Chief of Military History, Department of the Army, 1955

Hunnicutt R. P. *Sherman*. Belmont, CA: Taurus Enterprises, 1978

— *Firepower: A History of the American Heavy Tank*. Novato, CA: Presidio Press, 1988

— *Stuart*. Novato, CA: Presidio Press, 1992

— *Pershing: A History of the Medium Tank T20 Series*. Bellingham, WA: Feist Publications, Inc., 1996

— *Half-Track: A History of American Semi-Tracked Vehicles*. Novato, CA: Presidio Press, 2001

— *Armored Car: A History of American Wheeled Combat Vehicles*. Novato, CA: Presidio Press, 2002

Johnson, David E. *Fast Tanks and Heavy Bombers: Innovation in the U.S. Army, 1917–1945*. New York: Cornell University Press, 1998

Mayo, Lida. *The Ordnance Department: On Beachhead and Battlefront*. Washington, DC: Office of the Chief of Military History, U.S. Army, 1968

Mesko, Jim. *M24 Chaffee in Action*. Carrollton, TX: Squadron Signal Publications, 1988

— *M3 Lee/Grant in Action*. Carrollton, TX: Squadron Signal Publications, 1995

— *M3 Half-Track in Action*. Carrollton, TX: Squadron Signal Publications, 1996

— *U.S. Armored Cars in Action*. Carrollton, TX: Squadron Signal Publications, 1998

— *U.S. Tank Destroyers in Action*. Carrollton, TX: Squadron Signal Publications, 1998

— *Amtracs in Action*. Carrollton, TX: Squadron Signal Publications, 1999

— *U.S. Self-Propelled Guns in Action*. Carrollton, TX: Squadron Signal Publications, 1999

— *Walk Around M4 Sherman*. Carrollton, TX: Squadron Signal Publications, 2000

Thomson, Harry C. and Linda Mayo. *The Ordnance Department: Procurement and Supply*. Washington, DC: Office of the Chief of Military History, U.S. Army, 1960

Zaloga, Steven J. *The Sherman Tank in U.S. and Allied Service*. London: Osprey Publishing, 1982

— *M3 Infantry Half-Track 1940–73*. Oxford: Osprey Publishing, 1994

— *Amtracs U.S. Amphibious Assault Vehicles*. Oxford: Osprey Publishing, 1999

— *M3 & M5 Stuart Light Tank 1940–45*. Oxford: Osprey Publishing, 1999

— *M26/M46 Pershing Tank 1943–53*. Oxford: Osprey Publishing, 2000

— *M8 Greyhound Light Armored Car 1941–91*. Oxford: Osprey Publishing, 2002

— *M10 and M36 Tank Destroyers 1942–53*. Oxford: Osprey Publishing, 2002

— *M4 (76mm) Sherman Medium Tank 1943–65*. Oxford: Osprey Publishing, 2003

— *M18 Hellcat Tank Destroyer 1943–97*. Oxford: Osprey Publishing, 2004

— *U.S. Armored Divisions: The European Theater of Operations, 1944–45*. Oxford: Osprey Publishing, 2004

— *M3 Lee/Grant Medium Tank 1941–45*. Oxford: Osprey Publishing, 2005

— *U.S. Tanks and Tank Destroyer Battalions in the ETO, 1944–45*. Oxford: Osprey Publishing, 2005*U.S. Armored Units in the North African and Italian Campaigns, 1942–45*. Oxford: Osprey Publishing, 2006

GLOSSARY

AAR	After Action Report
AGF	Army Ground Forces
Alco	American Locomotive Company
AP	Armor-Piercing
APC	Armor-Piercing Capped
APC-T	Armor-Piercing Capped Tracer
APG	Aberdeen Proving Ground
AP-T	M72 Shot Tracer
ARV	Armored Recovery Vehicle
BARV	Beach Armored Recovery Vehicle
CDL	Canal Defense Light
CHA	Cast Homogeneous Armor
CWS	Chemical Warfare Section
DD	Duplex Drive
ETO	European Theater of Operations
FHA	Face-Hardened Armor
FMC	Food Machinery Corporation
GE	General Electric
GMC	Gun Motor Carriage
HE	High Explosive
HEAT	High-Explosive Antitank
HMC	Howitzer Motor Carriage
HVAP-T	Hypervelocity Armor-Piercing Tracer
HVSS	Horizontal Volute Spring Suspension
LVT	Landing Vehicle, Tracked
PD	Point-Detonating
PTO	Pacific Theater of Operations
RHA	Rolled Homogeneous Armor
SQ	Super-Quick
T-AC	Tank-Automotive Center
TO&E	Table of Organization and Equipment
VVSS	Vertical Volute Spring Suspension
WP	White Phosphorus

INDEX

References to images are in **bold**.

A

B

C

H

K

L

M

N

O

P

T

U

V

W